DIGITAL PANOPTICON

How America's Surveillance State Monitors Its Citizens

Larry Brower

FOREWARD

To the relentless pursuit of truth, the unwavering defense of privacy, and the enduring spirit of those who dare to question the edifice of power. This book is dedicated to the whistleblowers who have illuminated the shadows, the journalists who have bravely reported from the front lines of informational warfare, and every citizen who believes that a free society cannot exist without the fundamental right to be left alone. May their courage inspire us all to safeguard the digital public square and ensure that technology serves humanity, rather than controlling it. This work is also for the future generations, so that they may inherit a world where transparency triumphs over secrecy, and liberty is not a forgotten relic of the past, but a vibrant, protected reality of the present and future.

In the years following the September 11, 2001, terrorist attacks, a profound shift occurred in the relationship between the state and its citizens. The imperative of national security, amplified by public anxiety and political urgency, created fertile ground for the expansion of governmental powers, particularly in the realm of intelligence gathering and surveillance. What began as a targeted response to a specific threat has, over two decades, metastasized into a vast, interconnected apparatus that monitors populations on an unprecedented scale. "Digital Panopticon" embarks on a comprehensive investigation into this surveillance state, dissecting the laws, technologies, and policies that have enabled its growth. We will trace the genesis of this pervasive monitoring, from the immediate post-9/11 legislative changes like the USA PATRIOT Act, which dramatically broadened government authority, to the sophisticated technological advancements that now allow for the collection and analysis of immense datasets.

This book is meticulously crafted for an audience that values rigorous analysis and factual reporting: informed citizens, civil liberties advocates, policymakers, legal professionals, technologists, and academics. You are likely concerned with the delicate balance between security and freedom, the ethical implications of emerging technologies, and the potential for government overreach. We aim to resonate with your analytical minds by providing a clear, authoritative, and evidence-based account, eschewing sensationalism for precision and depth. Through narrative threads that weave together technical details, legal arguments, and the palpable human impact of constant observation, we will illustrate concepts like metadata collection with concrete examples, referencing seminal court cases like *Katz v. United States* to highlight evolving privacy expectations. We will delve into the operations of agencies like the NSA and FBI, drawing upon declassified reports and whistleblower accounts to illuminate their methods and scope. Furthermore, we will critically examine the indispensable role of private corporations, from telecommunications giants to social media platforms, in building and sustaining this surveillance infrastructure. Our journey will uncover the architecture of mass surveillance, the machinery of data analysis, and the often-overlooked human cost—the erosion of privacy, the chilling effect on free speech, and the potential for discrimination. We will scrutinize the limitations of oversight and accountability, explore the legal battles fought for privacy rights, and confront the complicity and profiteering of corporate entities. Finally, we will consider the global dimension of surveillance, emerging trends, and the urgent need for reform, ultimately empowering you, the reader, to understand your role in shaping a future where liberty and security can coexist.

1: THE GENESIS OF PERVASIVE SURVEILLANCE

The morning of September 11, 2001, was a day that irrevocably altered the American psyche and, in its wake, reshaped the very foundations of its governance. The coordinated terrorist attacks on the World Trade Center and the Pentagon were not merely acts of violence; they were a visceral assault on the nation's sense of security and invincibility. The images of collapsing towers, the plumes of smoke against a clear blue sky, and the sheer scale of loss—nearly 3,000 lives extinguished in a matter of hours—unleashed a wave of fear and grief that permeated every level of society. This collective trauma created an immediate and profound imperative: to prevent such an event from ever happening again. In the frantic aftermath, this imperative began to eclipse other deeply held values, including the traditional understanding of privacy and civil liberties.

The psychological impact on the American public was immediate and profound. The attacks shattered the illusion of physical distance from global conflicts and demonstrated a new, insidious form of warfare waged not against armies, but against civilian populations. This vulnerability, brought so starkly into focus, fostered a deep-seated anxiety. People grappled with a primal fear, a sense that the unthinkable had become thinkable, and that the familiar comforts of everyday life could be shattered without warning. This pervasive sense of insecurity was amplified by round-the-clock media coverage that replayed the horrors, keeping the images and the emotions raw. The feeling of being unsafe, of being under siege, created a national mood that was highly receptive to strong, decisive action.

This climate of fear and the urgent demand for security quickly translated into a political environment ripe for the expansion of government power. Leaders across the political spectrum recognized the public's overwhelming desire for reassurance and protection. The narrative that emerged was one of an existential threat, a clear and present danger that required extraordinary measures to confront. In this atmosphere, dissenting voices or concerns about the potential overreach of new security initiatives were often marginalized or dismissed as unpatriotic or naive. The urgency of the situation seemed to justify a suspension of the usual deliberative processes and a willingness to grant significant new authorities to intelligence and law enforcement agencies. The focus was singular: to identify and neutralize threats, regardless of the cost to established norms or individual privacy.

The perceived existential threat posed by terrorism was unprecedented in its nature and its public recognition. Unlike previous conflicts or security challenges that might have felt distant or abstract, the 9/11 attacks brought the danger home with brutal clarity. This direct experience of vulnerability created a unique political and social consensus. Suddenly, concepts that had been debated in academic circles or legal journals – the balance between security and liberty, the limits of government surveillance – became matters of urgent public concern. The government's role was redefined in the eyes of many: not just to protect citizens from external aggression, but from insidious, unseen enemies who could strike anywhere, at any time. This redefinition was crucial in laying the groundwork for the surveillance state that would emerge over the subsequent years.

This shift in national priorities manifested in a series of rapid legislative and executive actions. The singular focus on counter-terrorism provided the impetus for a dramatic expansion of intelligence gathering capabilities. Agencies that had previously operated with more constrained mandates were empowered with new tools and authorities, often justified by the argument that traditional legal and procedural hurdles could impede the fight against an adaptive and ruthless enemy. The need to "connect the dots," a phrase that became commonplace, implied a requirement for more data, more comprehensive monitoring, and more sophisticated analytical tools to identify potential threats before they materialized. This imperative set the stage for a profound transformation in how the United States approached national security, a transformation that would deeply and permanently alter the landscape of privacy and civil liberties. The immediate post-9/11 era was characterized by a potent cocktail of fear, a desire for decisive leadership, and a national willingness to accept increased government power in the name of collective safety. This potent combination created a fertile ground for the seeds of pervasive surveillance to be sown, not as a deliberate conspiracy, but as a seemingly logical, albeit ultimately transformative, response to an unprecedented crisis.

The psychological shockwaves of September 11, 2001, were not confined to the immediate hours and days following the attacks. They reverberated through American society, creating a national consciousness deeply unsettled by the realization of its vulnerability. The visual and emotional testimony of the attacks—the collapsing towers, the desperate final calls from those trapped, the images of first responders rushing into danger—instilled a profound sense of insecurity that lingered long after the dust settled. This pervasive anxiety fueled a widespread desire for reassurance and for tangible actions that would demonstrate the government's commitment to preventing future atrocities. It created an environment where the very notion of security became paramount, often eclipsing other considerations that, in more placid times, would have been central to public discourse.

The media played a crucial role in amplifying and sustaining this national mood. For weeks and months following the attacks, the events of that day were replayed endlessly. The graphic imagery and the human stories associated with the tragedy served to reinforce the sense of ongoing threat. This constant media saturation, while intended to inform and unite, also had the effect of embedding a deep-seated fear within the populace. The abstract concept of terrorism was transformed into a tangible, immediate danger that could, and did, strike at the heart of the nation. This collective trauma created a psychological vulnerability that political leaders could, and did, leverage. The public was, understandably, looking for solutions, for a sense of control in a world that suddenly felt unpredictable and dangerous.

In this atmosphere of heightened anxiety and a pressing need for reassurance, the political landscape shifted dramatically. A national consensus emerged, largely unifying the country behind the imperative to bolster national security. This consensus was not achieved through a slow, deliberative process, but rather in the urgent aftermath of a shocking event. The political will to act decisively was immense, and the perceived existential nature of the threat provided the justification for extraordinary measures. Concerns about civil liberties, which are often a central tenet of American political debate, were, in this immediate post-9/11 period, often framed as secondary to the overriding goal of preventing further attacks. Any suggestion of caution or critique regarding new security initiatives could be easily framed as a lack of commitment to the safety of the nation, a position that few politicians or citizens were willing to adopt.

The concept of an "existential threat" was central to the narrative that justified the rapid implementation of new security measures. Terrorism, as embodied by the 9/11 attacks, was presented not merely as a crime or a political act, but as a fundamental challenge to the very existence of the United States and its way of life. This framing elevated the stakes considerably. It moved the discussion from routine law enforcement to a broader, more existential struggle that demanded extraordinary responses. This rhetoric helped to normalize the idea that established norms and legal precedents might need to be re-evaluated, or even temporarily suspended, in the face of such a grave danger. The government was not just asked to protect its citizens; it was portrayed as being engaged in a war for survival, and in such a war, the traditional rules of engagement were implicitly understood to be different.

This imperative to prevent future attacks set the stage for a dramatic expansion of intelligence gathering capabilities. The attacks themselves highlighted perceived failures in the intelligence community's ability to "connect the dots"—to share information effectively between different agencies and to analyze disparate pieces of data in a way that could reveal an impending plot. The response was to equip these agencies with the tools and authorities they claimed were necessary to overcome these perceived deficiencies. This included a significant increase in funding, a relaxation of some interagency barriers, and, crucially, new legal authorities that would broaden the scope of surveillance. The assumption was that more information, collected more broadly and analyzed more comprehensively, would lead to better intelligence and, therefore, greater security.

The immediate aftermath of 9/11 witnessed the swift passage of legislation that dramatically expanded the government's powers. The USA PATRIOT Act, enacted just 45 days after the attacks, is the most prominent example. This act, rushed through Congress with minimal debate, significantly broadened the government's ability to conduct surveillance, access records, and collect information, often with reduced judicial oversight. Provisions that allowed for expanded wiretapping, access to library records, and the sharing of intelligence between law enforcement and intelligence agencies were implemented under the banner of national security. The speed and scope of this legislative response underscore the profound shift in priorities. What might have been debated for years in peacetime was enacted in weeks under the intense pressure of perceived national emergency. This legislation did not merely augment existing capabilities; it fundamentally altered the legal framework within which surveillance could operate, creating the initial scaffolding for the pervasive surveillance apparatus that would become a hallmark of the ensuing decades.

The justification for these expanded powers was rooted in the idea that intelligence agencies needed new tools to combat a new kind of enemy—one that operated in the shadows, communicated through encrypted channels, and could strike anywhere without warning. The traditional models of intelligence gathering, which often relied on specific leads or the investigation of suspected wrongdoing, were deemed insufficient. The prevailing theory was that by collecting vast amounts of data on communications, financial transactions, and movements, intelligence agencies could identify patterns, anomalies, and connections that might reveal a plot in its nascent stages. This led to a crucial transition in surveillance strategy: from a focus on specific, identifiable threats to a more generalized approach of mass data collection, a shift that would profoundly alter the relationship between the state and the individual.

The initial focus of the post-9/11 security apparatus was on identifying and tracking suspected terrorists. Agencies like the Central Intelligence Agency (CIA) and the National Security Agency (NSA) were tasked with disrupting plots, dismantling terrorist networks, and preventing future attacks. This involved developing and deploying new technologies for communications interception, data analysis, and surveillance, often operating in secrecy to maintain operational effectiveness. Early efforts included programs aimed at monitoring the communications of individuals believed to be connected to terrorist organizations, analyzing financial transactions for suspicious patterns, and tracking the movements of individuals entering or leaving the country. These early operations, while couched in the language of counter-terrorism, represented the nascent stages of a massive expansion in intelligence gathering capabilities. The technological infrastructure and the legal frameworks that would enable broader surveillance were being developed and tested during this formative period, driven by the urgent need to prevent another 9/11.

This era also saw the beginnings of a critical shift in the nature of surveillance itself. The initial focus on specific targets, individuals or groups suspected of engaging in terrorist activities, began to broaden. The understanding grew that identifying potential threats required not just looking for overt signs of wrongdoing, but also analyzing vast datasets to find subtle patterns and connections that might not be immediately apparent. This led to the development of programs designed to collect information on a much larger scale, often encompassing the communications of individuals who were not suspected of any wrongdoing. The rationale was that by casting a wider net, intelligence agencies could increase their chances of identifying the few individuals who might pose a threat. This transition from targeted surveillance, based on suspicion, to mass data collection, based on the assumption that all data might be relevant, was a fundamental turning point, and it set the stage for the pervasive surveillance that would characterize the following decades.

The technological advancements of the late 20th and early 21st centuries played a pivotal role in enabling this expansion of surveillance. The proliferation of the internet, mobile phones, and digital communication technologies created an unprecedented volume of data that could be collected and analyzed. Intelligence agencies recognized the potential of these technologies to provide a comprehensive picture of individuals' activities, relationships, and even thoughts. This led to a concerted effort to develop and deploy sophisticated tools for data mining, communications interception, and sophisticated analytical techniques. The pace of technological innovation often outstripped public understanding and the ability of lawmakers to adequately address its implications. This created a situation where surveillance capabilities could grow and evolve in ways that were often hidden from public view, operating behind a veil of classification and national security secrecy.

The role of private sector technology companies in developing and providing these tools also began to become apparent during this period. Telecommunications companies, internet service providers, and software developers were essential partners in the government's expanded surveillance efforts, often providing the infrastructure and access points necessary for data collection. This symbiotic relationship between government intelligence agencies and private industry would become a defining characteristic of the modern surveillance state, blurring the lines between public and private sector data collection and raising complex questions about accountability and oversight. The post-9/11 imperative, therefore, not only spurred legislative action and technological development but also fostered a new kind of partnership that would fundamentally reshape the landscape of privacy and security in the United States.

The ink was barely dry on the nation's collective grief when Congress, acting with an alacrity that stunned many observers, passed the USA PATRIOT Act. Enacted a mere 45 days after the September 11th attacks, this sweeping piece of legislation represented a seismic shift in the balance between governmental power and individual liberties. Its very title, the "Uniting and Strengthening America by Providing Appropriate Tools Required to Intercept and Obstruct Terrorism Act," signaled its singular purpose: to equip the national security apparatus with unprecedented capabilities to thwart future terrorist plots. The urgency with which it was drafted and passed, bypassing the usual deliberative legislative processes, underscored the palpable fear that gripped the nation and the widespread willingness to grant extraordinary powers in the name of collective safety. This was not merely an enhancement of existing tools; it was a fundamental recalibration of the legal framework governing surveillance and information gathering.

At its core, the PATRIOT Act was designed to dismantle perceived barriers that hindered intelligence and law enforcement agencies from effectively monitoring and interdicting potential threats. One of its most significant and controversial provisions was the expansion of Section 215 of the Foreign Intelligence Surveillance Act (FISA). Prior to the Act, obtaining business records, including those of libraries or communication providers, required a showing of relevance to an ongoing criminal investigation. Section

215, however, dramatically lowered this threshold. It allowed the FBI, with a court order from the Foreign Intelligence Surveillance Court (FISC), to obtain "any tangible things" relevant to an investigation to obtain foreign intelligence information or to protect against international terrorism or clandestine intelligence activities. This broad language was interpreted by government agencies to encompass a vast array of records, from phone logs and internet browsing histories to financial transactions and even, controversially, library records and book purchase histories. The implication was that any information held by a third party could potentially be accessed by the government without the subject's knowledge or consent, so long as it was deemed relevant to a national security investigation, however broadly defined.

Another critical aspect of the PATRIOT Act was its relaxation of the "wall" that had previously separated domestic law enforcement from foreign intelligence gathering. For decades, the legal framework had attempted to maintain a distinction between the investigatory powers of agencies like the FBI, which primarily dealt with domestic crimes, and those of agencies like the CIA, which focused on foreign intelligence. This separation was intended to prevent intelligence agencies from engaging in domestic surveillance or law enforcement activities that would infringe upon the civil liberties of U.S. citizens. The PATRIOT Act effectively blurred these lines, allowing for greater information sharing and, in some instances, joint operations between these agencies. This was justified by the argument that terrorism was a global phenomenon, and that the intelligence gathered abroad could be vital for domestic security, and vice versa. However, critics argued that this fusion of powers created opportunities for intelligence agencies to engage in domestic surveillance under the guise of foreign intelligence gathering, thereby circumventing constitutional protections afforded to American citizens.

The Act also introduced "sneak and peek" provisions, which allowed law enforcement to delay notifying individuals that their property had been searched. This meant that law enforcement could enter a premises, conduct a search, and seize evidence without the knowledge of the property owner. While proponents argued this was crucial for disrupting ongoing investigations and preventing suspects from destroying evidence, opponents decried it as a direct assault on the Fourth Amendment's protection against unreasonable searches and seizures, which traditionally included the requirement of notification. The rationale for delayed notification was that immediate disclosure might alert terrorist cells or hinder the collection of further intelligence by allowing suspects to destroy evidence or alter their communications. This provision, in effect, allowed the government to conduct surveillance in secret, undermining the fundamental principle of transparency in government actions.

Furthermore, the PATRIOT Act significantly expanded the government's ability to conduct roving wiretaps. Previously, wiretap orders were specific to a particular telephone number or communication facility. Under the new provisions, a single warrant could authorize the surveillance of any phone or communication device that a suspect might use, even if the specific device was unknown to law enforcement at the time the warrant was issued. This meant that a person could be under surveillance without their knowledge, even if they were using a different phone or device than the one initially identified in the warrant. This was framed as a necessary adaptation to the evolving methods of communication used by terrorists, who might switch phones or use public telephones to evade detection. However, it also raised concerns about the potential for overbroad surveillance and the erosion of privacy in personal communications, as the scope of surveillance could expand far beyond the initially identified communication method.

The national security letter (NSL) provisions of the PATRIOT Act also came under intense scrutiny. NSLs are administrative subpoenas used by the FBI to obtain certain types of information, such as subscriber information from telecommunications companies, without prior judicial approval. The Act expanded the scope of information that could be obtained via NSLs and crucially, prohibited the recipient of an NSL from disclosing that they had received one. This gag order provision meant that companies providing records or information could not inform their customers that the government had accessed their data, creating a system of secret surveillance facilitated by private entities. Critics argued that the unchecked use of NSLs and their accompanying gag orders bypassed judicial oversight and created a significant power imbalance, allowing the government to collect vast amounts of personal data without any transparency or accountability. The ability for the FBI to issue NSLs without a court order meant that the checks and balances inherent in the judicial process were circumvented, empowering the executive branch with a powerful tool for information gathering that was shielded from public and judicial scrutiny.

The legal challenges to the PATRIOT Act began almost immediately after its passage. Civil liberties organizations, including the American Civil Liberties Union (ACLU), filed lawsuits arguing that the Act violated fundamental constitutional rights, particularly those guaranteed by the First, Fourth, and Fifth Amendments. The core of these arguments centered on the broad interpretation of "relevant" in Section 215, the expanded use of NSLs with gag orders, the "sneak and peek" provisions, and the relaxation of the wall between domestic and foreign intelligence. While some of these legal challenges were ultimately unsuccessful, or their impact mitigated by subsequent legislative modifications, they succeeded in drawing public attention to the implications of the Act and in initiating a national debate about the appropriate limits of government surveillance in a post-9/11 world. The courts, grappling with the novel challenges posed by modern technology and the exigencies of national security, often found themselves navigating uncharted legal territory.

One of the most significant early legal battles involved the challenge to Section 215. In *Doe v. Ashcroft* (later *Doe v. Gonzales* and then *Doe v. Holder*), a challenge was brought against the constitutionality of Section 215, arguing that it allowed for the seizure of sensitive information without sufficient justification or oversight. While the initial ruling in *Doe v. Ashcroft* found Section 215 unconstitutional, this decision was later vacated. The subsequent legal proceedings highlighted the difficulty in challenging these provisions, particularly when information was classified and the specifics of the government's actions were not fully disclosed. The case underscored the secrecy surrounding the implementation of the PATRIOT Act, making it exceedingly difficult for individuals to know if their rights were being violated, let alone to seek redress.

The PATRIOT Act was not a static piece of legislation. It included sunset clauses for many of its most controversial provisions, meaning they were set to expire unless reauthorized by Congress. This created recurring moments of intense debate and political maneuvering. Each reauthorization cycle, particularly in 2005 and 2006, saw renewed arguments from civil liberties advocates for stricter limitations and from national security proponents for continued or expanded powers. During these debates, proponents of the Act often emphasized the successes attributed to its provisions in thwarting terrorist plots, though the specifics of these claims were often classified and difficult to independently verify. Conversely, opponents highlighted the potential for abuse and the chilling effect on free speech and association, citing instances where the Act's powers were allegedly misused or over-applied.

Despite the controversies, many of the Act's key provisions were eventually made permanent or were reauthorized with modifications that did not fundamentally curtail the government's expanded surveillance capabilities. The process of reauthorization often involved compromises that, in the eyes of critics, further diluted civil liberties protections. For instance, while some minor reforms were introduced, the core mechanisms for accessing information and conducting surveillance largely remained intact. The protracted debates over reauthorization, while offering moments of public discourse, ultimately failed to reverse the trajectory towards increased government surveillance powers established in the immediate post-9/11 environment. The legislative battles became a recurring feature of the American political landscape, each cycle reinforcing the established norms of expanded surveillance.

The aftermath of the PATRIOT Act extended beyond its legislative journey. It profoundly shaped the operational practices of intelligence and law enforcement agencies, embedding a culture of pervasive data collection. The infrastructure and legal authorities established by the Act became foundational for subsequent surveillance programs, some of which were later revealed to the public through whistleblowers. The success of the PATRIOT Act in achieving its stated goals of preventing terrorism remained a subject of debate, with supporters pointing to the absence of another major attack on American soil as evidence of its effectiveness, while critics argued that the vast expansion of surveillance powers came at an unacceptable cost to privacy and civil liberties, and that intelligence successes could have been achieved through less intrusive means. The broad interpretation and application of its provisions created a chilling effect, potentially discouraging legitimate dissent and the free exchange of ideas, as individuals became aware that their communications and activities might be monitored. The legacy of the PATRIOT Act is thus a complex one, marked by the enduring tension between the perceived necessities of national security and the fundamental rights and freedoms that define American democracy. It fundamentally altered the relationship between the state and the individual, ushering in an era where pervasive surveillance became not an exception, but an increasingly normalized aspect of governance. The very notion of privacy in the digital age was irrevocably reshaped by its passage and subsequent extensions, setting the stage for even more sophisticated and far-reaching surveillance capabilities in the years to come.

The immediate aftermath of September 11th, 2001, witnessed a furious acceleration in the operational tempo of American intelligence agencies. The shock of the attacks did not merely galvanize legislative action like the PATRIOT Act; it fundamentally reoriented the priorities and methodologies of the nation's intelligence apparatus. Agencies that had historically focused on foreign adversaries or specific, targeted threats were suddenly thrust into a dynamic and pervasive fight against a decentralized, networked enemy operating within and across borders. This pivot required not just new legal authorities, but a rapid, often ad hoc, expansion of existing capabilities and the hurried development of new ones, all under the immense pressure to prevent recurrence.

The Central Intelligence Agency (CIA) and the National Security Agency (NSA), as the primary foreign intelligence and signals intelligence entities respectively, found themselves at the vanguard of this intensified effort. Their mandates, already broad, were now interpreted with an urgency that pushed the boundaries of what was previously considered feasible or even permissible. The concept of "tracking suspected terrorists" evolved from identifying and monitoring known operatives to a much more ambitious and technologically intensive endeavor: identifying potential threats from vast, anonymized datasets. This involved a significant increase in the collection of communications metadata – who was talking to whom, when, and for how long – alongside a renewed focus on acquiring actual content where possible.

The early years were characterized by a desperate scramble to build the necessary technological infrastructure to support this expanded mission. Existing surveillance systems, designed for more traditional intelligence gathering, were rapidly upgraded and repurposed. New systems were conceived and deployed with unprecedented speed, often with minimal public disclosure or congressional oversight. The rationale was simple: the nation was at war, and the intelligence services needed the tools to fight it effectively. This often meant prioritizing operational capability over meticulous adherence to established protocols or the slow, deliberative process of building consensus around new technological deployments.

One of the most significant early operational shifts involved the NSA's role in collecting communications data. While the NSA had long been engaged in signals intelligence, the post-9/11 environment saw a dramatic expansion in the scope and scale of its activities, particularly concerning the communications of individuals within the United States or involving U.S. persons. The PATRIOT Act, through provisions like the revised Section 215 of FISA, provided a legal framework for the acquisition of what were termed "tangible things," which intelligence agencies interpreted to include vast troves of telecommunications records held by private companies. This led to the establishment of programs designed to ingest and analyze massive amounts of metadata, often in bulk, from domestic and international telecommunications providers. The objective was to create a comprehensive network map that could, in theory, reveal connections between known terrorists and their associates, or identify nascent communication patterns indicative of planning.

The operationalization of these data collection programs was a complex undertaking. It involved forging partnerships with telecommunication companies, negotiating agreements for data access, and developing the analytical tools capable of sifting through petabytes of information. These early efforts were often classified, and the precise nature of the data collected, the companies involved, and the legal justifications used were shrouded in secrecy. The NSA's role, in particular, became central to many of these early intelligence operations. Its capacity to intercept and process vast quantities of electronic communications, from phone calls and emails to internet traffic, was seen as indispensable. The agency's technological prowess, honed over decades of signals intelligence, was now directed towards the singular goal of counter-terrorism.

Beyond metadata, early operations also sought to gather intelligence on the ground, often in collaboration with foreign partners. The CIA, in particular, intensified its human intelligence (HUMINT) operations in regions known to harbor terrorist groups. This involved recruiting informants, deploying clandestine operatives, and working with allied intelligence services to track individuals, disrupt financing networks, and disrupt planned attacks. The insights gained from these human sources were often integrated with signals intelligence, creating a more holistic picture of terrorist organizations and their activities. The fusion of HUMINT with signals intelligence (SIGINT) became a hallmark of the post-9/11 intelligence strategy, with agencies striving to break down traditional silos to share and analyze information more effectively.

The development and deployment of new technologies were a constant feature of these early operations. Agencies invested heavily in sophisticated data mining and analytical software, aiming to identify patterns and anomalies that might signify terrorist activity. This included the use of algorithms to search for keywords, connections between individuals, and unusual communication patterns. The concept of "predictive analysis" began to gain traction, with the hope of identifying potential threats before they materialized. This involved developing models and profiles based on known terrorist behaviors and communication methods, and then applying these to the vast datasets being collected.

However, this rapid technological advancement and operational expansion often outpaced the existing oversight mechanisms. While Congress passed the PATRIOT Act and established procedures for the Foreign Intelligence Surveillance Court (FISC) to approve certain surveillance requests, the sheer volume and novelty of the data being collected created significant challenges. The FISC, designed to review targeted wiretap applications based on probable cause, was not necessarily equipped to handle broad requests for bulk data or to provide meaningful oversight on programs that collected information on millions of individuals incidentally. The secrecy surrounding these programs meant that even congressional oversight committees had limited visibility into the full scope of intelligence activities.

The ethical considerations of such pervasive data collection were also becoming increasingly apparent, even in these early years. While the primary objective was to prevent terrorist attacks, the potential for these tools to be used for other purposes, or to infringe upon the privacy of innocent citizens, was a growing concern among some policymakers and legal scholars. The debate over the balance between security and liberty, ignited by the PATRIOT Act, extended directly into the operational realm. How much data collection was too much? At what point did the incidental collection of information on U.S. persons constitute an illegal search? These questions were being grappled with, often in classified forums, as agencies sought to implement their new authorities.

The early implementation of Section 215 of FISA, for example, proved particularly contentious. The broad language allowing the FBI to obtain "any tangible things" relevant to foreign intelligence or counter-terrorism investigations was interpreted to permit the collection of vast quantities of telephone metadata. The legal justification for this broad interpretation, and for the FISC's approval of such requests, was based on the idea that such records were essential for tracing the communications of suspected terrorists. However, critics argued that this approach constituted a mass surveillance program that violated the Fourth Amendment, as it allowed the government to access the private communications of countless individuals without individualized suspicion. The secrecy surrounding these applications meant that the public, and indeed much of Congress, remained largely unaware of the scale of this data collection.

The post-9/11 period also saw an increased reliance on National Security Letters (NSLs), which are administrative subpoenas that do not require prior judicial approval. The PATRIOT Act expanded the types of information that could be obtained through NSLs and, crucially, allowed for gag orders that prohibited recipients from disclosing their receipt of an

NSL. This meant that telecommunications companies and other entities could be compelled to provide customer data to the FBI without being able to inform their customers. This tool was widely used in the early years to gather subscriber information, call detail records, and other data deemed relevant to counter-terrorism investigations. The ability to obtain this information quickly and without judicial oversight was seen by law enforcement as a significant advantage, but it also raised serious concerns about due process and the erosion of privacy.

The operationalization of intelligence capabilities in the years immediately following 9/11 was a period of rapid innovation and significant expansion, largely driven by a perceived existential threat. Agencies like the CIA and NSA were empowered and directed to find and disrupt terrorist plots, leading to the development and deployment of sophisticated technologies and methodologies for data collection and analysis. The initial focus was on building the capacity to "connect the dots," that is, to identify patterns and relationships within vast amounts of information that might indicate illicit activities. This involved not only technological infrastructure but also the development of new analytical techniques and the retraining of personnel to deal with the unique challenges of counter-terrorism in a networked world.

The NSA, in particular, played a crucial role in developing the capabilities to collect and analyze communications metadata on a massive scale. Programs were established to acquire detailed records of telephone calls, internet usage, and other electronic communications. The legal basis for much of this collection was found in Section 215 of FISA, as amended by the PATRIOT Act, which allowed the government to obtain "any tangible things" relevant to foreign intelligence and counter-terrorism investigations. This broad authority was interpreted to permit the collection of records in bulk, without requiring individualized suspicion for each person whose records were collected. The intention was to create a comprehensive dataset that could be searched and analyzed to identify potential threats.

The CIA, meanwhile, ramped up its human intelligence operations and its efforts to track individuals of interest. This involved working with foreign intelligence services, deploying operatives, and using advanced interrogation techniques (which later became a subject of significant controversy) to gather information on suspected terrorists and their networks. The agency also focused on disrupting terrorist financing and preventing the flow of weapons and other materials to terrorist organizations. The operational tempo for both agencies was exceptionally high, with a constant demand for actionable intelligence to prevent future attacks.

The technological infrastructure supporting these operations was a key focus. Significant investments were made in data storage, processing power, and analytical software. The goal was to enable agencies to collect, store, and analyze unprecedented volumes of data. This included the development of sophisticated search algorithms, pattern recognition tools, and visualization software designed to help analysts identify connections and trends that might otherwise be missed. The concept of data fusion, or the integration of information from various sources (SIGINT, HUMINT, open-source intelligence, etc.), became increasingly important, as agencies recognized that a more comprehensive understanding of threats required the synthesis of diverse datasets.

However, the rapid expansion of these capabilities also raised significant questions about oversight and the protection of civil liberties. The legal frameworks established by the PATRIOT Act, while intended to provide the necessary tools for counter-terrorism, were often interpreted broadly by the agencies themselves. The Foreign Intelligence Surveillance Court (FISC), which was tasked with reviewing surveillance requests, operated in secret and often approved the government's requests with little public scrutiny. Critics argued that the court was too deferential to the executive branch and that the secrecy surrounding its proceedings made it difficult to assess the legality and appropriateness of the government's surveillance activities.

The use of National Security Letters (NSLs) also expanded significantly in the early years. These administrative subpoenas allowed the FBI to obtain certain types of information, such as subscriber data and basic subscriber information, without prior judicial approval. The PATRIOT Act broadened the scope of information that could be obtained via NSLs and importantly, allowed for the imposition of gag orders, which prohibited the recipient of an NSL from disclosing its existence. This meant that companies providing data to the government could not inform their customers that their information had been accessed. This lack of transparency was seen by critics as a serious impediment to accountability and a significant threat to individual privacy.

The ethical considerations surrounding these early intelligence operations were substantial, though often overshadowed by the urgency of national security. The collection of vast amounts of data, much of it belonging to innocent citizens, raised concerns about the potential for misuse and the erosion of privacy. The lack of transparency surrounding many of these programs made it difficult for the public to understand the extent to which their communications and activities were being monitored. This created a "black box" effect, where the government possessed powerful surveillance tools, but the mechanisms for ensuring their lawful and ethical use were opaque.

In essence, the early years of post-9/11 intelligence

operations were characterized by a rapid build-up of capabilities, driven by the imperative to prevent further attacks. This involved a significant expansion of data collection, the development of new analytical tools, and a close working relationship between agencies like the CIA and NSA. However, this expansion also occurred within a context of limited transparency and evolving legal interpretations, setting the stage for ongoing debates about the balance between national security and individual liberties that would continue to shape surveillance practices in the years to come. The operational tempo was set by the perceived threat, and the infrastructure and legal frameworks were rapidly adapted, often in secret, to meet that challenge. This nascent era of pervasive surveillance, born out of necessity, would lay the groundwork for the complex and often controversial intelligence apparatus that emerged in the subsequent decades. The intelligence community was not merely reacting to a new threat; it was actively building the architecture of a new kind of warfare, one fought in the digital shadows and fueled by unprecedented data aggregation. The initial focus was on identifying known actors and their immediate networks, but the technological acceleration quickly pointed towards a broader, more encompassing approach to threat detection, where potential threats were inferred from patterns within massive datasets, even in the absence of direct suspicion. This fundamentally altered the relationship between the individual and the state, shifting the paradigm from targeted surveillance based on cause to pervasive monitoring in anticipation of potential harm.

The immediate aftermath of September 11th, 2001, undeniably catalyzed an unprecedented expansion of intelligence capabilities and a fundamental reorientation of surveillance strategies. As the previous section detailed, the initial shock and the perceived existential threat led to a rapid build-up of technological infrastructure and legal authorities, primarily focused on identifying and disrupting terrorist networks. Agencies like the NSA and CIA were empowered to collect and analyze data on a scale previously unimaginable, driven by the urgent need to "connect the dots" and prevent future attacks. However, the narrative of this expansion does not end with the identification of known threats. A crucial, and often less scrutinized, development during this period was the strategic shift from surveillance meticulously focused on specific, identified threats to a far broader, more generalized approach of mass data collection. This transition was not a sudden abandonment of original goals but rather an evolutionary repurposing of the very infrastructure and legal frameworks that had been so hastily erected.

The rationale behind this pivot was rooted in the evolving understanding of how modern terrorist organizations operated. The decentralized nature of groups like Al-Qaeda, their reliance on global communication networks, and the sheer difficulty in distinguishing genuine threats from legitimate activities necessitated a more comprehensive, data-driven approach. Intelligence analysts realized that simply tracking known operatives or intercepted communications was insufficient. The true challenge lay in identifying emergent threats, understanding the hidden connections within vast global networks, and anticipating plots before they materialized. This ambitious goal required a fundamental change in methodology: instead of casting a narrow net for specific individuals or communications, the objective became to collect and analyze immense volumes of data, believing that within this ocean of information, patterns indicative of illicit activity would inevitably surface. This marked a significant departure from traditional intelligence gathering, which had historically been more targeted, requiring at least some level of suspicion or specific intelligence to initiate surveillance.

The legal and technological scaffolding built in the wake of

9/11 proved remarkably adaptable to this broader mandate. Section 215 of the Foreign Intelligence Surveillance Act (FISA), for instance, which had initially been interpreted to allow the acquisition of "tangible things" relevant to terrorism investigations, was soon understood to encompass the collection of bulk metadata. This interpretation extended beyond communications directly linked to suspected terrorists. The belief was that by analyzing the calling patterns, email contacts, and internet activity of millions of ordinary citizens, analysts could identify anomalies—unusual communication hubs, sudden changes in contact patterns, or connections to known problematic individuals—that might flag nascent threats or reveal previously unknown associates of terrorist groups. The logic was that even if the initial collection swept up innocent individuals, the sheer scale of the data allowed for a process of filtering and analysis that could, in theory, isolate the relevant information without requiring individualized suspicion at the point of initial collection. This approach effectively shifted the burden of proof; instead of the government needing to demonstrate suspicion to collect data on an individual, individuals were subject to collection, and it was up to them, or rather the analytical process, to demonstrate they were *not* a threat.

The strategic imperative was to create what some in the intelligence community referred to as a "total information awareness" capability, or at least a foundational dataset that could be interrogated for potential links to terrorism. The collection of telephone metadata, for example, was not limited to call detail records of individuals on watchlists. Instead, programs were established to collect virtually all domestic and international telephone metadata, including numbers called, duration of calls, and time of calls, often from major telecommunications providers. This data was then stored in massive databases, making it accessible for analysis. The idea was to have a comprehensive record of who was communicating with whom, allowing analysts to map out communication networks and identify individuals who, while not themselves suspects, might be connected to known threat actors. The rationale was that a single, seemingly innocuous phone call between an innocent citizen and a suspected terrorist could, when viewed within the context of millions of other communications, reveal a crucial link.

This shift also extended to other forms of communication and digital activity. Internet metadata, email records, and even more granular data points began to be collected under broad interpretations of national security directives and FISA authorities. The ability of agencies like the NSA to tap into undersea cables and internet backbone junctions allowed for the interception of vast quantities of data flowing across the globe. While the legal justifications for intercepting communications of U.S. persons or communications originating or terminating within the United States remained subject to stricter FISA requirements, the sheer volume of data collected often resulted in the incidental collection of information pertaining to millions of Americans who had no connection to terrorism. The argument was that such incidental collection was unavoidable in the pursuit of foreign intelligence and that the privacy interests of innocent individuals were adequately protected by limitations on how this data could be accessed, searched, and retained.

The strategic appeal of this mass data collection approach lay in its potential to uncover the "unknown unknowns"— threats and actors that intelligence agencies were not even aware existed. By casting a wide net, the hope was to identify individuals who might be radicalized domestically, or foreign actors using seemingly legitimate communication channels to plan attacks. The analytical tools developed for this purpose were crucial. Sophisticated algorithms and data-mining techniques were employed to sift through petabytes of information, looking for specific keywords, unusual communication patterns, or connections between individuals that might warrant further investigation. This allowed analysts to move beyond simply identifying known threats to proactively searching for potential threats embedded within the broader population.

The repurposing of infrastructure and legal frameworks was a critical element in this transition. Existing partnerships with telecommunications companies, forged in the immediate post-9/11 era to facilitate targeted data requests, were leveraged to enable broader data acquisition programs. Similarly, the expanded authorities granted under the PATRIOT Act and the interpretations of FISA Section 215 provided the legal cover for collecting data on a scale that significantly outstripped any previous intelligence operation. The secrecy surrounding these programs, often justified by the need to protect national security and operational methods, meant that the public and even many members of Congress had limited visibility into the true scope of this generalized data collection. This opacity allowed the intelligence community to develop and implement these broad surveillance programs with minimal public debate or oversight concerning their implications for privacy and civil liberties.

The strategic rationale was clear: in an era where threats could emerge from anywhere and at any time, relying solely on pre-existing suspicion was a recipe for failure. The intelligence community argued that a proactive, data-driven approach was essential to identify emerging threats before they could manifest. This involved building comprehensive databases of communication records, travel patterns, financial transactions, and other digital footprints. The goal was to create a searchable repository of information that could be analyzed to reveal connections, identify anomalies, and predict potential future actions. This represented a profound shift in the nature of surveillance, moving from a reactive model based on specific intelligence to a proactive model based on comprehensive data analysis.

The concept of "connecting the dots" thus evolved from linking pieces of intelligence about known threats to building a vast, interconnected web of information about ordinary citizens, with the hope that patterns of illicit activity would emerge. This fundamentally changed the relationship between the state and the individual. Previously, government surveillance typically required a warrant based on probable cause, meaning there had to be a reasonable belief that a crime had been or was about to be committed. The mass data collection programs, however, operated on a different principle: collect as much data as possible on as many people as possible, and then use analytical tools to find the needles in the haystack. This approach, while potentially effective in uncovering hidden threats, also raised significant concerns about the erosion of privacy, the potential for misuse of data, and the implications of a society where virtually all digital interactions could be subject to government scrutiny, even in the absence of any wrongdoing. The transition from specific threats to broad data collection was not merely a technical adjustment; it was a strategic redefinition of intelligence gathering, driven by the perceived necessities of a new era of security.

The fervent pursuit of enhanced security capabilities in the wake of 9/11 did not merely expand existing methodologies; it ignited a technological arms race, a relentless drive to develop and deploy ever more sophisticated tools for observation and analysis. This surge in innovation was not confined to incremental improvements but represented a paradigm shift, introducing capabilities that fundamentally altered the scale and nature of government surveillance. At the heart of this transformation lay the burgeoning fields of data mining and advanced analytics. The sheer volume of information being collected, as previously discussed, was immense, but its utility was directly proportional to the ability to process and interpret it. Agencies poured resources into developing algorithms capable of sifting through petabytes of data, searching for patterns, anomalies, and connections that human analysts, however skilled, could never hope to identify alone. These were not simple keyword searches; they were complex algorithms designed to identify behavioral patterns, social networks, and even predictive indicators of future actions, all derived from seemingly innocuous digital footprints.

Communications interception, long a staple of intelligence gathering, also underwent a dramatic technological evolution. While the physical interception of phone calls and emails had always been possible, the digital age presented new challenges and opportunities. The development of sophisticated decryption techniques, advanced packet sniffing technologies, and the ability to tap into global internet backbone infrastructure allowed intelligence agencies to intercept and analyze digital communications on an unprecedented scale. This included not only the content of communications but also the metadata – who was communicating with whom, when, for how long, and from where. The ability to process this metadata in near real-time transformed it from a supplementary source of information into a primary tool for mapping relationships and identifying potential threats. The technical challenge was immense, requiring significant investment in hardware, software, and specialized personnel, but the perceived payoff in terms of enhanced situational awareness was deemed to be of paramount importance.

Facial recognition technology emerged as another transformative element in this technological surge. Initially developed for more specific applications, such as identifying known criminals or persons of interest, its capabilities rapidly advanced. High-resolution cameras, coupled with increasingly accurate algorithms capable of identifying individuals from even partial or low-quality images, began to be deployed in public spaces. The potential was clear: to create a persistent, automated identification system that could track individuals as they moved through the physical world, much like digital surveillance tracked them online. While the legal and ethical implications of such pervasive visual monitoring were only beginning to be debated, the technology itself was being rapidly refined and integrated into existing surveillance frameworks. The ability to link an individual's physical presence to their digital identity, and vice versa, created a powerful, integrated surveillance system.

Location tracking, a natural corollary to facial recognition and communication intercepts, also saw significant technological leaps. GPS technology, once primarily associated with navigation, became a ubiquitous tool for monitoring movement. Beyond dedicated tracking devices, the proliferation of smartphones meant that vast amounts of location data were being generated passively by ordinary citizens. Intelligence agencies developed the capacity to access and analyze this data, either through direct partnerships with telecommunications companies or through exploiting vulnerabilities in data flows. The ability to reconstruct an individual's movements, identify their frequent locations, and map their daily routines provided a highly granular layer of surveillance, offering insights into personal habits and associations that could be correlated with other collected data.

Crucially, this technological revolution was not solely the product of government research and development. The private sector played an indispensable, and often understated, role. Technology companies, driven by commercial imperatives and a desire to provide cutting-edge solutions, were at the forefront of developing many of these advanced surveillance tools. From the sophisticated algorithms used for data mining and pattern recognition to the hardware and software that enabled mass communications interception, private companies were the architects and engineers of the modern surveillance apparatus. These companies, operating in a rapidly evolving market, often developed capabilities before public understanding or legislative frameworks could adequately address their implications. This dynamic created a situation where powerful surveillance technologies were readily available, even as the societal debate about their use lagged significantly behind.

The integration of these disparate technologies created a synergy, a powerful ecosystem of observation and analysis. Data collected from one source could be cross-referenced and validated with data from another, creating a more complete and often intimate picture of individuals' lives. A communication intercept could be correlated with location data, which could then be linked to facial recognition matches from public surveillance cameras, all analyzed by sophisticated data-mining algorithms. This integrated approach meant that even seemingly insignificant pieces of information could become highly revealing when combined with other data points. The technological sophistication allowed for the automation of many of these processes, enabling the continuous monitoring and analysis of vast populations without the need for a commensurate increase in human resources.

The pace of technological advancement frequently outstripped the capacity of public discourse and legislative bodies to comprehend and regulate it. By the time a new surveillance capability was developed and deployed, the underlying technology had often evolved, presenting new challenges for oversight. This created a significant asymmetry of knowledge and power. Intelligence agencies and the private companies that supported them possessed a deep understanding of these technologies, while the public and many elected officials remained largely unaware of the full extent of their capabilities and deployment. This information gap allowed for the widespread implementation of these tools with minimal public scrutiny, fostering an environment where the expansion of surveillance was often a fait accompli rather than a subject of democratic deliberation.

The legal frameworks that were in place, while expanded in the post-9/11 era, were often ill-equipped to address the novel challenges posed by these new technologies. Concepts of privacy that had been developed in an analog age struggled to adapt to a world where every digital interaction and every movement in public could be recorded, analyzed, and stored. The legal justifications for surveillance, often rooted in national security and foreign intelligence, were applied to technologies that blurred the lines between domestic and foreign, and between targeted and indiscriminate collection. This legal ambiguity provided fertile ground for the expansion of surveillance, as agencies could operate within perceived legal gray areas, often relying on broad interpretations of existing statutes to deploy new technological tools.

The development of "all-source analysis" capabilities further underscored the technological arms race. This concept involved integrating intelligence from every available source – human intelligence, signals intelligence, imagery intelligence, and open-source information – into a unified analytical framework. The technological challenge lay in creating the infrastructure and analytical tools to manage and fuse such diverse datasets. Advanced databases, secure inter-agency communication networks, and sophisticated analytical software were all necessary components. The goal was to create a comprehensive, 360-degree view of potential threats, but this ambition inherently required the collection and integration of data on a scale that inevitably encompassed the activities of millions of ordinary citizens. The technological imperative to achieve "all-source analysis" directly fueled the demand for more data, thus driving the expansion of collection capabilities across the board.

Furthermore, the drive for speed and efficiency in intelligence operations necessitated the automation of many analytical processes. Machine learning and artificial intelligence began to be explored and implemented to assist in identifying patterns, flagging anomalies, and even generating predictive assessments. This shift from human-centric analysis to algorithm-driven insights was a direct consequence of the overwhelming volume of data. The ability of algorithms to process information at speeds far exceeding human capability was a crucial factor in making mass surveillance a potentially viable strategy. However, it also raised new questions about bias in algorithms, the transparency of analytical processes, and the potential for errors to have profound consequences for individuals wrongly identified or flagged by automated systems.

The symbiotic relationship between government intelligence agencies and the private technology sector became increasingly apparent. As agencies sought to acquire and deploy cutting-edge surveillance technologies, they often turned to private companies that possessed the expertise and resources to develop them. This led to a complex web of contracts, partnerships, and data-sharing agreements. While the private sector was often motivated by profit and innovation, its role in enabling government surveillance had profound implications for civil liberties. The very companies that provided the tools for mass observation also often held vast amounts of personal data, creating a scenario where commercial interests and national security objectives became intertwined, often with limited public awareness or oversight. This collaboration laid the groundwork for future discussions about the extent of corporate responsibility in the surveillance state.

The sophistication of data aggregation and correlation tools also played a pivotal role. Beyond simply collecting raw data, intelligence agencies developed the capability to aggregate and correlate information from disparate sources, creating comprehensive profiles of individuals. This involved linking phone records with internet browsing history, travel data with financial transactions, and social media activity with public records. The technological challenge was to build the databases and analytical engines capable of performing these complex cross-referencing operations. The result was the creation of detailed, often deeply personal, dossiers on millions of people, many of whom had no connection to any criminal or terrorist activity. The ability to correlate data in this manner transformed seemingly innocuous information into potentially incriminating evidence, highlighting the power and pervasiveness of the evolving surveillance apparatus. This era marked a profound technological acceleration, fundamentally altering the landscape of privacy and security, and setting the stage for the pervasive surveillance systems that would become increasingly visible in the years to come. The sheer ingenuity and rapid development of these tools created a potent, often unseen, mechanism for observing and understanding the populace, a reality that was only beginning to dawn on a public largely unequipped to grapple with its implications. The technological arms race was, in essence, a race against time, as capabilities were developed and deployed, often before the ethical and legal frameworks could catch up, creating a new paradigm of state-citizen interaction built on a foundation of pervasive technological observation.

2: THE ARCHITECTURE OF SURVEILLANCE INFRASTRUCTURE

The National Security Agency, or NSA, stands as a formidable and often enigmatic pillar within the architecture of the United States' surveillance infrastructure. Its mandate, rooted in the collection and decryption of foreign intelligence and the protection of U.S. national security information, has evolved dramatically, particularly in the digital age. While its primary focus is ostensibly external, the sheer scale and nature of its operations inevitably and extensively encompass the domestic realm, often blurring the lines that were once thought to clearly demarcate the boundaries between foreign intelligence gathering and the privacy rights of American citizens. The agency's role is not merely that of a data collector; it is a sophisticated processor, an analyst, and a key architect of the technological systems that enable pervasive electronic surveillance on a global scale.

The capabilities of the NSA are vast, built upon decades of investment in signals intelligence (SIGINT). This encompasses the interception, processing, and analysis of all forms of electronic communication, from traditional telephone conversations and faxes to the far more complex and voluminous digital traffic that constitutes the modern internet. The agency's technical prowess is legendary, built on a foundation of advanced cryptology, sophisticated analytical software, and an expansive global network of collection facilities. Following the events of September 11, 2001, these capabilities were significantly augmented and applied with a broadened scope, driven by a perceived need to identify and disrupt terrorist plots before they could materialize. This intensification of effort led to the development and deployment of programs that could, in theory, collect and analyze nearly every form of electronic communication originating from or passing through U.S. networks, as well as those connected to global communication infrastructure.

At the heart of the NSA's operational capacity lies its ability to acquire and process enormous quantities of data. Whistleblower disclosures, most notably those by former contractor Edward Snowden, have provided unprecedented, albeit controversial, insights into the sheer magnitude of this data collection. Programs like PRISM, revealed to be a direct collection from major U.S. technology companies, allowed the NSA access to vast troves of emails, chat logs, videos, photos, and other digital information belonging to foreign nationals, but critically, also incidentally capturing the communications of American citizens. Similarly, the collection of telephone metadata, famously exposed by Snowden's leaks concerning Verizon, demonstrated the NSA's capacity to gather billions of phone records annually, detailing who called whom, when, for how long, and from where. While the NSA maintained that this metadata was used only to identify connections to foreign targets, the sheer volume of data collected meant that a significant portion of American domestic communications was being cataloged and stored, creating a searchable database of intimate personal connections.

The infrastructure supporting these operations is nothing short of monumental. The NSA operates a global network of listening posts, utilizing a combination of satellite interception, undersea cable taps, and partnerships with telecommunications providers. Facilities like the Bluffdale Data Center in Utah, a massive complex designed for storing and processing colossal amounts of intercepted data, underscore the agency's commitment to quantitative analysis. This infrastructure is not merely about capturing raw signals; it is about creating a persistent, accessible repository of digital life. The ability to store and search through this data, often for extended periods, transforms ephemeral electronic communications into a permanent record, accessible for retrospective analysis. The technical architecture is designed to handle the exponential growth of digital information, employing advanced processing techniques and artificial intelligence to sift through petabytes of data in near real-time, identifying patterns, connections, and anomalies that might indicate a threat.

Furthermore, the NSA's reach extends to the internet's most fundamental infrastructure. Through programs like MUSCULAR, jointly operated with the British Government Communications Headquarters (GCHQ), the agency gained direct access to the internal networks of major technology companies, intercepting vast amounts of unencrypted data as it flowed between data centers. This demonstrated a capacity to bypass individual user-level protections and tap directly into the arteries of the internet. The agency's technical expertise in breaking encryption, or developing methods to circumvent it, is a core component of its mission. While much of this effort is directed at foreign adversaries, the pervasive nature of digital communication means that the tools and techniques developed for foreign intelligence gathering can, and have, inadvertently or intentionally, swept up the communications of domestic individuals.

The legal frameworks under which the NSA operates have been a subject of intense scrutiny and debate. The Foreign Intelligence Surveillance Act (FISA), originally enacted to govern electronic surveillance for foreign intelligence purposes, has been amended multiple times to accommodate new technologies and perceived threats. Section 702 of FISA, for example, authorizes the surveillance of non-U.S. persons located outside the United States, but its implementation has raised significant concerns about the incidental collection of American citizens' communications. The legal justifications often hinge on the distinction between targeting foreign individuals and the inevitable collection of domestic data as a byproduct. However, the technical reality of mass data collection often makes this distinction practically meaningless, as data flows are not neatly compartmentalized.

The agency's role also involves collaboration with other intelligence agencies, both domestic and international. Within the U.S., the NSA works closely with agencies like the FBI and the CIA, sharing intelligence and analytical capabilities. Its partnerships with foreign intelligence agencies, particularly those within the Five Eyes alliance (comprising the United States, the United Kingdom, Canada, Australia, and New Zealand), amplify its collection and analytical reach across the globe. These alliances facilitate a reciprocal sharing of intercepted data, creating a global surveillance network where each member nation's capabilities are leveraged to the benefit of all. This international cooperation, while framed as essential for combating global threats, further entrenches the NSA's position as a central node in a vast, interconnected surveillance apparatus that spans continents and oceans.

The technical challenges faced by the NSA are immense, not least of which is the sheer volume and complexity of electronic communications. The agency must constantly adapt to new communication technologies, encryption methods, and the evolving nature of online activity. Its mission necessitates continuous investment in research and development, seeking out novel methods for intercepting, decrypting, and analyzing data. This relentless pursuit of technological advantage places the NSA at the forefront of advancements in fields such as artificial intelligence, machine learning, and big data analytics, often pushing the boundaries of what is technically feasible. However, this innovation also creates a dynamic where the agency's capabilities can outpace public understanding and legislative oversight, leading to a persistent tension between security imperatives and civil liberties concerns.

The NSA's operational posture is characterized by a high degree of secrecy. Its programs, budgets, and activities are largely classified, making independent oversight and public accountability exceptionally difficult. While legislative bodies and internal inspector generals provide some level of oversight, the classified nature of its work means that the full extent of its capabilities and operations remains hidden from public view. This opacity is often defended on national security grounds, but it also creates an environment where potential abuses of power or overreach can occur with limited checks and balances. The revelations of whistleblower Edward Snowden fundamentally challenged this secrecy, bringing to light the profound implications of the NSA's surveillance activities for privacy and democratic freedoms worldwide.

In essence, the National Security Agency is not merely a participant in the surveillance infrastructure; it is one of its primary architects and enablers. Its technical expertise, global reach, and unparalleled capacity for data collection and analysis place it at the apex of the state's ability to observe and understand electronic communications. The agency's mission, while ostensibly focused on foreign intelligence, has demonstrably resulted in the widespread collection and analysis of data that includes the communications of American citizens, raising fundamental questions about privacy, civil liberties, and the balance between security and freedom in the digital age. The sheer scale of its operations, powered by cutting-edge technology and extensive infrastructure, underscores the pervasive nature of modern surveillance and the central role the NSA plays in its implementation.

The agency's commitment to signals intelligence is not static; it is a dynamic and ever-evolving endeavor. The NSA invests heavily in research and development to stay ahead of advancements in telecommunications, encryption, and cybersecurity. This proactive approach means that as new ways of communicating emerge, the NSA is simultaneously developing the means to intercept and understand them. For instance, the proliferation of encrypted messaging applications, which use end-to-end encryption to secure communications, presents a significant challenge. While direct interception of encrypted content may be difficult, the NSA's strategies often involve targeting the metadata surrounding these communications, the endpoints of the communication, or seeking vulnerabilities in the systems themselves. This ongoing technological arms race necessitates a constant refinement of tools and techniques, ensuring that the agency's capabilities remain relevant in the face of evolving digital security practices.

Furthermore, the NSA's role extends to the development and deployment of specialized hardware and software designed for intelligence collection. This includes sophisticated network intrusion tools, decryption algorithms, and data analysis platforms. These are not off-the-shelf solutions; they are custom-built, highly specialized systems designed for specific intelligence requirements. The agency often works in concert with private sector contractors to develop and maintain these advanced technologies, creating a complex ecosystem of public-private collaboration in the realm of surveillance. This partnership allows the NSA to leverage the innovation and expertise of the commercial technology sector, while also potentially outsourcing some of the more sensitive aspects of its operations.

The concept of "all-source analysis" is critical to understanding the NSA's operational philosophy. This approach involves integrating intelligence from all available sources – human intelligence (HUMINT), signals intelligence (SIGINT), imagery intelligence (IMINT), geospatial intelligence (GEOINT), and open-source intelligence (OSINT) – into a unified analytical framework. The NSA plays a pivotal role in processing and analyzing the SIGINT component of this intelligence, but its capabilities are also applied to fuse this data with information gathered by other agencies. This integration allows for a more comprehensive understanding of targets, threats, and geopolitical landscapes. For example, a communication intercept might be correlated with satellite imagery of a particular location or with information from a human source, creating a richer, more contextualized intelligence product. The technological infrastructure required for such all-source analysis is immense, involving massive databases, secure networks, and sophisticated analytical software capable of managing and cross-referencing vast and disparate datasets.

The Snowden revelations also brought to light programs that specifically targeted the communications of individuals, even if those individuals were not primarily foreign adversaries. For instance, the "Upstream" collection, which involves tapping into internet backbone infrastructure, allows the NSA to intercept communications as they transit through global networks. While the stated purpose is to collect foreign intelligence, the technical reality is that these taps capture a wide range of data, including that of U.S. persons, often before it can be adequately protected by domestic legal safeguards. The justification for this collection often relies on the argument that the data is being collected from foreign infrastructure or that the individuals are communicating with foreign targets, but critics argue that this approach represents an erosion of privacy for ordinary citizens.

The NSA's involvement in cybersecurity, while seemingly counter-intuitive to surveillance, is deeply intertwined. The agency is responsible for defending U.S. government networks against cyberattacks and for developing offensive cyber capabilities. These offensive capabilities, often developed in secret, can be used to infiltrate foreign networks, gather intelligence, and disrupt enemy operations. The tools and techniques developed for these offensive operations can also be adapted for surveillance purposes, further expanding the NSA's reach. The dual nature of these capabilities – defensive and offensive – highlights the complex and often blurred lines within the intelligence community's technological endeavors.

The agency's reliance on partnerships with telecommunications companies and internet service providers (ISPs) is another crucial aspect of its operational architecture. While some of these partnerships are mandated by law, others may be voluntary or based on less transparent agreements. These collaborations provide the NSA with access to crucial infrastructure and customer data, enabling the large-scale collection that underpins its operations. The legal justifications for these requests often fall under FISA, with specific court orders authorizing the collection of certain types of data. However, the sheer volume and scope of these requests, as revealed in declassified documents, indicate a systemic reliance on private sector infrastructure for intelligence gathering.

The global network of collection points is another significant element of the NSA's infrastructure. Beyond its primary bases in the United States, the agency operates or has access to facilities worldwide, strategically positioned to intercept communications traffic. These sites leverage various technologies, including satellite dishes, ground stations, and clandestine tapping operations, to capture signals intelligence. The international nature of communication networks means that intelligence gathering often requires a global presence, and the NSA's extensive network of facilities and partnerships is a testament to this reality. This global footprint allows the agency to intercept a vast array of electronic communications, from international phone calls and emails to satellite transmissions and data flowing through undersea cables.

The analytical side of the NSA is as critical as its collection capabilities. Raw data, no matter how vast, is useless without sophisticated analytical tools and skilled personnel to interpret it. The agency employs advanced algorithms, artificial intelligence, and machine learning to identify patterns, anomalies, and connections within the collected data. These analytical capabilities are designed to sift through terabytes of information to identify potential threats, track individuals of interest, and provide insights into the activities of foreign adversaries. The development of these analytical tools is a continuous process, driven by the need to process increasingly large volumes of data and to identify more subtle indicators of malicious activity.

The legal and ethical debates surrounding the NSA's activities are ongoing and complex. While proponents argue that its surveillance powers are essential for national security, critics raise serious concerns about privacy, civil liberties, and the potential for government overreach. The incidental collection of U.S. persons' data, the broad interpretation of national security justifications, and the lack of transparency in many of the agency's operations are all points of contention. The revelations of the past decade have intensified these debates, leading to calls for greater oversight, accountability, and reform of the legal frameworks governing intelligence collection. The NSA, therefore, operates not only within a technological and operational framework but also within a complex legal and societal landscape, where the pursuit of security is constantly weighed against the protection of fundamental rights. Its role as a central pillar of the surveillance infrastructure is thus intrinsically linked to these ongoing societal dialogues about the appropriate limits of state power in the digital age. The sheer breadth of its mandate and the technological sophistication it employs mean that understanding the NSA is crucial to understanding the architecture of modern surveillance.

The architecture of the United States' surveillance infrastructure is not a monolithic entity, but rather a complex, interconnected ecosystem where various agencies and their capabilities are interwoven. Central to this operational framework is the principle of interagency cooperation and the seamless flow of data between entities charged with different facets of national security and law enforcement. This intricate web of collaboration is designed to leverage the unique strengths and mandates of each organization, creating a synergistic effect that amplifies the collective intelligence-gathering and analytical power of the U.S. government. The National Security Agency (NSA), with its unparalleled capabilities in signals intelligence (SIGINT), often acts as a foundational data provider, but its contributions are amplified and contextualized through shared intelligence with a multitude of other government bodies.

The Federal Bureau of Investigation (FBI), for instance, plays a crucial role in domestic investigations and counterintelligence operations. While the NSA's primary focus is on foreign intelligence, the FBI operates with a domestic jurisdiction, tasked with investigating federal crimes, including terrorism, espionage, and cybercrime within the United States. The synergy between the NSA and the FBI is particularly evident in counterterrorism efforts. Information gathered by the NSA through electronic surveillance, such as communications intercepts or metadata analysis related to suspected foreign terrorist organizations, can be crucial for the FBI's investigations into potential domestic links or planned attacks on U.S. soil. Conversely, the FBI's extensive network of informants, its ability to conduct physical surveillance, and its domestic investigative powers can provide valuable context or leads that might inform the NSA's foreign intelligence collection priorities. This information sharing is typically facilitated through formal channels and agreements, ensuring that data is disseminated to the agencies best positioned to act upon it, while also attempting to adhere to legal and policy guidelines that delineate the boundaries between foreign intelligence and domestic law enforcement.

Similarly, the Central Intelligence Agency (CIA), primarily responsible for foreign intelligence gathering and covert operations, also engages in significant data sharing with the NSA. The CIA's human intelligence (HUMINT) operations, which involve clandestine sources and methods to collect information on foreign governments, organizations, and individuals, can provide critical insights that complement the NSA's signals intelligence. For example, a human source within a foreign government might provide information about a planned cyberattack, which the NSA could then attempt to corroborate or track through intercepted communications or network activity. The NSA, in turn, might identify communication patterns or networks associated with this threat, which could then be used by the CIA to identify potential human sources or to refine its intelligence collection strategy. This cross-pollination of intelligence disciplines – SIGINT from the NSA and HUMINT from the CIA – is fundamental to creating a comprehensive understanding of complex global threats.

The establishment of specific programs and frameworks has been instrumental in facilitating this interagency data flow. Following the September 11, 2001, terrorist attacks, a significant impetus was placed on breaking down traditional inter-agency stovepipes. The creation of the Department of Homeland Security (DHS) in 2003, consolidating numerous disparate agencies and their functions, was a direct response to this perceived need for greater coordination. Within DHS, various components, such as Immigration and Customs Enforcement (ICE), Customs and Border Protection (CBP), and the Transportation Security Administration (TSA), all collect and generate vast amounts of data related to individuals and their movements. This data, ranging from travel manifests and border crossing records to passenger screening information, can be valuable for other intelligence agencies. For example, information indicating that a person of interest to the NSA or FBI has traveled to or from a region of concern might be flagged and shared, providing a crucial piece of the intelligence puzzle.

A key mechanism for integrating intelligence from multiple sources and agencies is the concept of "fusion centers." These are state and major urban area fusion centers, established and operated by state and local law enforcement agencies, often with support from federal agencies like the FBI and DHS. Their primary purpose is to serve as central hubs for the collection, analysis, and dissemination of intelligence and information related to a wide range of threats, including terrorism, criminal activity, and public safety concerns. These centers are designed to bring together personnel from various federal, state, local, and tribal agencies, as well as private sector partners, to foster collaboration and information sharing. The NSA, through its intelligence feeds and analytical products, can contribute to the intelligence picture being developed at these fusion centers. Conversely, intelligence gathered by state and local law enforcement through routine policing or specific investigations can be uploaded to these centers, potentially leading to the identification of connections to federal or international threats that fall within the purview of agencies like the NSA.

The legal and operational frameworks that govern this data sharing are often complex and have evolved significantly over time. The Foreign Intelligence Surveillance Act (FISA), as amended, provides the statutory basis for certain types of electronic surveillance conducted by the U.S. government. While FISA's primary focus is on foreign intelligence collection, its application and the sharing of information obtained under its authority with domestic law enforcement agencies have been subjects of intense debate and legal scrutiny. Section 702 of FISA, for example, authorizes the U.S. government to collect foreign intelligence information by targeting non-U.S. persons reasonably believed to be located outside the United States. However, the nature of internet communications means that the communications of U.S. persons can be "incidentally" collected during such targeted surveillance. Regulations and procedures exist to govern how this incidentally collected information can be accessed, used, and disseminated by domestic agencies like the FBI. This process, often referred to as "backdoor searches," allows the FBI to query data collected by the NSA under Section 702, provided that certain criteria are met and that the query is not intended to circumvent domestic law enforcement authorities.

The intelligence community's ability to share information is also facilitated by overarching intelligence-sharing agreements and directives, such as Presidential Policy

Directive-28 (PPD-28), which governs the signaling intelligence activities of the U.S. government. PPD-28, issued under the Obama administration, aimed to provide greater transparency and accountability for SIGINT activities, while also recognizing the necessity of international cooperation and the collection of intelligence. It stipulated that U.S. persons' communications that are incidentally collected should be protected and that access to such information should be minimized and subject to appropriate oversight. However, the interpretation and implementation of these directives are crucial. The broad definition of what constitutes a "foreign intelligence purpose" or a "national security" threat can, critics argue, create loopholes that allow for a wider sharing of data than originally intended.

The establishment of the Terrorist End User Identification (TEUID) database, or similar systems, exemplifies the push for integrated information sharing. These databases aim to consolidate information about individuals suspected of involvement in terrorism, drawing data from various sources across the intelligence and law enforcement communities. The NSA's SIGINT collection would contribute to populating such databases, providing technical intelligence on communication patterns, network affiliations, or expressed intentions. This information, when combined with data from other sources, such as travel records, financial transactions, or known associates, creates a more robust profile of individuals and networks of concern.

The sharing of technical capabilities and analytical methodologies between agencies is another vital aspect of interagency cooperation. The NSA, for instance, develops highly sophisticated tools and techniques for cryptanalysis, network exploitation, and data parsing. These capabilities, refined through years of dedicated research and development, can be shared or adapted for use by other agencies. The FBI might receive specialized software developed by the NSA to analyze encrypted communications encountered in domestic investigations, or the CIA might leverage NSA's expertise in network mapping to understand the infrastructure of a foreign cyber adversary. This technological transfer ensures that the entire intelligence apparatus benefits from the cutting edge of SIGINT capabilities, even if direct access to NSA's raw collection is restricted.

The role of the Director of National Intelligence (DNI) is also central to fostering this interagency cooperation. The DNI, as the head of the U.S. Intelligence Community (IC), is responsible for overseeing and integrating the intelligence efforts of its 17 member agencies. This includes ensuring that intelligence is collected, analyzed, and disseminated effectively and that agencies work collaboratively rather than in silos. The DNI's office often issues directives and guidance on information sharing, data standards, and the development of common intelligence platforms, all aimed at improving the coherence and effectiveness of the overall intelligence enterprise.

However, this extensive interagency cooperation and data sharing, while ostensibly designed to enhance national security, raises significant questions regarding oversight and accountability. When information gathered by one agency, potentially under different legal authorities or with varying privacy protections, is shared with another, the chain of accountability can become convoluted. For instance, if the NSA collects data under Section 702 of FISA and then shares it with the FBI for a domestic investigation, who is primarily accountable if the initial collection or subsequent access to that data is found to be improper? The blurriness of these lines of responsibility can make it challenging to pinpoint where a violation occurred and to ensure that appropriate redress is available.

The sheer volume of data being shared across agencies

also presents a significant challenge for oversight bodies, such as congressional committees or internal inspector generals. Effectively reviewing the legality and propriety of every data transfer and access request across multiple agencies, each with its own set of operational procedures and legal justifications, is a monumental task. The classified nature of many of these operations further complicates oversight, as those tasked with reviewing the legality of the activities often do not have access to the full context or the granular details of the data being shared.

Furthermore, the emphasis on "all-source analysis" — the practice of integrating intelligence from all available sources — inherently requires robust interagency data sharing. The NSA's SIGINT is a critical component of this, but it must be fused with HUMINT from the CIA, imagery intelligence (IMINT) from the National Geospatial-Intelligence Agency (NGA), open-source intelligence (OSINT) from various entities, and other forms of intelligence. The success of all-source analysis depends directly on the willingness and ability of agencies to share their data and analytical products. This creates a dynamic where the imperative for comprehensive intelligence may, in practice, outweigh stricter adherence to the compartmentalization of data based on original collection authorities or privacy concerns.

The evolution of technology also plays a significant role in shaping interagency cooperation and data sharing. Cloud computing, advanced data analytics, and artificial intelligence are enabling agencies to store, process, and analyze vast datasets more efficiently. These technological advancements often facilitate the integration of data from disparate sources, making it easier to create unified intelligence platforms. However, they also introduce new challenges related to data security, privacy, and the potential for unintended data aggregation. As agencies adopt these shared technological infrastructure, the lines between their individual data holdings can become even more blurred, necessitating a constant recalibration of oversight and accountability mechanisms.

The reliance on private sector partners, particularly technology companies and telecommunications providers, further complicates the landscape of interagency cooperation. These companies are often crucial conduits for data, either through lawful intercept requests, data-sharing agreements, or as providers of the infrastructure upon which much of the nation's communications flow. Information gathered by agencies from these private entities may then be shared with other government bodies, creating multi-layered data flows that are difficult to trace and oversee comprehensively. The legal authorities and contractual agreements governing these relationships are therefore critical elements in understanding the architecture of surveillance and the mechanisms of interagency data sharing.

In conclusion, the intricate web of interagency cooperation and data sharing is a foundational element of the United States' surveillance infrastructure. Agencies like the NSA, FBI, CIA, and DHS, along with numerous other entities, contribute their unique capabilities and data to a collective intelligence enterprise. Fusion centers, specific legal frameworks like FISA, and overarching directives all serve to facilitate this data flow. While this cooperation is presented as essential for national security, it also presents significant challenges for oversight and accountability, raising complex questions about privacy and the potential for governmental overreach as information is passed between organizations with differing mandates and authorities. The ongoing technological advancements and the reliance on private sector partners further accentuate the need for robust and adaptable oversight mechanisms to ensure that the pursuit of security does not undermine fundamental rights and civil liberties.

The architecture of surveillance, as we've begun to understand, is deeply intertwined with the legal frameworks that authorize and govern its operations. These statutes act as the scaffolding upon which intelligence agencies build their capabilities, dictating what information can be collected, how it can be obtained, and to whom it can be disseminated. Central to this legal architecture, particularly for foreign intelligence and counterterrorism efforts, is the Foreign Intelligence Surveillance Act of 1978, commonly known as FISA. Enacted in the wake of revelations about abuses of government surveillance powers during the Vietnam War and Watergate era, FISA was intended to provide a statutory basis for electronic surveillance and other forms of intelligence gathering conducted for foreign intelligence purposes. It established a framework for obtaining warrants from a specialized court, the Foreign Intelligence Surveillance Court (FISC), before conducting surveillance on individuals within the United States suspected of being agents of a foreign power. This court, composed of federal judges appointed by the Chief Justice of the United States, operates in secret, hearing applications for surveillance orders submitted by agencies like the FBI, often based on information provided by the NSA.

FISA's original intent was to strike a balance between national security needs and the privacy rights of individuals, particularly U.S. persons. However, the evolving nature of technology and the persistent threat of terrorism have led to significant amendments and interpretations of the Act, expanding its reach and scope. One of the most consequential of these amendments came in the form of Section 702 of the FISA Amendments Act of 2008. This section dramatically altered the landscape of foreign intelligence surveillance by authorizing the U.S. government to target non-U.S. persons reasonably believed to be located outside the United

States for the purpose of acquiring foreign intelligence information. Crucially, Section 702 allows for the collection of communications content and metadata from a broad range of targets without individual court orders, provided that the targets are not U.S. persons located within the U.S. The NSA is the primary agency utilizing this authority, often through "ticking clock" or "dragnet" collection programs aimed at identifying and disrupting foreign threats.

The critical, and often controversial, aspect of Section 702 collection arises when the communications of U.S. persons are "incidentally" swept up in the surveillance of foreign targets. This occurs because communications often travel through networks that are subject to lawful interception, and the distinction between a foreign target and a U.S. person communicating with that target can be blurred. FISA regulations, particularly as they have been interpreted and refined over time, address this "incidental collection." While Section 702 does not permit the direct targeting of U.S. persons, it does allow for the querying of the collected data to find communications involving U.S. persons, provided that these queries are conducted for legitimate foreign intelligence or national security purposes and are not intended to circumvent domestic law enforcement prohibitions. This process, frequently referred to as "backdoor searches" by critics, allows agencies like the FBI to access U.S. person data that was initially collected under the foreign intelligence authorities of Section 702. The FBI's ability to conduct such queries has been a significant point of contention, with concerns raised about the potential for bypassing Fourth Amendment protections against unreasonable searches and seizures. The Attorney General must approve the query procedures, and the FISC reviews them for legality. However, the broad scope of foreign intelligence purposes, coupled with the sheer volume of data collected, has fueled ongoing debates about the adequacy of these safeguards and the transparency of the querying process.

Beyond FISA and its amendments, other legal tools have been instrumental in the government's ability to gather information. National Security Letters (NSLs), for instance, are a powerful administrative subpoena used by the FBI and other agencies to compel the disclosure of certain business records and other information from third parties, such as telecommunications companies and financial institutions. Unlike warrants, NSLs do not require prior judicial approval. They are issued directly by FBI field offices under specific statutory authorities, often with the sole signature of an FBI official. The information obtained through NSLs can include subscriber information, toll billing records, and electronic communication transactional records. While the government argues that NSLs are essential for rapid intelligence gathering and are only used when specific statutory criteria are met, critics have raised concerns about their potential for abuse, the lack of independent judicial oversight at the point of issuance, and the gag orders that typically accompany them, preventing recipients from disclosing that they have received an NSL. The scope of information that can be obtained via NSLs has also been a subject of legal challenge and legislative adjustment over the years.

Another significant legal provision that emerged from the post-9/11 era is Section 215 of the USA PATRIOT Act. This section amended Section 215 of Title 18 of the U.S. Code, authorizing the FBI, with the approval of the FISC, to obtain from any business or person records relevant to an investigation into international terrorism or clandestine intelligence activities. The scope of "business" and "records" was interpreted broadly, leading to the government's collection of vast amounts of metadata, including bulk telephony metadata. Under this authority, the government could order telecommunications companies to provide it with records of all calls made or received by individuals within a certain geographic area or connected to certain communication networks. The government's explanation for this collection was that it allowed them to map communication networks and identify potential links to terrorist organizations. However, the interpretation of "relevance" and the authorization for bulk collection under Section 215 were vigorously debated, culminating in a landmark decision by the Second Circuit Court of Appeals, which found the program to be unauthorized by the statute. Following this ruling and subsequent legislative action, the bulk telephony metadata collection program under Section 215 was significantly curtailed, with more targeted approaches now being employed.

The legal framework surrounding government surveillance is a dynamic and often opaque area. While FISA, Section 702, NSLs, and Section 215 represent some of the most prominent legal authorities, numerous other statutes, executive orders, and interagency agreements contribute to the complex web of legal justifications for intelligence gathering. For example, the Electronic Communications Privacy Act (ECPA) governs the privacy of electronic communications and provides the legal basis for law enforcement to obtain various types of data from electronic service providers, although its provisions are distinct from those used for foreign intelligence. Executive Order 12333, issued in 1981 and subsequently amended, provides broad authority for U.S. intelligence activities conducted outside the territorial limits of the United States, often without the need for court orders, and forms a significant basis for NSA's extraterritorial surveillance.

The secrecy surrounding many of these legal authorities and their implementation makes public understanding and oversight exceptionally challenging. The FISC, by its very nature, operates in closed sessions, and its opinions and orders are often classified. This lack of transparency makes it difficult for civil liberties advocates, legal scholars, and the public to assess the full extent to which these legal frameworks are being utilized and whether they are operating within constitutional bounds. While reforms have been introduced over time, such as increased congressional oversight and mandatory declassification reviews for certain FISC opinions, the fundamental opacity of the system persists. The debate over the legal basis for surveillance is not merely academic; it has profound implications for the privacy rights of individuals, the balance of power between the executive branch and other branches of government, and the fundamental principles of a democratic society. The constant tension between the perceived necessity of robust intelligence gathering for national security and the imperative to protect civil liberties underscores the need for continued scrutiny and public discourse regarding these powerful legal tools.

The legal framework is not static; it is constantly being shaped by new technologies, evolving threats, and judicial interpretations. For instance, the rise of encrypted communications, the use of anonymizing technologies like Tor, and the global nature of the internet present ongoing challenges for intelligence agencies and, consequently, for the legal authorities designed to govern their activities. The very definition of what constitutes "foreign intelligence" or an "agent of a foreign power" can be subject to interpretation in ways that expand or contract the scope of surveillance. Moreover, the sharing of information across agencies, as discussed previously, raises complex legal questions about the proper application of authorities and the preservation of privacy protections when data moves from one legal jurisdiction or set of rules to another. The legal architecture, therefore, is not merely a set of statutes, but a complex ecosystem of laws, regulations, court decisions, and policy directives that collectively define the boundaries and capabilities of the surveillance state. The ongoing efforts to balance security and liberty, often through legislative action and judicial review, demonstrate the enduring significance of these legal underpinnings in shaping the nation's approach to intelligence gathering in the digital age. The ongoing legal battles and legislative debates surrounding these authorities highlight the profound societal implications of how these laws are written, interpreted, and applied, and they underscore the critical role of the legal framework in defining the relationship between the government and its citizens in an era of pervasive data collection.

The intricate tapestry of modern surveillance infrastructure is not woven solely by government agencies; it is, to a significant extent, a collaborative creation, powered and enabled by the private sector. The digital arteries of our global communication networks, the vast data repositories held by internet giants, and the very hardware that connects us all – these are the domains of private corporations. Without their infrastructure, services, and the data they collect and store, the capabilities of government intelligence agencies would be vastly curtailed, if not rendered practically inert in the digital age. This symbiotic, and at times coercive, relationship forms a foundational pillar of the surveillance state, transforming companies that provide essential services into unwitting, or perhaps sometimes willing, partners in intelligence gathering.

Telecommunications giants, from the colossal providers of mobile and broadband internet to the companies that manage the undersea cables and satellite networks crisscrossing the globe, represent the most fundamental layer of this infrastructure. They are the conduits through which the vast majority of digital communications flow. Legal frameworks, often rooted in national security imperatives and facilitated by the very statutes previously discussed, empower governments to compel these companies to provide access to customer data and communications content. While explicit warrants are typically required for domestic law enforcement surveillance, directives under national security laws, such as those derived from FISA or its related amendments, can place significant obligations on these companies concerning foreign intelligence information. These obligations may include the retention of data for specified periods, the provision of technical capabilities to facilitate lawful interception, and the disclosure of subscriber information and metadata. The sheer scale of their operations means that even targeted requests, when aggregated, can yield an immense volume of information about user behavior, location, and social connections.

Internet service providers (ISPs), whether they be global behemoths or smaller regional players, are equally critical. They are the gatekeepers to the internet for billions of users, managing the flow of data packets and maintaining extensive records of user activity. Under various legal authorities, ISPs can be compelled to disclose information such as IP addresses associated with user accounts, connection logs, and even, in some circumstances, the content of communications that pass through their networks. The legal mechanisms for obtaining this data can range from traditional warrants based on probable cause to less stringent administrative subpoenas or national security letters, depending on the legal context and the nature of the information sought. The infrastructure they manage is not merely passive; it actively routes, caches, and often logs vast quantities of data, making it an indispensable resource for intelligence agencies seeking to monitor online activities.

The rise of social media platforms and other large online service providers has introduced another, perhaps even more pervasive, layer to the surveillance architecture. These companies collect and analyze an unprecedented amount of personal data, encompassing not only communications content but also social graphs, user preferences, demographic information, and behavioral patterns. The terms of service that users agree to, often without deep understanding, grant these platforms broad rights to collect, process, and utilize this data, typically for advertising and service improvement. However, these same data troves are of immense interest to intelligence agencies. Legal instruments like Section 702 of FISA, which permits the targeting of non-U.S. persons outside the United States for foreign intelligence purposes, have proven particularly effective in accessing data held by these global platforms. When a targeted foreign individual communicates with individuals in the United States, or uses services hosted by companies operating within the U.S. jurisdiction, their communications and associated metadata can be collected. The subsequent querying of this aggregated data for information pertaining to U.S. persons, as discussed previously, highlights the complex interplay between foreign intelligence collection and domestic privacy concerns, with these platforms acting as crucial repositories.

Beyond these core infrastructure providers, the ecosystem of data brokers and analytics companies represents a more opaque, yet equally significant, dimension of the surveillance landscape. These companies specialize in aggregating, processing, and selling vast datasets derived from a multitude of sources, including public records, commercial transactions, online activities, and even data purchased from other companies. While their primary business model is often commercial, the data they amass – often anonymized or pseudonymized, but frequently re-identifiable – provides intelligence agencies with rich profiles of individuals, both domestically and internationally. The legal basis for government access to this data can be more varied, sometimes relying on contractual agreements, public domain information, or specific statutory authorities that permit the acquisition of commercially available data for national security purposes. The line between legitimate commercial data aggregation and intelligence gathering can become exceedingly blurred in this sector, as agencies can leverage the commercial market to acquire data that might otherwise be protected or difficult to obtain directly.

The relationship between government intelligence needs and private corporations is not solely one of compelled compliance. Financial incentives, contractual agreements, and even co-development of technologies often play a significant role. Government contracts, particularly those related to defense and national security, can provide substantial revenue streams for technology companies. This can create a vested interest in maintaining cooperative relationships and developing capabilities that align with intelligence agency requirements. Furthermore, the development of new surveillance technologies, from sophisticated data analysis software to specialized hardware for signal interception, is frequently driven by innovation within the private sector. Intelligence agencies may then procure these technologies or collaborate with companies on their development, effectively outsourcing certain aspects of surveillance capability building. This can lead to a situation where private companies are not merely providing access to existing infrastructure but are actively designing and implementing new tools for government surveillance.

The legal frameworks that govern this relationship are designed to facilitate cooperation, but they also create a complex web of obligations and potential liabilities for corporations. Companies are often subject to strict confidentiality requirements, preventing them from disclosing the extent of their cooperation with government agencies. This silence, mandated by law, makes it difficult for the public and even for shareholders to fully understand the role these companies play in the surveillance infrastructure. The legal justifications for data requests can be broad, and the interpretation of terms like "foreign intelligence information" or "national security" can evolve, potentially expanding the scope of data that companies are obligated to provide. This creates a dynamic environment where corporate compliance is a continuous process of navigating evolving legal demands and technological capabilities.

Moreover, the global nature of the technology sector means that data is often collected, processed, and stored across multiple jurisdictions with varying legal protections. A U.S. intelligence agency's request for data held by a company with servers in Europe, for instance, can trigger complex legal negotiations and jurisdictional challenges. While U.S. national security laws might provide a basis for such requests, data privacy regulations in other countries, such as the GDPR, may impose different obligations and restrictions. This international dimension adds another layer of complexity to the architecture of surveillance, as it involves not just the direct relationship between a government and a corporation, but also the interplay of international law, bilateral agreements, and the extraterritorial reach of national statutes. Companies operating on a global scale must therefore contend with a patchwork of legal requirements, making compliance a significant operational challenge.

The financial incentives for cooperation can also manifest in more subtle ways. The demand for data analytics, cybersecurity, and communication services from government entities creates a lucrative market for technology firms. Companies that can demonstrate a willingness and ability to meet the unique needs of intelligence agencies, including the provision of data access and analytical tools, may find themselves in a more advantageous competitive position. This can create a subtle pressure to align business practices with perceived government expectations, even in the absence of direct legal compulsion. The ongoing evolution of data-centric business models, where data itself is a primary commodity, further intertwines corporate interests with the data-gathering ambitions of states.

The process by which governments obtain data from private entities is often characterized by a degree of opacity. While some legal frameworks, like the warrant process for domestic law enforcement, involve judicial oversight, many national security-related requests operate with less public visibility. National Security Letters (NSLs), as mentioned earlier, are administrative subpoenas that do not require judicial approval at the point of issuance, and their use is often accompanied by gag orders. Similarly, directives issued under authorities like Section 702 can be based on certifications from the Attorney General and reviews by the Foreign Intelligence Surveillance Court (FISC), but the operational details and the full extent of data acquisition remain classified. This lack of transparency means that the public has limited insight into the precise nature and scale of the data that private corporations are providing to government intelligence agencies.

The reliance on private sector infrastructure also raises questions about accountability and oversight. When data is collected and analyzed by private companies, even under government direction, the lines of responsibility can become blurred. Who is ultimately accountable for privacy violations or erroneous data collection – the government agency, the contracting company, or the employees of either? The complex contractual relationships, the use of subcontractors, and the proprietary nature of much of the technology involved can create a diffusion of accountability, making it challenging to pinpoint responsibility when things go wrong. This diffusion of accountability is a critical concern for civil liberties advocates, who argue that the outsourcing of surveillance functions to private entities can, in turn, outsource accountability.

Furthermore, the commercialization of data collection and analysis by private firms can sometimes outpace or circumvent existing legal and ethical safeguards. As companies develop increasingly sophisticated methods for collecting, aggregating, and monetizing data, governments can readily leverage these capabilities for intelligence purposes. This can lead to a situation where the technology itself drives the expansion of surveillance, with legal frameworks struggling to keep pace. The continuous innovation within the private technology sector means that the infrastructure of surveillance is constantly evolving, presenting ongoing challenges for oversight and regulation. The sheer volume and variety of data being generated and collected by private entities – from browsing history and location data to biometric information and social media interactions – provide a rich and readily accessible resource for intelligence agencies, transforming the private sector into a de facto data-gathering arm of the state.

The legal frameworks that compel or facilitate this cooperation are diverse and often interlinked. While FISA and its amendments provide a significant basis for foreign intelligence collection, other statutes and authorities may also be invoked, depending on the nature of the data and the context of the request. For instance, the Stored Communications Act (SCA), part of the Electronic Communications Privacy Act (ECPA), governs government access to data held by electronic communication service providers, setting out different standards and procedures depending on the type of data and its location (e.g., whether it is stored or in transit). The application of these different legal tools to data held by private entities highlights the complexity of the legal landscape and the varying levels of protection afforded to different types of information.

The financial motivations of these corporations are also a crucial factor. While some may comply with government requests out of a sense of civic duty or legal obligation, the economic realities of the technology sector cannot be ignored. Government contracts and partnerships can represent significant sources of revenue, incentivizing companies to develop and maintain capabilities that align with intelligence agency needs. Moreover, companies that can effectively manage and secure large volumes of data, and provide access to it in a structured and timely manner, may be seen as more valuable partners by government entities. This creates a powerful economic incentive for cooperation, further embedding private corporations within the surveillance architecture.

The debate over the role of private corporations in surveillance infrastructure is therefore multifaceted, encompassing legal, economic, and ethical dimensions. It questions the extent to which essential public services have become conduits for state surveillance, the adequacy of transparency and accountability mechanisms, and the fundamental implications for privacy and civil liberties when the private sector's data-gathering capabilities are leveraged for national security purposes. The very nature of the digital economy, driven by data collection and analysis, has inadvertently or deliberately created the fertile ground upon which modern surveillance states can flourish, making the actions and obligations of private corporations a central, and often controversial, element of this evolving landscape.

The intricate web of modern surveillance is not confined by national borders; it is a globally interconnected system, orchestrated and utilized by intelligence agencies operating far beyond their domestic territories. While previous discussions have illuminated the foundational role of domestic private sector infrastructure in enabling surveillance, this section pivots to the extraterritorial reach of these capabilities, examining how governments, most notably the United States through agencies like the National Security Agency (NSA), engage in the systematic collection of data that originates in, transits through, or pertains to individuals and entities in other nations. This global architecture of surveillance is built upon a sophisticated understanding of the world's digital infrastructure – the undersea cables, the internet backbone, the satellite networks, and the data centers that form the nervous system of global communication.

Central to this global data collection is the concept of "upstream" collection. This method involves tapping into the physical infrastructure of the internet itself, specifically the high-capacity fiber optic cables and other conduits that carry vast amounts of internet traffic between countries. Agencies like the NSA, through programs that have been revealed over time, have developed the capacity to intercept this traffic at crucial chokepoints. These chokepoints are often located at facilities that act as major transit points for internet data, such as internet exchange points (IXPs) or the landing stations where undersea cables emerge from the ocean floor. By gaining access to these locations, often through partnerships or covert means, intelligence agencies can collect a broad spectrum of communications data, including emails, instant messages, web browsing activity, and voice-over-IP communications, as they traverse international boundaries. This is not merely about targeted intercepts; it is about capturing a significant portion of the data flowing across these critical arteries. The sheer volume of information processed at these points means that even a fraction of it, when collected systematically, can yield an immense trove of intelligence.

The Prism program, famously brought to light by Edward

Snowden, offered a stark illustration of how these global networks are leveraged. Prism is designed to obtain data directly from the servers of major U.S. internet companies, including those that operate globally and hold data on users worldwide. While Prism primarily targets communications of non-U.S. persons located outside the United States for foreign intelligence purposes, the architecture it taps into has profound implications for individuals in other nations. The companies involved are global giants, with user bases that span continents. When these companies store or process data pertaining to individuals in various countries, that data becomes accessible under the provisions of programs like Prism, provided the legal criteria are met. The program's ability to query this vast repository of communications data, stored by companies with a global presence, means that the surveillance net can extend far beyond the immediate geographical scope of the targeting authority. It transforms global service providers into unintentional conduits for intelligence gathering on a worldwide scale.

The legal justifications for such extraterritorial collection are often rooted in national security imperatives and specific legal frameworks, such as Section 702 of the Foreign Intelligence Surveillance Act (FISA) in the United States. Section 702 allows for the acquisition of foreign intelligence information from non-U.S. persons who are reasonably believed to be located outside the United States. However, the practical application of this law, particularly in conjunction with upstream collection and the use of global platforms, creates a complex situation. When a communication involves individuals in different countries, or when data is hosted on servers located in the U.S. but pertains to individuals abroad, the lines of jurisdiction and privacy become blurred. The potential for incidental collection of communications of U.S. persons, or individuals in allied nations, through broad data sweeps is a significant concern.

The global nature of the internet means that data is not static; it moves across continents, is stored in multiple locations, and is routed through various networks. This inherent mobility of digital information makes it susceptible to collection at numerous points. For instance, a seemingly innocuous email sent between two individuals in France might transit through a server in the United States or be routed via an undersea cable that is monitored. The legal basis for intercepting this communication in such a scenario can become a contentious issue, particularly when it involves multiple jurisdictions with differing legal standards for privacy and surveillance. The extraterritorial application of domestic surveillance laws, while often justified on grounds of national security and the need to gather intelligence on foreign threats, raises fundamental questions about national sovereignty and the digital rights of individuals worldwide.

International agreements and collaborations play a crucial role in enabling and facilitating these global surveillance networks. Intelligence-sharing alliances, such as the Five Eyes (comprising the United States, the United Kingdom, Canada, Australia, and New Zealand), have established frameworks for cooperation in intelligence gathering and analysis. These alliances often involve reciprocal sharing of intercepted communications data, effectively creating a shared intelligence commons. While these arrangements are primarily focused on mutual security interests, they also mean that data collected by one member nation, often utilizing its own technological capabilities and legal authorities, can be shared with and utilized by other member nations. This can result in a situation where a nation's citizens might be subjected to surveillance by a foreign intelligence agency, even if that agency does not have direct authority over them within their own borders, all facilitated by these collaborative agreements.

The legal architecture supporting these global operations is multifaceted. Beyond FISA, other domestic laws and international agreements can empower intelligence agencies to collect data across borders. For example, bilateral agreements concerning law enforcement and intelligence cooperation can create pathways for data sharing and joint operations. However, the scope and transparency of these agreements vary significantly. The growing reliance on cloud computing further complicates the jurisdictional landscape. Data stored by global cloud providers might reside in data centers located in countries different from where the user is located or where the cloud provider is headquartered. This creates a complex legal matrix where access requests can involve multiple national laws, data protection regulations (such as the GDPR in Europe), and international treaties.

The implications of these global data collection networks for the privacy of individuals worldwide are profound. For non-U.S. persons, the direct impact is clear: their communications and online activities can be intercepted and analyzed by a foreign government. For U.S. citizens, the situation is more nuanced. While U.S. law generally protects the communications of U.S. persons, the extraterritorial collection of data, particularly through upstream collection or by accessing data held by global companies, can result in the incidental or even programmatic acquisition of information pertaining to U.S. citizens when their communications interact with foreign systems or individuals. The "backdoor search" loophole, where data collected under foreign intelligence authorities is subsequently queried for information about U.S. persons without a warrant, highlights this concern. When intelligence agencies collect vast amounts of data globally, the likelihood of capturing data related to their own citizens, even if not the primary target, increases significantly.

The development of specialized technologies by defense contractors and technology firms further fuels this global surveillance infrastructure. Companies that design and build network infrastructure, encryption technologies, and data analytics platforms often collaborate with government intelligence agencies, either through direct contracts or by developing products with built-in capabilities that can be leveraged for intelligence purposes. The very design of the internet and its global backbone, while enabling unprecedented connectivity, also presents vulnerabilities that can be exploited for surveillance. The commercial imperatives driving innovation in this sector, coupled with national security demands, create a powerful synergy that supports the expansion of global data collection capabilities.

The secrecy surrounding many of these global operations makes it exceedingly difficult to ascertain their full scope and impact. While revelations from whistleblowers have provided glimpses into these vast networks, much remains classified. This lack of transparency hinders public debate and oversight, making it challenging to assess whether these surveillance activities are proportionate, necessary, and in compliance with evolving international norms and expectations regarding privacy and human rights. The global nature of the digital economy means that data flows seamlessly across borders, but the legal and ethical frameworks governing surveillance often struggle to keep pace with this reality.

Moreover, the reliance on upstream collection and partnerships with telecommunications and internet infrastructure providers in various countries highlights issues of national sovereignty. When a foreign intelligence agency can effectively tap into a nation's communication infrastructure, it raises questions about a nation's ability to protect the privacy of its own citizens and control the flow of information within its borders. While intelligence-sharing agreements are often presented as tools for mutual security, they can also create dependencies and empower foreign intelligence entities to operate within or upon the digital territories of other states. This can lead to a perception, and indeed a reality, of intrusive extraterritorial surveillance, potentially undermining trust between nations and fueling concerns about a global surveillance state.

The technological underpinnings of global surveillance are constantly evolving. As encryption methods become more sophisticated, intelligence agencies develop new techniques and tools to circumvent or exploit them. The increasing use of mobile devices, the Internet of Things (IoT), and the vast amounts of data generated by these connected devices present new frontiers for data collection. These technologies, often developed and deployed by global corporations, operate across national boundaries, creating a complex and ever-shifting landscape for surveillance. The challenge lies in understanding how these global technological trends intersect with the legal and policy frameworks governing data collection and privacy, particularly in an era where national security concerns often drive the expansion of surveillance capabilities. The architecture of global surveillance is, therefore, a dynamic interplay of technology, law, international relations, and economic interests, all converging to create a pervasive system of data harvesting that spans the planet.

3: THE MACHINERY OF MASS SURVEILLANCE

The digital realm, often perceived as a mere conduit for communication, is in reality a vast repository of information, much of which is meticulously cataloged and stored. Beyond the explicit content of our emails, messages, and calls lies a layer of data so ubiquitous and revealing that it has become the cornerstone of modern surveillance operations: metadata. This is not the substance of our conversations, but rather the contextual information that surrounds them, forming a detailed blueprint of our interactions and, by extension, our lives. Understanding metadata collection is pivotal to grasping the sheer scale and intimate invasiveness of contemporary surveillance machinery.

At its core, metadata is simply "data about data." In the context of telecommunications and digital communications, this translates into information that describes the attributes of a communication event. Consider a simple phone call. The content is the spoken words exchanged between two individuals. The metadata, however, would include precisely who initiated the call, the number they called, the date and time of the call, its duration, and potentially the location from which the call was made. Similarly, for an email, metadata would encompass sender and recipient addresses, subject lines, timestamps for when it was sent and received, and the IP addresses involved in its transit. For internet browsing, it includes the websites visited, the duration of each visit, the IP addresses of the servers accessed, and the times of access. This information, seemingly dry and technical, forms the bedrock upon which extensive analytical capabilities are built.

The sheer volume of this data is staggering. Every digital interaction, from a text message to a social media post, from a website visit to a GPS location ping from a mobile device, generates metadata. This data is collected and aggregated by telecommunications companies, internet service providers (ISPs), social media platforms, and a host of other entities that facilitate our online and offline lives. These companies, often compelled by legal obligations or voluntary cooperation with government agencies, are essentially the initial collectors and custodians of this massive metadata stream. They possess the infrastructure to capture, store, and process this information on an industrial scale, often for their own operational needs, such as billing, network management, and service improvement, but increasingly as a direct resource for intelligence agencies.

The significance of metadata for surveillance agencies lies in its ability to paint a comprehensive picture of an individual's activities, relationships, and habits without ever needing to access the actual content of their communications. This is the essence of its power and its danger. By analyzing the metadata of a person's communications, investigators can discern patterns that might otherwise remain hidden. For instance, knowing who a person communicates with, how frequently, at what times, and for how long can reveal the nature of their relationships—whether they are close friends, professional contacts, or casual acquaintances. The timing and duration of calls or messages can indicate daily routines, work schedules, or even periods of intense personal engagement. Location data, often embedded within mobile phone metadata, can map out a person's movements, revealing where they live, work, socialize, and travel.

This relational and temporal mapping can be incredibly revealing. Imagine a scenario where a person communicates extensively with a known dissident at odd hours and in specific locations. This metadata, even without knowing the content of their conversations, can raise red flags and warrant further scrutiny. Conversely, a complete lack of communication with certain individuals or groups, coupled with a regular pattern of communication with others, can also reveal significant social and political affiliations. Metadata can expose clandestine meetings, identify key contacts within a network, and trace the flow of information, even if the information itself is encrypted or otherwise inaccessible. It provides the "who, what, when, where, and how often" of our digital lives, creating a searchable and analyzable tapestry of our existence.

The techniques for collecting this metadata are as varied as the sources. In many jurisdictions, laws permit telecommunication providers and ISPs to collect and retain customer data, including metadata, for a specified period. This retention can be mandated by government regulation, often under the guise of facilitating law enforcement investigations or ensuring national security. Historically, this data was primarily used for billing and network management. However, with the advent of mass surveillance programs, this stored metadata has become a primary target for intelligence gathering. Agencies can then issue warrants or subpoenas, or in some cases, access this data through broader legal authorities that permit bulk collection or programmatic access.

Programs that focus on "upstream" collection, as previously discussed, are not solely concerned with the content of data packets. They also meticulously capture the metadata associated with that traffic. Even if the content of a communication is encrypted, the metadata—the source and destination IP addresses, the timestamps, the size of the data packet—is typically unencrypted and readily available. This unencrypted metadata can still reveal a great deal. For example, knowing that a particular IP address, associated with a specific individual, communicated with a server known to host illicit content, or with another IP address linked to known criminal activity, is highly valuable intelligence. The metadata provides the investigative leads, even if the content requires further effort to obtain or is unobtainable.

The aggregation and analysis of metadata are where its true power for surveillance is unleashed. Sophisticated algorithms and analytical tools are employed to sift through the immense volumes of data, identifying patterns, connections, and anomalies. Link analysis, for instance, is a technique that visualizes relationships between entities based on their communication metadata. This can reveal hidden connections within terrorist networks, criminal organizations, or even broader social movements. By plotting who communicates with whom, these analyses can identify central figures, key facilitators, and the overall structure of a group.

Consider the implications of this data for individual privacy. A person's metadata can reveal a pattern of visiting certain health-related websites, indicating a potential medical condition they may wish to keep private. It can show frequent communication with a particular religious or political group, revealing their affiliations and beliefs. It can map out their movements to a mosque, a church, a political rally, or a doctor's office, each piece of information contributing to a detailed, often intimate, personal profile. The aggregation of these seemingly innocuous data points, over time and across different communication platforms, can create a far more comprehensive and revealing picture than any single piece of information could provide on its own.

The value of metadata is also amplified by its persistence and its ability to be correlated with other datasets. Unlike ephemeral conversations, metadata is often stored for extended periods, allowing for retrospective analysis and the identification of long-term patterns. Furthermore, metadata from telecommunications can be combined with metadata from other sources—financial transactions, social media activity, location data from apps, travel records—to build an even more robust and invasive profile. This cross-referencing capability is a hallmark of sophisticated surveillance operations, transforming disparate pieces of information into a coherent, detailed narrative of an individual's life.

The legal frameworks surrounding metadata collection vary by jurisdiction, but the trend in many countries has been towards enabling broader access for intelligence and law enforcement agencies. In some nations, bulk metadata collection is permissible, where agencies can acquire large volumes of data without needing to demonstrate probable cause for suspicion against specific individuals. This contrasts with traditional surveillance, which typically requires a warrant based on specific evidence of wrongdoing. The argument often put forward for such broad collection is that it allows agencies to identify threats and investigate them efficiently in the digital age, where individuals can operate anonymously and communicate across borders instantaneously.

The metadata itself can be incredibly revealing regarding an individual's social graph. Who are their closest contacts? Who do they reach out to during moments of crisis or celebration? Who are the people they communicate with least frequently, but perhaps for the longest durations? These questions, answerable through metadata analysis, speak volumes about an individual's personal and professional life. It can reveal family connections, romantic relationships, friendships, and even tentative acquaintanceships. The absence of communication with certain individuals can be as telling as the presence of communication.

Moreover, the temporal aspect of metadata is crucial. Knowing when communications occur can provide insights into routines and habits. Are there regular late-night calls to a particular number? Does a person frequently communicate with international contacts during specific hours? This temporal mapping can reveal work schedules, personal habits, or even patterns associated with specific activities or events. A sudden change in communication patterns— a cessation of contact with certain individuals, an increase in communication with others, or a shift in the times of communication—can be a significant indicator of a change in an individual's circumstances or activities.

The collection of location metadata from mobile devices, often mandated or requested from telecommunications providers, adds another critical dimension. This data can track an individual's movements over time, creating a detailed geographical log of their daily life. Knowing that someone visited a specific political organization's headquarters, a place of worship, a doctor's office, or the residence of a known activist can provide powerful context to their communications and activities. The combination of who they communicate with, when, and where they are during those communications creates a deeply granular understanding of their interactions and associations.

The impact of this pervasive metadata collection on civil liberties is a subject of ongoing debate. Critics argue that the ability of governments to amass such detailed records of individuals' communications and movements, even without accessing the content, constitutes a significant invasion of privacy. They contend that this constant surveillance can have a chilling effect on freedom of expression and association, as individuals may self-censor their communications and activities for fear of being monitored or misinterpreted. The potential for misuse of this data, whether for political targeting, discrimination, or other nefarious purposes, is also a significant concern.

The evolution of communication technologies continuously creates new avenues for metadata collection. The proliferation of smartphones, the Internet of Things (IoT) devices, and various messaging applications all generate metadata that can be harvested. Smart home devices, wearable technology, and connected vehicles all produce data streams that, while not always directly communication-related, often contain contextual information that functions similarly to metadata. The challenge for privacy advocates is to keep pace with the relentless technological advancements that expand the reach and detail of this data collection.

In essence, metadata collection represents a form of surveillance that is both pervasive and deeply personal. It operates by meticulously documenting the outer shell of our digital interactions, providing a wealth of information about our social connections, routines, movements, and affiliations. While it may not always reveal the spoken or written word, the patterns and associations exposed by metadata analysis can be just as revealing, if not more so, in painting a comprehensive and intimate portrait of an individual's life. This digital footprint, inadvertently or deliberately left by every interaction, has become an indispensable tool for intelligence agencies and a profound challenge to the concept of privacy in the 21st century. The sheer scale and granular detail of this collected information underscore the transformative power of metadata in the machinery of mass surveillance.

While metadata provides a skeletal outline of our digital

lives, the ultimate prize for surveillance agencies is the content itself—the actual words spoken on a phone call, the text of an email, the body of an instant message, or the posts on a social media platform. This direct interception of content represents a more invasive and, arguably, more revealing facet of mass surveillance, moving beyond the "who, what, when, and where" of metadata to the "what" in its most granular form. The capability to access and analyze the substance of private communications has been a long-standing objective of intelligence agencies, and the digital age has provided unprecedented opportunities, as well as formidable challenges, in achieving this goal.

The technical infrastructure for content interception is vast and often operates in complex partnership with the very companies that provide our communication services. Telecommunications companies and internet service providers (ISPs) are the gatekeepers to vast networks of data flow. For them to facilitate content interception, intelligence agencies typically require legal authorization, which can take various forms depending on the jurisdiction and the nature of the targeted communication. In many countries, this involves a court order or a warrant, which generally requires demonstrating probable cause that specific illegal activity is occurring or is about to occur, and that the information sought is critical to an investigation. However, the legal frameworks are often a complex and evolving landscape, with different standards for different types of communication and different types of requests.

For instance, the interception of phone calls has historically been a core component of intelligence gathering. Through agreements and legal mandates with telecommunications providers, agencies can gain access to the live voice streams of conversations. This often involves specialized equipment that connects directly to the telecommunication network infrastructure, allowing for the real-time monitoring of specific phone lines. The process is generally covert, with the telecommunications provider facilitating the diversion of the targeted call traffic to the surveillance agency. Similarly, for internet communications, such as emails and instant messages, interception can occur at various points within the network. This might involve tapping into internet backbone infrastructure, gaining access to data stored by email providers, or monitoring traffic flowing through specific networks. The legal basis for such interceptions often stems from legislation designed to allow law enforcement and intelligence agencies to conduct surveillance in the face of evolving communication technologies.

The legal justifications for content interception are often rooted in national security and the prevention of serious crime. In the United States, for example, the Foreign Intelligence Surveillance Act (FISA) allows for the electronic surveillance of individuals suspected of being agents of foreign powers, without necessarily requiring the same level of probable cause as domestic law enforcement. This distinction is crucial for intelligence agencies operating abroad or targeting foreign individuals. However, when such surveillance incidentally sweeps up the communications of U.S. citizens or residents, complex legal questions arise regarding privacy protections. The Communications Assistance for Law Enforcement Act (CALEA) in the U.S., for instance, mandates that telecommunications carriers and manufacturers of communication equipment provide assistance to law enforcement and intelligence agencies in intercepting communications. This legislation effectively codifies the obligation of these companies to build "backdoors" or facilitate access for lawful interception.

Beyond traditional phone calls and emails, the interception of online communications presents a more complex and evolving challenge. Social media platforms, messaging apps, and Voice over IP (VoIP) services all generate vast amounts of data that can be subject to interception. The technical means for this often involve working directly with the service providers. These companies, holding the vast databases of user communications, are often compelled by court orders or national security directives to provide access to specific user data. This can range from retrieving historical message logs to conducting real-time monitoring of active conversations. The scope and nature of these requests are often kept secret, making it difficult to ascertain the full extent of content interception occurring across various platforms.

One of the significant challenges faced by intelligence agencies in content surveillance is encryption. As communication technologies have advanced, so too have the methods used to secure them. End-to-end encryption, for example, ensures that only the sender and the intended recipient can read a message, with intermediate parties, including the service provider, unable to access the content. This poses a substantial obstacle to direct content interception. When communications are end-to-end encrypted, even if an agency can intercept the data packets, the content will be unreadable without the decryption key, which is held by the users. This has led to a continuous technological arms race, with intelligence agencies seeking ways to circumvent or weaken encryption, while privacy advocates and technology companies work to strengthen it.

Efforts to overcome encryption have taken several forms. One approach involves targeting the devices themselves. If an agency can gain unauthorized access to a user's device —whether through malware, physical access, or exploiting software vulnerabilities—they can potentially decrypt communications before they are encrypted or after they are decrypted on the device. This often involves sophisticated hacking techniques and zero-day exploits. Another strategy involves leveraging vulnerabilities in the implementation of encryption protocols or the surrounding ecosystem. For example, if a service provider that offers encrypted communications is compelled to hand over its decryption keys, or if weaknesses are found in the servers used to manage those keys, this can compromise the security of the encrypted communications.

Furthermore, intelligence agencies often seek to exploit the metadata, as discussed previously, to identify individuals and communications that warrant further, more intrusive surveillance. Even if the content is encrypted, the metadata can reveal who is communicating with whom, when, and from where. This information can then be used to justify targeted requests for content interception, either by compelling service providers to decrypt communications on their end (if not end-to-end encrypted) or by seeking other means to access the content. The debate over "lawful access" or "backdoors" in encrypted communications is a significant area of contention. Proponents argue that such mechanisms are necessary for combating terrorism and serious crime, while critics contend that creating such vulnerabilities compromises the security of all users and opens the door to potential abuse by governments or malicious actors.

The legal and ethical considerations surrounding content interception are profound. Unlike metadata, which can be seen as a record of communication, content is the direct expression of thought, opinion, and personal life. The ability of the state to access and analyze this content without the explicit consent or knowledge of the individuals involved raises fundamental questions about privacy, freedom of expression, and the balance of power between the individual and the state. The potential for misuse is immense; content surveillance can be used to suppress dissent, monitor political opponents, or even coerce individuals through the revelation of sensitive personal information.

The scale of content interception, while often cloaked in secrecy, is believed to be significant. Programs like PRISM, revealed by Edward Snowden, demonstrated how intelligence agencies could compel U.S. technology companies to provide access to user data, including the content of emails, instant messages, and other communications, under FISA Section 702. While these programs are ostensibly targeted at foreign intelligence gathering, the sheer volume of data collected means that communications of individuals outside the scope of initial suspicion can be incidentally acquired. The legal justifications for such incidental collection are often complex, and the oversight mechanisms in place are frequently criticized as being insufficient to protect civil liberties.

The technical means of interception are constantly evolving. As new communication platforms emerge, intelligence agencies must adapt their methods. The rise of ephemeral messaging apps, where messages disappear after a short period, presents a particular challenge, as there may be no stored content to retrieve. This necessitates real-time interception capabilities. Similarly, the increasing use of peer-to-peer communication systems and decentralized networks can make it harder to identify central points for interception.

The operational procedures for content surveillance are highly sophisticated. Agencies employ specialized analysts who sift through vast quantities of intercepted communications, looking for patterns, keywords, and indicators of illegal activity or national security threats. The sheer volume of data collected means that automated tools and artificial intelligence are increasingly used to help filter and analyze the content, identifying potential leads that human analysts can then investigate further. This reliance on algorithms and automated systems also raises concerns about potential biases and errors, which could lead to misidentification or the unjust targeting of individuals.

The legal frameworks governing content interception are often a patchwork of statutes and court interpretations that have struggled to keep pace with technological change. While some jurisdictions have robust legal safeguards and oversight mechanisms, others have more permissive laws that allow for broader surveillance powers. The extraterritorial reach of surveillance also presents a significant challenge, with agencies in one country potentially intercepting the communications of individuals in another country, often with limited legal recourse for the affected individuals.

In conclusion, the direct interception of communication content is a critical component of mass surveillance. It moves beyond the contextual clues of metadata to access the substance of our private lives. While legal authorizations and technical partnerships with service providers are typically involved, the scale and scope of these operations, coupled with the ongoing challenges posed by encryption, highlight the pervasive nature of contemporary surveillance and its profound implications for individual privacy and civil liberties. The ongoing efforts to circumvent encryption and the debates surrounding lawful access underscore the continuous struggle to balance national security imperatives with fundamental human rights.

The preceding discussion illuminated the interception of digital communications, delving into how intelligence agencies can access the content of our conversations and messages. However, the operational reality of modern surveillance extends far beyond targeted interception. A foundational element, often preceding or running parallel to direct content acquisition, is the practice of *bulk data harvesting*. This is not about picking a specific conversation to listen in on; it is about casting an enormous net across the digital ocean, collecting everything that swims within it, regardless of immediate relevance or suspicion. The sheer volume of data generated daily by billions of individuals engaging in countless online activities creates a treasure trove for those seeking to monitor populations. This subsection will dissect the mechanics, infrastructure, and profound implications of this indiscriminate data collection and storage.

At its core, bulk data harvesting is the systematic and indiscriminate collection of digital information on a massive scale. Unlike targeted surveillance, which focuses on specific individuals or communications based on predefined criteria, bulk harvesting operates on a principle of inclusivity. The objective is to acquire as much raw data as possible, with the expectation that patterns, connections, and significant information will emerge through subsequent analysis. This data can encompass a staggering array of digital footprints: the aforementioned metadata, but also the content of communications, browsing histories, location data, social media activity, financial transactions, and even biometric information, all collected and stored in vast repositories. The rationale often presented is that in an era where threats can emerge from unexpected quarters and evolve rapidly, having a comprehensive historical record of digital activity allows agencies to investigate past events, identify precursors to future threats, and understand the broader networks within which individuals operate.

The technological infrastructure required to support such widespread data acquisition is monumental, representing a significant investment in computing power, storage capacity, and network bandwidth. Data centers, often referred to as "data lakes" or "data warehouses," are the physical embodiment of this infrastructure. These are not your typical server rooms; they are sprawling complexes housing hundreds of thousands, if not millions, of servers, interconnected by high-speed networks. The energy consumption of these facilities is staggering, often rivaling that of small cities. Cooling systems must operate continuously to prevent the vast arrays of processors from overheating, and redundant power supplies ensure uninterrupted operation. Within these data centers, specialized hardware and software are deployed to ingest, process, and store the incoming data streams.

Ingestion refers to the process of collecting data from various sources and feeding it into the storage system. This often involves setting up collection points at critical junctures in the global telecommunications and internet infrastructure. For example, agreements with internet service providers (ISPs), telecommunications companies, and even major online platforms allow intelligence agencies to tap into the flow of data. This can be achieved through physical taps on fiber optic cables, access to network switches, or direct data-sharing agreements with service providers, often compelled by law or facilitated through partnerships. The diversity of data sources necessitates flexible ingestion tools capable of handling different data formats, protocols, and velocities. Imagine a system designed to pull in streaming video, email traffic, social media posts, and financial transaction logs simultaneously – each requiring specific handling and formatting.

Once ingested, the data needs to be processed and made searchable. This is where sophisticated algorithms and powerful computing resources come into play. Raw data, often in unstructured or semi-structured formats, is transformed into a more organized and analyzable state. This processing can involve cleaning the data (removing duplicates or corrupted entries), categorizing it, and indexing it for efficient retrieval. For instance, text data might be parsed to extract keywords, entities (names, places, organizations), and sentiment. Metadata associated with communications might be linked to user accounts or devices. The goal is to create a structured representation of the vast sea of raw data, enabling analysts to query it effectively.

The storage aspect of bulk data harvesting is perhaps the most defining characteristic. Unlike targeted surveillance where data is often collected with a specific investigative purpose and retained for a limited period, bulk harvesting entails the long-term retention of massive datasets. This is driven by the principle that the value of data increases over time, as it can be correlated with future events or used to establish historical patterns. The sheer scale means that exabytes (one exabyte is a billion gigabytes) of data are routinely stored. This necessitates the development of advanced storage technologies, including distributed file systems, object storage, and highly scalable databases. The cost of maintaining these vast digital archives is immense, encompassing not only hardware and electricity but also the software licenses, maintenance, and the skilled personnel required to manage them.

The retrospective search capability is the ultimate purpose of this massive data accumulation. With the data stored and processed, analysts can then query these digital archives to identify connections, anomalies, or specific pieces of information that might be relevant to ongoing investigations or future intelligence requirements. These queries can be incredibly complex, involving the cross-referencing of data from multiple sources over extended periods. For example, an analyst might search for all communications between two individuals over a five-year period, cross-referenced with their online activities, travel records, and financial transactions, all within minutes. This capability transforms static data into a dynamic tool for intelligence gathering, allowing for a depth of analysis that was previously unimaginable.

The implications of such indiscriminate data collection

are far-reaching. From a civil liberties perspective, the knowledge that virtually all our digital interactions are being recorded and stored indefinitely raises profound concerns. The potential for misuse, whether intentional or accidental, is significant. Data breaches at these massive storage facilities could expose an unprecedented amount of personal information, leading to identity theft, blackmail, or reputational damage on a massive scale. Furthermore, the mere existence of such comprehensive surveillance infrastructure can have a chilling effect on freedom of expression and association, as individuals may self-censor their communications and activities for fear of being misinterpreted or flagged by automated systems.

The legal justifications for bulk data harvesting often revolve around national security and the need to stay ahead of evolving threats. In many jurisdictions, legislation has been enacted or interpreted to permit the collection and retention of data that is deemed "necessary and proportionate" for such purposes. However, the definition of what constitutes "necessary and proportionate" in the context of mass surveillance is a subject of intense debate. Critics argue that the indiscriminate nature of bulk harvesting inherently violates privacy principles, as it collects data on individuals who pose no threat. The argument is often made that if intelligence agencies have specific reasons to suspect someone, they should seek targeted warrants or authorizations rather than collecting data on entire populations and then sifting through it.

The practical implementation of bulk data harvesting involves a complex interplay of technology, law, and policy. The algorithms used for data processing and analysis are not neutral; they are designed with specific objectives and can embed biases. For instance, algorithms trained on data that disproportionately represents certain demographic groups might unfairly flag individuals from those groups. The development and deployment of these analytical tools are often shrouded in secrecy, making it difficult to assess their accuracy, fairness, and potential for error.

Furthermore, the notion of "incidental collection" is a critical aspect of bulk data harvesting. When vast datasets are collected, it is inevitable that the communications and activities of individuals who are not the primary targets of surveillance will be captured. The legal frameworks governing how this incidentally collected data can be handled, retained, and analyzed are often complex and vary significantly between jurisdictions. In some cases, there are strict rules about identifying and purging data pertaining to citizens or residents of the collecting country, while in others, the threshold for retaining such data may be lower.

The sheer scale of data storage also presents challenges related to data management and governance. Ensuring the integrity and security of these massive datasets over long periods requires robust systems and protocols. Metadata about the data itself – its origin, collection date, format, and any processing that has occurred – becomes crucial for understanding and managing these archives. As technology evolves, older storage formats may become obsolete, requiring ongoing efforts to migrate data to newer, more accessible systems.

The role of third-party contractors and vendors in the operation of these data harvesting and storage systems is also significant. The private sector often plays a crucial role in developing and maintaining the hardware and software infrastructure required for mass surveillance. This reliance on private companies raises questions about accountability, oversight, and the potential for commercial interests to influence government surveillance practices. The outsourcing of data management and analysis can create additional layers of complexity in ensuring that data is handled responsibly and in accordance with legal and ethical standards.

The evolution of data harvesting practices is closely tied to advancements in artificial intelligence and machine learning. These technologies enable the automated analysis of vast datasets, allowing for the identification of patterns and anomalies that might be missed by human analysts. Machine learning algorithms can be trained to recognize specific behaviors, detect deviations from normal patterns, and even predict future actions. While these tools offer significant advantages in managing and extracting insights from enormous volumes of data, they also introduce new challenges related to transparency, explainability, and the potential for algorithmic bias. The "black box" nature of some AI systems means that it can be difficult to understand precisely why a particular piece of data was flagged or why a certain conclusion was reached, raising concerns about due process and accountability.

In essence, bulk data harvesting represents the technological and strategic foundation upon which much of modern mass surveillance is built. It is the creation of a digital memory, a vast repository of information that can be accessed and analyzed retrospectively. The infrastructure required is immense, the process is indiscriminate, and the implications for privacy and civil liberties are profound. The continuous accumulation and storage of data, facilitated by ever-advancing technology, underscore the pervasive and enduring nature of state-level surveillance capabilities in the digital age, creating a persistent state of data collection that touches almost every aspect of our online lives. The challenge lies not just in the collection itself, but in the long-term stewardship and responsible use of these unprecedented archives of human activity.

The preceding discussion illuminated the interception

of digital communications, delving into how intelligence agencies can access the content of our conversations and messages. However, the operational reality of modern surveillance extends far beyond targeted interception. A foundational element, often preceding or running parallel to direct content acquisition, is the practice of *bulk data harvesting.* This is not about picking a specific conversation to listen in on; it is about casting an enormous net across the digital ocean, collecting everything that swims within it, regardless of immediate relevance or suspicion. The sheer volume of data generated daily by billions of individuals engaging in countless online activities creates a treasure trove for those seeking to monitor populations. This subsection will dissect the mechanics, infrastructure, and profound implications of this indiscriminate data collection and storage.

At its core, bulk data harvesting is the systematic and indiscriminate collection of digital information on a massive scale. Unlike targeted surveillance, which focuses on specific individuals or communications based on predefined criteria, bulk harvesting operates on a principle of inclusivity. The objective is to acquire as much raw data as possible, with the expectation that patterns, connections, and significant information will emerge through subsequent analysis. This data can encompass a staggering array of digital footprints: the aforementioned metadata, but also the content of communications, browsing histories, location data, social media activity, financial transactions, and even biometric information, all collected and stored in vast repositories. The rationale often presented is that in an era where threats can emerge from unexpected quarters and evolve rapidly, having a comprehensive historical record of digital activity allows agencies to investigate past events, identify precursors to future threats, and understand the broader networks within which individuals operate.

The technological infrastructure required to support such widespread data acquisition is monumental, representing a significant investment in computing power, storage capacity, and network bandwidth. Data centers, often referred to as "data lakes" or "data warehouses," are the physical embodiment of this infrastructure. These are not your typical server rooms; they are sprawling complexes housing hundreds of thousands, if not millions, of servers, interconnected by high-speed networks. The energy consumption of these facilities is staggering, often rivaling that of small cities. Cooling systems must operate continuously to prevent the vast arrays of processors from overheating, and redundant power supplies ensure uninterrupted operation. Within these data centers, specialized hardware and software are deployed to ingest, process, and store the incoming data streams.

Ingestion refers to the process of collecting data from various sources and feeding it into the storage system. This often involves setting up collection points at critical junctures in the global telecommunications and internet infrastructure. For example, agreements with internet service providers (ISPs), telecommunications companies, and even major online platforms allow intelligence agencies to tap into the flow of data. This can be achieved through physical taps on fiber optic cables, access to network switches, or direct data-sharing agreements with service providers, often compelled by law or facilitated through partnerships. The diversity of data sources necessitates flexible ingestion tools capable of handling different data formats, protocols, and velocities. Imagine a system designed to pull in streaming video, email traffic, social media posts, and financial transaction logs simultaneously – each requiring specific handling and formatting.

Once ingested, the data needs to be processed and made searchable. This is where sophisticated algorithms and powerful computing resources come into play. Raw data, often in unstructured or semi-structured formats, is transformed into a more organized and analyzable state. This processing can involve cleaning the data (removing duplicates or corrupted entries), categorizing it, and indexing it for efficient retrieval. For instance, text data might be parsed to extract keywords, entities (names, places, organizations), and sentiment. Metadata associated with communications might be linked to user accounts or devices. The goal is to create a structured representation of the vast sea of raw data, enabling analysts to query it effectively.

The storage aspect of bulk data harvesting is perhaps the most defining characteristic. Unlike targeted surveillance where data is often collected with a specific investigative purpose and retained for a limited period, bulk harvesting entails the long-term retention of massive datasets. This is driven by the principle that the value of data increases over time, as it can be correlated with future events or used to establish historical patterns. The sheer scale means that exabytes (one exabyte is a billion gigabytes) of data are routinely stored. This necessitates the development of advanced storage technologies, including distributed file systems, object storage, and highly scalable databases. The cost of maintaining these vast digital archives is immense, encompassing not only hardware and electricity but also the software licenses, maintenance, and the skilled personnel required to manage them.

The retrospective search capability is the ultimate purpose of this massive data accumulation. With the data stored and processed, analysts can then query these digital archives to identify connections, anomalies, or specific pieces of information that might be relevant to ongoing investigations or future intelligence requirements. These queries can be incredibly complex, involving the cross-referencing of data from multiple sources over extended periods. For example, an analyst might search for all communications between two individuals over a five-year period, cross-referenced with their online activities, travel records, and financial transactions, all within minutes. This capability transforms static data into a dynamic tool for intelligence gathering, allowing for a depth of analysis that was previously unimaginable.

The implications of such indiscriminate data collection are far-reaching. From a civil liberties perspective, the knowledge that virtually all our digital interactions are being recorded and stored indefinitely raises profound concerns. The potential for misuse, whether intentional or accidental, is significant. Data breaches at these massive storage facilities could expose an unprecedented amount of personal information, leading to identity theft, blackmail, or reputational damage on a massive scale. Furthermore, the mere existence of such comprehensive surveillance infrastructure can have a chilling effect on freedom of expression and association, as individuals may self-censor their communications and activities for fear of being misinterpreted or flagged by automated systems.

The legal justifications for bulk data harvesting often revolve around national security and the need to stay ahead of evolving threats. In many jurisdictions, legislation has been enacted or interpreted to permit the collection and retention of data that is deemed "necessary and proportionate" for such purposes. However, the definition of what constitutes "necessary and proportionate" in the context of mass surveillance is a subject of intense debate. Critics argue that the indiscriminate nature of bulk harvesting inherently violates privacy principles, as it collects data on individuals who pose no threat. The argument is often made that if intelligence agencies have specific reasons to suspect someone, they should seek targeted warrants or authorizations rather than collecting data on entire populations and then sifting through it.

The practical implementation of bulk data harvesting involves a complex interplay of technology, law, and policy. The algorithms used for data processing and analysis are not neutral; they are designed with specific objectives and can embed biases. For instance, algorithms trained on data that disproportionately represents certain demographic groups might unfairly flag individuals from those groups. The development and deployment of these analytical tools are often shrouded in secrecy, making it difficult to assess their accuracy, fairness, and potential for error.

Furthermore, the notion of "incidental collection" is a critical aspect of bulk data harvesting. When vast datasets are collected, it is inevitable that the communications and activities of individuals who are not the primary targets of surveillance will be captured. The legal frameworks governing how this incidentally collected data can be handled, retained, and analyzed are often complex and vary significantly between jurisdictions. In some cases, there are strict rules about identifying and purging data pertaining to citizens or residents of the collecting country, while in others, the threshold for retaining such data may be lower.

The sheer scale of data storage also presents challenges related to data management and governance. Ensuring the integrity and security of these massive datasets over long periods requires robust systems and protocols. Metadata about the data itself – its origin, collection date, format, and any processing that has occurred – becomes crucial for understanding and managing these archives. As technology evolves, older storage formats may become obsolete, requiring ongoing efforts to migrate data to newer, more accessible systems.

The role of third-party contractors and vendors in the

operation of these data harvesting and storage systems is also significant. The private sector often plays a crucial role in developing and maintaining the hardware and software infrastructure required for mass surveillance. This reliance on private companies raises questions about accountability, oversight, and the potential for commercial interests to influence government surveillance practices. The outsourcing of data management and analysis can create additional layers of complexity in ensuring that data is handled responsibly and in accordance with legal and ethical standards.

The evolution of data harvesting practices is closely tied to advancements in artificial intelligence and machine learning. These technologies enable the automated analysis of vast datasets, allowing for the identification of patterns and anomalies that might be missed by human analysts. Machine learning algorithms can be trained to recognize specific behaviors, detect deviations from normal patterns, and even predict future actions. While these tools offer significant advantages in managing and extracting insights from enormous volumes of data, they also introduce new challenges related to transparency, explainability, and the potential for algorithmic bias. The "black box" nature of some AI systems means that it can be difficult to understand precisely why a particular piece of data was flagged or why a certain conclusion was reached, raising concerns about due process and accountability.

In essence, bulk data harvesting represents the technological and strategic foundation upon which much of modern mass surveillance is built. It is the creation of a digital memory, a vast repository of information that can be accessed and analyzed retrospectively. The infrastructure required is immense, the process is indiscriminate, and the implications for privacy and civil liberties are profound. The continuous accumulation and storage of data, facilitated by ever-advancing technology, underscore the pervasive and enduring nature of state-level surveillance capabilities in the digital age, creating a persistent state of data collection that touches almost every aspect of our online lives. The challenge lies not just in the collection itself, but in the long-term stewardship and responsible use of these unprecedented archives of human activity.

Moving beyond the mere collection and storage of data, the true power of mass surveillance lies in its sophisticated analytical capabilities. Once the vast oceans of digital information have been harvested, the critical task becomes sifting through this data to extract meaningful intelligence. This is where advanced analytical techniques, often powered by artificial intelligence and machine learning, come into play. These algorithms are designed to identify patterns, connections, and anomalies within the data that might indicate potential threats, criminal activity, or simply behaviors that deviate from a perceived norm. The goal is often not just to understand what has happened, but to predict what *might* happen.

One of the most prominent and controversial applications of these analytical techniques is in the realm of *predictive policing*. The premise behind predictive policing is to use historical crime data, combined with a wide array of other datasets, to forecast where and when crimes are likely to occur. Algorithms analyze factors such as past crime locations, times, socioeconomic indicators, weather patterns, and even social media sentiment to generate "hotspot" maps that direct police patrols to areas deemed at higher risk. The underlying logic is that by concentrating resources in these predicted high-crime areas, law enforcement can deter crime before it happens or respond more rapidly when it does.

However, the implementation of predictive policing algorithms is fraught with significant challenges and ethical concerns. A primary issue is the inherent bias that can be embedded within the data itself. Historical crime data often reflects existing patterns of policing, which may disproportionately target certain neighborhoods or demographic groups, particularly those in lower-income or minority communities. When algorithms are trained on this biased data, they can perpetuate and even amplify these existing disparities. This can lead to a feedback loop where increased police presence in already over-policed areas generates more data, which in turn reinforces the prediction that these areas are inherently more prone to crime, irrespective of the actual underlying reality. The result is often a cycle of surveillance and enforcement that unfairly targets marginalized populations, leading to increased arrests for minor offenses and further entrenching systemic inequalities.

For instance, consider an algorithm trained on arrest data that shows a higher incidence of drug-related arrests in a particular urban neighborhood. If this neighborhood has historically been subjected to more aggressive stop-and-frisk policies or more intensive surveillance due to socioeconomic factors, the algorithm may learn to associate that neighborhood with drug activity. Consequently, it might direct more police resources to that area, leading to more stops and arrests, even if the actual prevalence of drug use is similar in other, less-policed neighborhoods. This creates a self-fulfilling prophecy, where the prediction of crime becomes the driver of the policing that generates the data confirming the prediction. The accuracy of these predictions is also a significant concern. While proponents argue that these tools can improve efficiency, critics point to studies that have found their predictive power to be questionable, especially when accounting for the underlying biases.

Beyond policing, similar analytical techniques are deployed in counter-terrorism efforts and cybersecurity. In counter-terrorism, vast datasets from communications intercepts, travel records, financial transactions, and online activities are analyzed to identify potential threats or radicalization pathways. Algorithms might be used to flag individuals exhibiting certain communication patterns, visiting specific websites, or associating with known individuals of interest. The sheer volume of data necessitates automated analysis, but the risk of false positives – flagging innocent individuals as potential threats – is substantial. A misinterpreted social media post, a seemingly innocuous online search, or a connection to someone who is later deemed suspicious can all trigger automated alerts, potentially leading to individuals being placed on watchlists or subjected to increased scrutiny without any evidence of wrongdoing.

The challenge of defining "threat" or "suspicious behavior" is also a complex one. These are not objective, easily quantifiable metrics. They are often socially and politically constructed. When algorithms are tasked with identifying these nebulous concepts, they rely on proxies and correlations derived from the data. These proxies may not accurately reflect the actual intent or behavior of an individual. For example, an algorithm might identify a pattern of encrypted communication as suspicious, even though encrypted communication is a common tool for privacy-conscious individuals, journalists, or whistleblowers. The risk is that the systems designed to identify genuine threats inadvertently criminalize legitimate activities or associations.

In cybersecurity, analytical techniques are used to detect malicious activity on networks, identify phishing attempts, and predict potential vulnerabilities. Machine learning models can be trained on vast logs of network traffic and user behavior to distinguish between normal and anomalous activities. This can be highly effective in identifying sophisticated attacks that might evade traditional signature-based detection methods.

However, even in this domain, the potential for bias and the interpretation of data remain critical. As cybersecurity threats evolve, so too must the analytical models. The continuous need to update and retrain these models introduces challenges related to data freshness, the introduction of new biases, and the potential for adversaries to manipulate the data used for training.

The ethical implications of using data analysis to make probabilistic judgments about individuals are profound. When we allow algorithms to assess risk or predict future behavior, we are essentially outsourcing judgment to systems that may not fully comprehend the nuances of human action or the context in which it occurs. This can lead to a form of "algorithmic determinism," where individuals are preemptively categorized and treated based on statistical probabilities rather than concrete evidence of wrongdoing. This raises fundamental questions about due process, presumption of innocence, and the right to be judged as an individual rather than as a data point within a larger statistical model.

Furthermore, the transparency and explainability of these analytical systems are often severely lacking. Many advanced machine learning models, particularly deep learning networks, operate as "black boxes." It can be incredibly difficult, even for the developers, to fully understand why a particular output was generated. When these systems are used in law enforcement or national security contexts, this lack of transparency becomes a significant issue. How can an individual challenge a decision made by an algorithm if the logic behind that decision cannot be articulated or understood? This opacity makes it difficult to identify and rectify errors, challenge biases, or ensure accountability. Without knowing *how* a decision was reached, it is nearly impossible to argue that the decision was flawed, unfair, or discriminatory.

The concept of "pre-crime" – identifying and intervening in potential criminal activity before it occurs – is a long-standing ambition of law enforcement. Predictive policing and broader data analytics offer the technological means to pursue this ambition on an unprecedented scale. However, the ethical and legal boundaries of intervening in potential future actions are far from settled. When does a statistical anomaly become a sufficient basis for intervention? What level of certainty is required? And crucially, what are the consequences for individual liberty when the state begins to police not just actions, but perceived intentions or probabilities of future actions?

The aggregation of diverse datasets – from communication metadata to location tracking, from financial transactions to social media interactions – creates a rich tapestry of information that can be interwoven by sophisticated analytical tools. This allows for the construction of detailed profiles of individuals, mapping out their social networks, habits, and potentially, their vulnerabilities. While such profiling can be useful in targeted investigations, its application in mass surveillance contexts, where the purpose is often to scan entire populations for deviations from the norm, raises significant concerns about privacy invasion and the potential for misuse. An individual's complete digital history, analyzed and interpreted by algorithms, can reveal intimate details of their lives that they may not wish to share and that may not be relevant to any legitimate security concern.

The continuous refinement of these analytical techniques, driven by advances in computing power and algorithmic sophistication, means that the capacity of surveillance states to monitor and predict behavior will only increase. As more data sources become available and more sophisticated analytical models are developed, the potential for pervasive, granular surveillance grows. The challenge for civil liberties advocates, policymakers, and the public is to grapple with the profound societal implications of these technologies. It requires a critical examination of the trade-offs between security and privacy, a rigorous assessment of the effectiveness and fairness of these analytical tools, and a robust debate about the ethical limits of using data to predict and shape human behavior. The machinery of mass surveillance, fueled by bulk data harvesting, is increasingly powered by an equally formidable engine of data analysis, capable of transforming mountains of raw information into actionable intelligence, and in doing so, reshaping the very nature of privacy, freedom, and justice in the digital age.

The ever-expanding infrastructure of mass surveillance is not confined to the silent harvesting of digital communications or the exhaustive cataloging of metadata. A parallel, and increasingly formidable, front in this technological arms race involves the deployment of technologies that can identify and track individuals based on their inherent physical characteristics: facial recognition and other biometric surveillance methods. These tools move surveillance from the realm of digital footprints to the tangible, observable reality of our physical presence in the world, transforming public spaces into vast, interconnected surveillance grids. The capacity to identify individuals in real-time, without their knowledge or consent, and to link them to extensive databases, represents a profound shift in the power dynamics between the state and the individual, fundamentally altering the nature of public life and personal liberty.

At its core, facial recognition technology operates by capturing an image of a face, analyzing key features – the distance between eyes, the shape of the nose, the contour of the jawline – and converting these into a unique numerical code or "faceprint." This digital signature is then compared against vast databases of known individuals. These databases can comprise a variety of sources, including government identification records (like driver's license photos or passport images), mugshots from law enforcement databases, images scraped from social media platforms, or even surveillance footage collected over time. The sophistication of the algorithms has advanced considerably, enabling systems to operate in varied lighting conditions, from different angles, and even with partial face occlusions, though the accuracy in such challenging environments remains a significant point of contention and ongoing development.

The deployment of facial recognition surveillance spans a diverse range of environments. It is increasingly common to find cameras equipped with this technology in public transportation hubs, airports, train stations, shopping malls, city streets, and even at private events. Law enforcement agencies are at the forefront of integrating these systems into their operations, using them to identify suspects wanted for crimes, to monitor individuals on watchlists, or to track the movements of specific persons of interest within a crowd. Beyond direct law enforcement applications, the technology is also being adopted by private entities for security purposes, retail analytics, and even for personalized marketing, further entrenching its presence in everyday life. This pervasive deployment means that individuals are, often unknowingly, subjected to constant biometric identification and tracking simply by existing and moving through public spaces.

The capabilities of these systems are, in theory, extensive. Advanced facial recognition can facilitate real-time identification, allowing authorities to flag an individual as they pass by a camera. It can also be used for retrospective analysis, enabling investigators to scan hours of surveillance footage to identify everyone who appeared in a particular location at a specific time. Furthermore, it can be integrated with other surveillance systems, such as license plate readers or gait analysis tools, to create a more comprehensive picture of an individual's movements and associations. The allure for security agencies lies in the promise of enhanced efficiency, the ability to proactively identify threats, and the creation of a digital trail that can trace an individual's presence across multiple locations and over extended periods.

However, the effectiveness and reliability of facial recognition technology are far from absolute, and its limitations are critical to understanding its true impact. Accuracy rates, while improving, are still highly variable and depend heavily on a multitude of factors, including the quality of the captured image, the angle of the face, lighting conditions, and the specific algorithm used. Crucially, these systems have repeatedly demonstrated a propensity for higher error rates when identifying women, people of color, and younger or older individuals, often due to biases in the datasets on which the algorithms were trained. These datasets, frequently drawn from predominately white male populations, can lead to a disproportionately high rate of false positives (incorrectly identifying an innocent person as a match) and false negatives (failing to identify a known individual).

The consequences of such misidentifications can be severe and life-altering. An innocent person could be wrongly detained, questioned, or placed on a watchlist simply because a facial recognition system made an error. This not only causes significant personal distress and inconvenience but can also lead to unwarranted scrutiny, reputational damage, and even legal repercussions. For communities already disproportionately targeted by law enforcement, the heightened risk of misidentification due to algorithmic bias exacerbates existing inequalities and can deepen mistrust between the public and authorities. The assumption of accuracy often inherent in the deployment of such technologies can lead to a dangerous overreliance, where algorithmic output is treated as definitive truth rather than as a probabilistic suggestion that requires further human verification.

The privacy implications of pervasive biometric surveillance are profound and deeply concerning. Unlike traditional forms of surveillance that might capture conversations or digital data, facial recognition directly targets an individual's identity and physical presence. The ability to track where people go, who they meet, and what events they attend, all without their consent, erodes the fundamental right to privacy and anonymity in public life. When every face in a crowd can be scanned, identified, and logged, the very notion of a private moment in a public space begins to dissolve. This constant, unobtrusive observation can create a chilling effect on freedom of assembly and expression. Knowing that one's presence at a protest, a political rally, a religious gathering, or even a sensitive medical appointment could be recorded and potentially linked to their identity might deter individuals from participating in legitimate civic activities or seeking support for personal matters.

This ubiquitous surveillance fosters an environment of constant observation, akin to living under a digital panopticon. Individuals may alter their behavior, self-censor their communications, or avoid certain places or associations simply out of fear of being monitored, misinterpreted, or flagged by automated systems. The normalization of such pervasive tracking can lead to a society where a baseline level of suspicion is applied to everyone, shifting the burden of proof of innocence onto the individual. The aggregation of biometric data over time allows for the creation of detailed behavioral profiles, mapping out an individual's routines, social connections, and daily habits. This granular insight into personal lives, when held by state or corporate entities, represents an unprecedented concentration of power.

Furthermore, the integration of facial recognition with other data streams amplifies its surveillance capabilities. When a faceprint can be linked to a mobile device's location data, financial transaction history, or online browsing activity, the picture that emerges is incredibly comprehensive and intrusive. This creates a powerful nexus of information that can be exploited for a variety of purposes, ranging from legitimate law enforcement investigations to more opaque forms of social control or commercial exploitation. The potential for mission creep is also significant; technologies initially deployed for serious crimes could gradually be expanded to monitor minor infractions, enforce social norms, or even track political dissent.

The legal and regulatory frameworks governing the use of facial recognition technology are still in their nascent stages, and there is considerable debate about how these powerful tools should be managed. Some jurisdictions have implemented outright bans or moratoriums on government use of facial recognition, citing privacy and civil liberties concerns. Others have adopted regulations that require transparency, independent oversight, and strict limitations on data retention and sharing. However, in many places, the deployment of these technologies has outpaced legislative action, leading to a patchwork of inconsistent rules and significant gaps in oversight. The question of whether facial recognition constitutes a "search" under existing legal doctrines, and thus requires a warrant, remains a contentious issue, with courts grappling to apply old legal principles to new technological realities.

The debate surrounding facial recognition surveillance

highlights a fundamental tension in modern society: the perceived need for enhanced security versus the protection of individual freedoms. Proponents argue that these technologies are essential tools for crime prevention, counter-terrorism, and maintaining public order, enabling authorities to identify threats more effectively and efficiently. They often emphasize the potential for real-time identification to prevent attacks or apprehend suspects before they can do further harm. The ability to quickly identify individuals in a crowd who may pose a risk, or to find missing persons, is presented as a clear benefit to public safety.

However, critics counter that the risks to civil liberties are too great and that the effectiveness of facial recognition is often overstated, particularly given its documented biases. They argue that the societal costs of creating a surveillance state, where every individual's movement and identity can be tracked, outweigh the purported security benefits. The potential for error, the chilling effect on democratic participation, and the erosion of anonymity are seen as fundamental threats to the fabric of a free society. Moreover, the increasing accuracy of AI-powered surveillance technologies raises the specter of a future where dissent can be easily identified and suppressed, and where personal freedoms are curtailed in the name of security, often without due process or public accountability. The development and deployment of these technologies represent a critical juncture, demanding careful consideration of their long-term societal implications and the establishment of robust safeguards to protect fundamental rights. The power to visually identify and track individuals across vast networks of cameras, linked to comprehensive databases, creates a chilling efficiency for surveillance states, transforming the very nature of public space into an arena of perpetual, automated scrutiny. The expansion of these biometric capabilities into everyday life signifies a gradual but significant erosion of the expectation of privacy and anonymity, fundamentally altering the relationship between the individual and the observing authority.

The growing adoption of facial recognition technology by both government agencies and private entities marks a significant escalation in the capabilities of mass surveillance. Unlike previous forms of observation that might have relied on human monitoring or the analysis of indirect digital traces, facial recognition offers a direct, automated, and highly scalable method of identifying and tracking individuals based on their unique biological features. This technology, when deployed on a wide scale, effectively transforms public spaces into areas of constant, indiscriminate biometric identification. Every individual moving through a monitored area, whether a busy street, an airport terminal, or a concert venue, can potentially have their identity captured, analyzed, and logged.

The technical underpinnings of facial recognition involve complex algorithms that analyze distinctive facial landmarks, such as the distance between the eyes, the shape of the nose, the prominence of cheekbones, and the contour of the jawline. These measurements are used to create a unique biometric template, often referred to as a "faceprint." This faceprint is then compared against large databases containing images of individuals. These databases can originate from a multitude of sources, including government-issued identification photographs (like driver's licenses and passports), criminal mugshots, passport control records, and increasingly, images scraped from social media platforms and other publicly available online sources. The technology's ability to perform these comparisons rapidly, often in real-time, allows for immediate identification or flagging of individuals against watchlists or databases of persons of interest.

The scope of deployment is broad and expanding. Law enforcement agencies frequently utilize facial recognition systems to identify suspects in criminal investigations, to scan crowds for known fugitives, or to monitor individuals on terror watchlists. Airports and border control agencies employ it for passenger identification and screening. Retailers are experimenting with it for loss prevention, customer analytics, and even personalized advertising. Public transportation systems, sports arenas, and university campuses are also increasingly incorporating this technology into their security infrastructure. This widespread adoption means that individuals are frequently subjected to biometric surveillance without their explicit consent or even their awareness, simply by virtue of moving through public or semi-public spaces.

A crucial aspect of this technological deployment is its integration with existing surveillance networks. Facial recognition cameras are often networked with vast arrays of other surveillance cameras, creating a comprehensive web of visual monitoring. This allows for the tracking of an individual's movements across multiple locations and over extended periods. When combined with other data sources, such as mobile phone location data, vehicle license plate recognition, and transaction records, facial recognition becomes a powerful tool for constructing detailed personal profiles and mapping out an individual's entire life. This aggregation of data points paints an increasingly intimate picture of an individual's habits, associations, and daily routines.

However, the capabilities of facial recognition technology are tempered by significant limitations, particularly concerning accuracy and bias. While advancements have been made, the technology's performance can vary dramatically depending on factors such as image quality, lighting conditions, viewing angle, and the presence of occlusions like masks or hats. Critically, numerous studies have consistently shown that facial recognition algorithms exhibit higher error rates when identifying individuals from certain demographic groups. Specifically, systems have demonstrated a greater propensity for false positives (incorrectly identifying an innocent person as a match) and false negatives (failing to identify a known individual) when analyzing the faces of women, people of color, and individuals of different age groups compared to white men. These biases often stem from the datasets used to train the algorithms, which may not adequately represent the diversity of the global population.

The consequences of these inaccuracies and biases are far-reaching and can have severe real-world impacts. An individual might be misidentified as a suspect in a crime, leading to unwarranted stops, interrogations, or even arrest. For communities already facing over-policing or racial profiling, the disproportionate error rates of facial recognition technology can exacerbate existing injustices and further erode trust in authorities. The reliance on these systems, often presented as objective and infallible, can mean that algorithmic errors are not adequately challenged, leading to potentially life-altering consequences for innocent individuals caught in the system. The very act of being flagged by such a system, regardless of accuracy, can result in increased scrutiny and a persistent digital record of suspicion.

The implications for privacy and civil liberties are profound. The ability to identify and track individuals in real-time and retrospectively, without their knowledge or consent, fundamentally undermines the expectation of anonymity in public life. This constant, pervasive surveillance can create a chilling effect on freedom of assembly and expression. Individuals may hesitate to participate in protests, attend political rallies, engage in religious practices, or seek sensitive medical care if they know their presence could be recorded, identified, and potentially used against them in the future. The knowledge that one's identity can be linked to their physical location and activities at any given moment can foster a sense of self-censorship and inhibit legitimate forms of civic engagement.

This pervasive biometric surveillance contributes to the creation of what some scholars describe as a "surveillance society," where a significant portion of the population is subject to continuous monitoring. The normalization of being constantly watched can lead to a behavioral shift, where individuals modify their actions and associations out of fear of being flagged or misinterpreted by automated systems. This shift can stifle spontaneity, creativity, and the open exchange of ideas that are crucial for a vibrant democratic society. The potential for "mission creep" is also a significant concern; technologies initially deployed for serious national security threats could be repurposed for monitoring minor offenses, enforcing social norms, or even identifying and suppressing political dissent.

The legal and regulatory landscape surrounding facial recognition technology remains complex and often lags behind the pace of technological development. Debates are ongoing regarding the legality of its use, particularly whether it constitutes a "search" requiring a warrant under constitutional protections. Some jurisdictions have enacted outright bans or moratoriums on government use of the technology, citing privacy and civil liberties concerns, while others have introduced regulations that mandate transparency, accountability, and limitations on data usage and retention. The lack of a consistent, comprehensive legal framework across different regions creates a situation where the application of this powerful surveillance tool can vary significantly, leaving individuals with uncertain protections for their fundamental rights.

The integration of facial recognition with other sophisticated surveillance tools, such as AI-powered data analysis, location tracking, and behavioral prediction algorithms, creates a formidable infrastructure for social control. This convergence of technologies allows for the creation of incredibly detailed profiles of individuals, identifying patterns of behavior, social networks, and potential vulnerabilities. While proponents argue that these tools are essential for enhancing security and preventing crime, critics contend that the potential for misuse, the erosion of privacy, and the inherent biases in the technology pose significant threats to individual liberty and democratic values. The challenge lies in finding a balance between legitimate security concerns and the fundamental rights to privacy, anonymity, and freedom of assembly and expression in an increasingly surveilled world. The technology itself is neutral, but its deployment within a surveillance framework, particularly one lacking robust oversight and clear ethical boundaries, transforms it into a powerful instrument that can reshape the very nature of public life and individual freedom.

4: THE HUMAN COST OF CONSTANT OBSERVATION

The ceaseless hum of surveillance, a ubiquitous backdrop to modern existence, quietly yet profoundly reshapes the contours of our personal lives, primarily through the insidious erosion of privacy and the curtailment of individual autonomy. When the awareness of being perpetually observed settles in, a subtle but powerful mechanism of self-regulation kicks in. This isn't a conscious decision to adhere to the law, but rather an ingrained reflex to conform, to avoid any behavior that might be misconstrued or flagged by the invisible, ever-watchful gaze of automated systems and their human overseers. The knowledge that every digital interaction, every physical movement, and increasingly, every facial expression can be logged, analyzed, and stored, fosters a climate of caution that can stifle genuine expression and free thought.

This pervasive sense of being watched exerts a chilling effect on fundamental freedoms, most notably on freedom of speech and association. Imagine attending a political rally, a protest against a government policy, or even a quiet gathering of like-minded individuals discussing sensitive social or political issues. If the knowledge that your presence, your interactions, and potentially your very identity are being recorded and associated with these activities is ever-present, the willingness to participate freely and vociferously diminishes. The risk of being added to a watchlist, subjected to increased scrutiny, or having one's dissent documented can be enough to deter even the most committed activists or concerned citizens. This isn't merely hypothetical; anecdotal evidence from various sectors suggests that individuals may self-censor their conversations, both online and offline, and shy away from participating in public discourse or social movements for fear of adverse consequences, however unlikely they might seem. The very possibility of being identified and linked to potentially controversial activities can lead to a prudent, yet ultimately limiting, retreat from public engagement.

Furthermore, the constant collection and aggregation of vast swathes of personal data directly attack the bedrock of individual autonomy: control over one's own information. Our personal data – our preferences, our health concerns, our social connections, our purchasing habits, our movements through space and time – is increasingly harvested, not with our informed consent, but as a byproduct of our engagement with the digital and physical world. This data is then curated, analyzed, and often commodified, shaping everything from the advertisements we see to the credit scores we are assigned, and even, in some instances, influencing our access to opportunities. When individuals lose control over who collects their data, how it is used, and for what purpose, their ability to shape their own narrative and direct their own lives is significantly compromised. Privacy, in this context, transforms from an inherent right into a privilege, a fragile state that is constantly under threat from technological advancement and the insatiable appetite for data. The loss of this control extends beyond mere inconvenience; it impacts our ability to define ourselves, to experiment with ideas and identities without permanent digital records, and to maintain the boundaries that protect our inner lives from external intrusion.

The impact on personal identity is multifaceted and deeply unsettling. Our sense of self is intrinsically linked to our experiences, our choices, and our ability to navigate the world without undue external influence or judgment. When every decision, every interaction, and every nascent thought is potentially cataloged and analyzed, there is a subtle pressure to conform to perceived norms and expectations. The fear of being misunderstood, of having a momentary lapse in judgment or an unconventional opinion permanently logged, can lead to a curated self-presentation that deviates from one's authentic identity. This can result in a gradual disconnect between the individual's inner life and their outward persona, leading to a sense of alienation and a loss of self-expression. The very freedom to be imperfect, to explore, and to evolve without constant scrutiny is a casualty of pervasive surveillance.

The concept of autonomy is also intrinsically tied to the freedom from undue interference in one's personal affairs. In an era of constant observation, this interference takes on new forms. It is not necessarily overt coercion, but rather the subtle influence exerted by the knowledge of being monitored. This can manifest as a reluctance to explore unconventional ideas, to engage in activities that might be perceived as unusual or potentially problematic, or even to seek help or support for personal issues for fear of the information being recorded and potentially used against them in the future. For instance, an individual struggling with mental health issues might hesitate to consult a therapist or attend support groups if they fear this information will be logged in a database accessible by potential employers or insurers. Similarly, someone exploring a new political ideology or engaging in research on sensitive topics might curtail their activities, fearing that their digital footprint will mark them as a person of interest. This self-imposed limitation on personal exploration and self-discovery represents a profound erosion of autonomy, as individuals begin to govern their own actions based on the perceived expectations of unseen observers.

Moreover, the normalization of surveillance can lead to a gradual desensitization, where the constant observation becomes so ingrained that it is accepted as an unavoidable aspect of modern life. This acceptance, however, does not negate its impact. It can foster a passive resignation, a sense that individual privacy and autonomy are simply too costly to reclaim in the face of perceived security benefits or technological inevitability. This resignation can further entrench surveillance practices, as it diminishes public resistance and facilitates the expansion of monitoring technologies into ever more intimate aspects of our lives. The struggle to reclaim privacy and autonomy is therefore not just a legal or technological challenge, but also a deeply psychological one, requiring a conscious effort to resist the normalizing pressures of constant observation and to advocate for the fundamental rights that underpin a free and self-determining society. The digital breadcrumbs we leave behind are not merely data points; they are fragments of our lives, our thoughts, and our very identities, and when these fragments are collected and controlled by external forces without our consent, our ability to be fully human, to be truly free, is irrevocably diminished.

The economic implications of this data-driven surveillance economy also contribute to the erosion of autonomy. Our personal information has become a valuable commodity, fueling industries that thrive on targeted advertising, personalized services, and predictive analytics. In this model, individuals are often incentivized to surrender their privacy in exchange for convenience or access to services, creating a Faustian bargain where immediate benefits are traded for long-term control over personal data. This creates a subtle form of coercion, as opting out of data collection often means opting out of participation in essential services or economic opportunities. As more aspects of life become digitized and data-dependent, the ability to navigate society without contributing to this surveillance apparatus becomes increasingly difficult, further limiting individual choice and reinforcing the power of data-gathering entities.

Consider, for example, the rise of "smart" cities, where sensors and interconnected devices are deployed to optimize traffic flow, manage resources, and enhance public safety. While these initiatives promise efficiency, they also generate an unprecedented volume of granular data about the daily lives of citizens. Information about where people travel, when they travel, and even their patterns of consumption can be collected and analyzed. While ostensibly for public good, this data can also be used to profile individuals, predict their behavior, and potentially influence their choices in ways that are not transparent or subject to individual consent. The very infrastructure designed to improve urban living can become an instrument of pervasive monitoring, blurring the lines between public service and intrusive surveillance.

The impact is not abstract; it has tangible consequences for social and political life. When individuals fear that their associations, their communications, or their online research could lead to negative repercussions, the vibrancy of civil society is inevitably dampened. Dissent becomes riskier, investigative journalism faces greater obstacles in gathering information, and the free exchange of ideas – the lifeblood of a democratic society – is curtailed. The ability to organize, to advocate for change, and to hold power accountable relies on a degree of privacy and anonymity that is systematically undermined by mass surveillance. Without these safeguards, the space for genuine public deliberation and democratic participation shrinks, leading to a populace that is more compliant, more cautious, and ultimately, less free.

The psychological toll of living in such an environment should not be underestimated. The constant awareness of being watched can lead to increased stress, anxiety, and a generalized sense of unease. This is not simply a matter of having "nothing to hide"; it is about the fundamental human need for private spaces where one can be ungucted, experiment with identity, and experience vulnerability without fear of judgment or exploitation. When these private spaces are eroded, the constant performative aspect of life can become exhausting, leading to burnout and a withdrawal from social and civic engagement. The very essence of individual freedom is rooted in the capacity to make choices and live one's life according to one's own values, unburdened by the oppressive weight of perpetual scrutiny.

Ultimately, the erosion of privacy and autonomy through pervasive surveillance is not merely a technological issue; it is a fundamental challenge to the principles of human dignity and self-determination. It represents a shift in the balance of power from the individual to the state and corporate entities, transforming citizens into data points to be managed, predicted, and influenced. Reclaiming these fundamental rights requires not only robust legal and technological safeguards but also a conscious societal effort to resist the normalization of surveillance and to reaffirm the intrinsic value of privacy and the freedom to live one's life without constant, intrusive observation. The ability to control one's own narrative, to choose what to reveal and what to keep private, is not a luxury but a prerequisite for a truly free and flourishing society.

The pervasive architecture of constant observation, while often justified under the banner of security and convenience, casts a long and undeniable shadow over the very foundations of a free society: freedom of speech and freedom of association. When the awareness that one's every utterance, every digital interaction, and every physical gathering can be logged, analyzed, and potentially stored indefinitely, takes root, it fosters an environment ripe for self-censorship and a gradual withdrawal from public life. This phenomenon, widely known as the "chilling effect," describes the disinclination to exercise fundamental rights due to the fear of reprisal, scrutiny, or unintended negative consequences stemming from the monitoring of one's activities.

Consider the individual contemplating participation in a political demonstration. The prospect of their presence being recorded – perhaps through facial recognition technology at a public square, CCTV footage along protest routes, or even by cell phone location data – can inject a significant element of risk. This risk isn't necessarily about imminent arrest for lawful assembly, but rather the more insidious possibility of their attendance being noted, cataloged, and linked to their broader digital footprint. This association might be with a particular political faction, a controversial cause, or even simply with a group perceived as critical of established authorities. The potential ramifications, however speculative, can loom large: a black mark on an employment record, increased scrutiny from law enforcement or intelligence agencies, or even subtle social ostracism if such information were to be leaked or misused. Faced with these potential deterrents, a person who might otherwise vocally advocate for change may choose to remain silent, to stay home, or to express their views only in the most guarded and private of forums, if at all. This quiet disengagement, multiplied across a population, erodes the collective voice necessary for a vibrant democracy and for holding power accountable.

The chilling effect extends beyond overt acts of protest to the everyday expression of ideas and opinions. In online spaces, where surveillance is particularly pervasive, individuals may self-edit their posts, comments, and even private messages. A nuanced discussion on a sensitive social issue might be watered down to avoid being misinterpreted by algorithmic content moderation or human analysts. A person researching a controversial topic, seeking to understand different perspectives or to gather information for a personal project, might hesitate to perform searches or visit certain websites for fear that their curiosity will be flagged as suspicious behavior. This reluctance to explore ideas freely, to engage in intellectual exploration without the constant pressure of being judged or categorized, fundamentally hinders the process of learning, critical thinking, and the formation of independent opinions. The marketplace of ideas, which thrives on open and unfettered discourse, becomes distorted and impoverished when individuals are afraid to voice their true thoughts or to explore challenging concepts.

Freedom of association, a cornerstone of civil society, is equally vulnerable. The right to gather with others, to form groups, and to collectively pursue shared interests or advocate for common goals, is severely hampered when such associations are under constant surveillance. Imagine a support group for individuals with a particular health condition, a book club discussing politically charged literature, or even a faith-based community. If participants believe their attendance, their conversations, and their membership in such groups are being monitored, the trust and openness necessary for genuine connection and support can be undermined. The fear that being seen with certain individuals or attending specific meetings could lead to unwanted attention or negative consequences can deter people from seeking out the solidarity and collective strength that associations provide. This can isolate individuals, weaken community bonds, and impede the formation of grassroots movements that are essential for social progress and for advocating for the rights of marginalized groups.

The erosion of these freedoms is not always overt or explicitly stated. Often, it operates through a diffused sense of unease, a subtle understanding that one is being watched, and that deviation from perceived norms could lead to complications. This internalized surveillance can be more insidious than direct censorship because it is self-imposed. Individuals become their own censors, anticipating the gaze of the watchers and modifying their behavior accordingly. This can lead to a society where conformity is implicitly rewarded and where dissent, even if expressed through entirely lawful means, is subtly discouraged. The vibrant cacophony of diverse opinions and the robust debate necessary for a healthy democracy can thus be muted, replaced by a more homogenized and compliant public discourse.

The implications for activism and political engagement are particularly stark. Grassroots movements, by their very nature, often challenge existing power structures and established norms. Their effectiveness relies on the ability of individuals to organize, to mobilize, and to express their grievances and demands publicly. When surveillance capabilities are ubiquitous, the very act of organizing becomes fraught with risk. The digital communications used to plan events, the geographical data of attendees, and the publicly visible expressions of solidarity can all be captured and used to identify, track, and potentially disrupt these movements. This can create a significant barrier to entry for potential participants, particularly those who may already be in vulnerable positions or who have less to gain and more to lose from engaging in activism. The fear of being labeled a troublemaker, of being placed on a list, or of having one's associations used against them can dissuade many from lending their energy and voice to causes they believe in.

Consider the case of journalists and whistleblowers, whose

work is critical to transparency and accountability. The ability to gather information, to protect sources, and to report on sensitive issues relies heavily on a degree of privacy. When surveillance technologies make it easier to identify and track communications between journalists and their sources, the flow of vital information can be severely curtailed. Sources may become unwilling to come forward for fear of being exposed, and journalists may face increased obstacles in verifying information and holding powerful entities accountable. This not only impacts the public's right to know but also weakens the checks and balances inherent in a democratic system.

Furthermore, the chilling effect is not confined to political activities. It can permeate artistic expression, academic research, and even personal relationships. Artists may self-censor their work to avoid controversy, fearing that their creations could be misinterpreted or used as evidence of disloyalty or deviance. Academics might steer clear of controversial research topics or express their findings in more cautious language if they believe their work is being monitored and could lead to professional repercussions or public backlash. Even intimate conversations between friends or family members might be guarded if there is a pervasive belief that these interactions are not truly private. This constant awareness of being potentially observed can lead to a profound sense of isolation and a stifling of creativity and intellectual exploration.

The concept of the "chilling effect" is not merely an abstract theoretical construct; it has tangible and measurable consequences for the health of democratic societies and the well-being of individuals. When people are afraid to speak out, to associate freely, or to explore challenging ideas, the collective capacity of society to address its problems, to innovate, and to adapt is diminished. The space for dissent, which is vital for the evolution of ideas and the correction of societal wrongs, shrinks. Moreover, the psychological burden of living under constant potential observation can lead to increased stress, anxiety, and a pervasive sense of unease, impacting individual mental health and overall quality of life.

The normalization of surveillance, as discussed previously, plays a crucial role in amplifying these chilling effects. When constant monitoring becomes an accepted, almost mundane, aspect of everyday life, the perceived threat of being watched can still exert a powerful influence on behavior, even if individuals don't consciously dwell on it. This passive acceptance can lead to a gradual but profound shift in societal norms, where open expression and bold association are replaced by a more cautious and conformist approach to public life. The very fabric of civil society, woven from the threads of open communication, voluntary association, and the fearless pursuit of knowledge, begins to fray when the invisible hand of surveillance creates a climate of apprehension. The potential for surveillance, whether realized or merely feared, acts as a constant deterrent, shaping behavior in ways that favor caution over courage, silence over speech, and isolation over association. This subtle yet pervasive influence poses a significant threat to the democratic ideals of an informed, engaged, and free citizenry.

The digital panopticon, a term increasingly used to describe the pervasive nature of modern surveillance, is not a neutral observer. Its gaze, far from being objective, is often shaped by the very societal inequalities it claims to combat. When algorithms are fed data that reflects a history of systemic discrimination, they inevitably learn and perpetuate those biases, transforming data points into pretexts for differential treatment. This is the insidious mechanism through which surveillance technologies can become powerful engines of discrimination and profiling, disproportionately impacting already marginalized communities.

At the heart of this issue lies the concept of algorithmic bias. Machine learning models, the sophisticated tools that power many surveillance systems – from facial recognition to predictive policing – are trained on vast datasets. If these datasets are derived from historical practices that have unfairly targeted certain groups, the algorithms will absorb and amplify these discriminatory patterns. For instance, if historical crime data shows a higher arrest rate for a particular racial minority in a specific neighborhood, an algorithm trained on this data might predict a higher likelihood of future criminal activity in that same demographic or area, regardless of individual behavior. This creates a feedback loop: increased surveillance and policing in these communities, leading to more arrests, which further reinforces the algorithm's biased predictions. The result is a self-fulfilling prophecy of suspicion, where individuals are profiled and scrutinized not based on their actions, but on the prejudiced patterns embedded in the data that underpins the surveillance systems.

This has been demonstrably observed in the realm of predictive policing. While often presented as a scientific approach to crime prevention, many such systems have been criticized for their reliance on data that reflects existing racial disparities in policing. Communities of color, often subjected to heavier police presence and more aggressive law enforcement tactics, naturally generate more data points related to stops, searches, and arrests. Predictive policing algorithms, when fed this skewed data, can direct police resources towards these same communities, leading to a higher rate of police interactions, even for minor infractions. This can result in increased arrests for low-level offenses, further cementing the community's status as a "high-crime" area in the eyes of the system. The individuals within these communities, therefore, face a heightened risk of being flagged, stopped, and questioned simply by virtue of their location and demographic profile, rather than any concrete suspicion of wrongdoing. This cycle of over-policing and algorithmic bias creates an environment of constant anxiety and distrust, eroding any semblance of equal treatment under the law.

Facial recognition technology, another ubiquitous surveillance tool, has also been shown to exhibit significant racial and gender biases. Studies have repeatedly indicated that these systems are less accurate when identifying individuals with darker skin tones, women, and certain ethnic groups. This inaccuracy can have serious consequences. In law enforcement contexts, misidentification can lead to wrongful stops, interrogations, or even arrests. Imagine a scenario where a person of color is wrongly identified by a faulty facial recognition system at a public gathering or in connection with a minor incident. The ensuing scrutiny, questioning, and potential detention can be deeply traumatizing and can have lasting repercussions on their reputation and freedom. The technology, intended to enhance security, instead becomes a tool for harassment and unjustified suspicion, particularly for those whose faces are less readily recognized by the underlying algorithms. The very people who might most need protection from unwarranted state intrusion are often the ones most vulnerable to the failures and biases of these technologies.

The religious dimension of profiling is equally concerning. In the post-9/11 era, intelligence agencies and law enforcement bodies have, in some instances, developed practices that disproportionately target individuals of Muslim faith or those perceived as having Middle Eastern or South Asian heritage. This can manifest in various ways, from increased scrutiny at airports and border crossings to surveillance of community institutions and online activities. The assumption that adherence to a particular faith or cultural background automatically correlates with a higher risk of engaging in extremist activities is a form of collective punishment that violates fundamental principles of individual liberty and due process. When surveillance efforts are guided by such generalized suspicion, innocent individuals are subjected to unwarranted attention, their privacy is invaded, and their sense of belonging and security is undermined. The fear that a particular religious practice or association could invite the watchful eye of surveillance agencies can lead to self-censorship and a reluctance to engage in community life, further isolating already scrutinized populations.

The erosion of trust between communities and law enforcement or intelligence agencies is a significant human cost of this discriminatory profiling. When certain groups perceive themselves as being consistently targeted and unfairly scrutinized, it breeds resentment and alienation. This can make it harder for law enforcement to gain cooperation and intelligence from these communities, ironically undermining the very security objectives these surveillance measures are meant to achieve. Instead of fostering a sense of shared security, discriminatory surveillance practices can create an adversarial relationship, pushing individuals further away from legitimate authorities and potentially driving them towards illicit or underground networks. The societal damage of such fractured trust can be profound, impacting community cohesion, civic engagement, and the overall legitimacy of governing institutions.

Beyond direct profiling by law enforcement, the vast repositories of data collected through constant observation can also be exploited by other entities, including private companies and malicious actors, to further discriminatory ends. While not directly government surveillance, the aggregation of personal data – browsing history, location data, social media interactions, purchase records – can be used to create detailed profiles that are then used for targeted advertising, credit scoring, or even employment screening. If these profiles are built upon biased assumptions or incomplete data, they can lead to individuals being unfairly denied opportunities, charged higher prices for goods and services, or subjected to intrusive marketing campaigns based on sensitive personal characteristics. For instance, someone with a perceived health condition might be shown only ads for expensive treatments, or an individual associated with certain online communities might be shown only certain types of political advertisements, subtly manipulating their perceptions and limiting their exposure to diverse viewpoints.

The normalization of such practices, where constant data collection and algorithmic analysis become the default for many aspects of life, creates a pervasive sense of vulnerability. Individuals from minority groups, who have historically faced discrimination, are acutely aware of the potential for surveillance to be weaponized against them. This awareness can lead to a heightened sense of fear and a cautious self-management of online and offline activities, a constant internal calculus of risk associated with expressing certain opinions, associating with particular people, or even accessing certain information. This internalized surveillance, driven by the very real threat of discriminatory profiling, can have a profound psychological impact, leading to stress, anxiety, and a diminished sense of personal agency.

Documented instances of discriminatory surveillance are crucial for understanding the tangible impact of these technologies. In various countries, surveillance programs have been accused of disproportionately monitoring minority religious or ethnic groups, leading to investigations, interrogations, and the creation of extensive databases on individuals who pose no threat. These practices often operate under broad justifications of national security, but critics argue they are rooted in prejudice and contribute to the stigmatization and marginalization of entire communities. The lack of transparency and accountability surrounding many surveillance operations makes it difficult to challenge these discriminatory patterns, further entrenching the problem. When the mechanisms of observation are opaque and their application is uneven, the potential for abuse and discrimination is significantly amplified.

The challenge lies not only in the explicit intent to discriminate, but also in the unintended consequences of systems designed without adequate consideration for the historical and societal context of inequality. Even if the designers of surveillance technology do not harbor malicious intent, the data on which these systems are trained can carry the weight of centuries of prejudice. Without rigorous auditing, bias detection, and a commitment to fairness in data selection and algorithm design, these technologies will continue to reproduce and exacerbate existing societal divides. This requires a multi-faceted approach: demanding transparency in how surveillance technologies are developed and deployed, advocating for robust legal and ethical frameworks that prohibit discriminatory profiling, and empowering communities to challenge the misuse of surveillance in ways that target them unfairly. The fight against discriminatory profiling through surveillance is, therefore, a critical front in the broader struggle for civil liberties and social justice, demanding vigilance and active resistance against the insidious ways technology can be used to entrench inequality.

The chilling effect of constant observation extends deeply into the bedrock of democratic societies: the practice of journalism. Investigative journalism, in particular, thrives on access to information, the protection of sources, and the freedom to pursue stories without undue interference. However, the pervasive surveillance capabilities wielded by governments and increasingly sophisticated private entities pose a direct and existential threat to these foundational elements. When journalists operate under the shadow of omnipresent digital monitoring, their ability to uncover and report on wrongdoing is severely compromised, creating a fertile ground for secrecy and unaccountability.

The fundamental challenge lies in the ability of surveillance apparatuses to trace and identify the sources of leaked information. In an era where almost every digital interaction leaves a trace – emails, encrypted messages, phone calls, metadata, even the location history of devices – tracking the flow of information from a whistleblower to a journalist has become a significantly easier task for well-resourced state actors. The assumption that digital communications are inherently private has long been eroded, replaced by the stark reality that many government agencies possess the technical means and legal authority to intercept, store, and analyze vast quantities of communication data. This capability transforms the act of leaking sensitive information, which is often crucial for exposing corruption, abuse of power, or significant societal harms, into an act fraught with peril for the source. The risk of exposure, leading to professional ruin, legal prosecution, or even imprisonment, becomes an almost unavoidable consequence for anyone daring enough to share information with the press.

This heightened risk directly impacts the willingness of potential whistleblowers to come forward. Individuals who witness illegal activities, unethical practices, or dangers to public safety within government or corporate structures often rely on the confidentiality of their disclosures. They may be motivated by a strong sense of civic duty, a moral imperative, or a desire to prevent harm, but their actions are intrinsically linked to the perceived safety of revealing such information. When surveillance capabilities make it highly probable that their identity will be uncovered, even through seemingly innocuous digital breadcrumbs, the calculus of speaking out shifts dramatically. The fear of retribution, whether it be job loss, reputational damage, or legal penalties, acts as a powerful deterrent. Consequently, vital information that could inform the public and hold powerful entities accountable remains buried, shielded not by legitimate secrecy, but by the fear of digital exposure.

Journalists, in turn, find their investigative efforts hampered by this chilling effect on sources. Without the assurance that their sources can speak to them safely and anonymously, journalists are often unable to access the critical information needed to pursue complex and important stories. They may be forced to rely on publicly available information, which is often sanitized or incomplete, or on sources who are willing to take extreme personal risks. This can lead to a scenario where only the bravest, or perhaps the most reckless, individuals are willing to engage with the press, potentially skewing the nature of the information received. Moreover, the very act of receiving sensitive information can place journalists under scrutiny. If authorities suspect a journalist of possessing classified or incriminating material, they may attempt to identify the source through the journalist's own digital footprint, metadata associated with shared files, or even by subpoenaing communication records.

The legal frameworks designed to protect journalistic activities and shield sources are increasingly being tested and, in some jurisdictions, circumvented by expansive surveillance powers. While many countries have laws intended to safeguard journalistic privilege, these protections are often riddled with loopholes or are subject to interpretation that favors state security over transparency. For example, national security exceptions can allow law enforcement to bypass standard procedures for obtaining journalist-related data, or to compel testimony that would reveal a source. The ability of governments to access metadata – the information about communications, such as who communicated with whom, when, and for how long – without necessarily reading the content of the communication, can be equally damaging. This metadata can be used to reconstruct a source's interactions with a journalist, providing crucial clues for identification.

The practical implications for investigative journalism are profound. Projects that require deep dives into sensitive government operations, corporate malfeasance, or intelligence failures become significantly more difficult and dangerous to undertake. Journalists might have to resort to less secure communication methods, employ extensive counter-surveillance techniques, or even meet sources in person in ways that carry their own risks. The cost and complexity of conducting thorough investigations are thus amplified, potentially limiting the scope of journalistic inquiry to stories that are less impactful or easier to prove through public records alone. This creates an uneven playing field, where powerful institutions can leverage their technological and legal advantages to operate with a greater degree of impunity, shielded from the critical gaze of the press.

Furthermore, the mere knowledge that one is being monitored can alter behavior, a phenomenon known as the "chilling effect." For journalists, this means a potential reluctance to pursue certain lines of inquiry that might draw unwanted attention from surveillance agencies. They might self-censor, avoiding topics or questions that could put them or their sources at risk. This internalized inhibition is a subtle but powerful form of control, capable of shaping the news agenda and limiting the range of public discourse. When journalists begin to second-guess what they can report or who they can speak to, the public's right to know is directly curtailed. The watchdog function of the press, its ability to hold power accountable by shedding light on hidden activities, is weakened when the very tools of observation are used to stifle the flow of information that would fuel such scrutiny.

The targeting of journalists and their sources through surveillance not only impacts specific investigations but also cultivates an environment of fear and distrust. When journalists are perceived as agents who can be monitored and whose sources are vulnerable, their credibility can be undermined in the eyes of both the public and the authorities. This can lead to a breakdown in the symbiotic, albeit sometimes adversarial, relationship between the press and government institutions, a relationship that is essential for a functioning democracy. The ability of journalists to act as intermediaries, bringing important information from internal dissenters to public awareness, is paramount. If that intermediary role is rendered too risky, the channels of communication and accountability begin to close.

The legal landscape surrounding digital surveillance and its impact on journalism is constantly evolving. Debates rage over the balance between national security and freedom of the press, and the interpretation of laws such as those pertaining to espionage or unauthorized disclosure of information. In many Western democracies, there is a tension between the public interest in transparency, as championed by the press, and the state's interest in maintaining secrecy for security or operational reasons. When surveillance technologies provide governments with unprecedented insight into journalistic activities and source networks, this tension often tilts in favor of increased governmental control and reduced transparency. The legal protections for journalists, which may have seemed robust in a pre-digital age, often prove insufficient against the sophisticated data-gathering capabilities of modern states.

Beyond direct tracking, governments can also use surveillance powers to intimidate journalists and their organizations. The threat of investigation, the disruption of communication, or the public disclosure of sensitive journalistic activities can serve as a warning to others. This can create a climate where media outlets become more cautious, less willing to challenge authority, and more susceptible to external pressures. The financial and human resources required to navigate legal challenges, defend against accusations, or simply to maintain secure communication channels can also divert essential resources away from actual reporting. This creates a significant barrier to entry and sustainability for investigative journalism, particularly for smaller or independent outlets that lack the extensive legal and security infrastructure of larger media corporations.

The international dimension of this issue is also critical. Journalists often collaborate across borders, and sources may be located in different jurisdictions from the journalist and the publication. This creates complex legal and ethical challenges, especially when surveillance laws and protections vary significantly from country to country. A journalist in a country with strong press protections might find their sources exposed if those sources communicate from a country with weaker privacy laws and more extensive state surveillance. This global reach of surveillance capabilities means that the threat to journalistic freedom is not confined by national borders.

The concept of "chilling effect" extends beyond the immediate fear of being caught. It also encompasses the broader societal impact of a less informed public. When investigative journalism is suppressed or significantly hindered by surveillance, the public loses access to critical information that could influence political decisions, public opinion, and civic engagement. Issues of corruption, malfeasance, and governmental overreach may go unreported or be reported in a diluted form, leading to a populace that is less equipped to make informed choices or to hold its leaders accountable. This erosion of informed citizenry is one of the most profound human costs of constant observation, as it undermines the very foundations of democratic participation and self-governance.

The role of technology companies in this ecosystem is also multifaceted. While many technology firms emphasize user privacy and security, their platforms are often the very conduits through which surveillance occurs. The vast troves of data collected by these companies, often aggregated and analyzed for commercial purposes, can also be accessed by governments through legal means, sometimes with limited oversight. This creates a complex web of responsibility, where the infrastructure of communication is simultaneously a tool for commerce, civic discourse, and state surveillance. The opacity surrounding data sharing agreements between tech companies and governments further exacerbates the challenges faced by journalists seeking to protect their sources and their work.

In essence, the surveillance state, with its insatiable appetite for data, casts a long shadow over the practice of investigative journalism and the courageous individuals who act as whistleblowers. It transforms the pursuit of truth into a high-stakes game of cat and mouse, where the odds are increasingly stacked against those who seek to expose wrongdoing. The ability of governments to track, monitor, and potentially expose sources not only endangers individuals but also fundamentally undermines the public's right to know and the press's capacity to serve as a bulwark against unchecked power. The corrosive effect of this constant observation is a silent but potent threat to democratic accountability, creating a more secretive, less transparent, and ultimately less free society. The fight to protect journalistic freedom and the space for whistleblowing is, therefore, intrinsically linked to the broader struggle for civil liberties in the digital age, demanding constant vigilance and robust advocacy.

The pervasive reach of modern surveillance, extending far beyond the investigative halls of journalism, infiltrates the very fabric of individual lives, weaving a complex tapestry of psychological and social consequences. When the invisible eyes of monitoring systems, whether governmental or corporate, become a perceived constant, the psychological landscape of individuals can shift dramatically. The knowledge, or even the suspicion, that one's digital communications, online activities, physical movements, and even personal habits are being recorded and analyzed can breed a gnawing sense of anxiety. This is not merely a hypothetical concern; it is a palpable undercurrent for many in an age where data trails are ubiquitous. The subconscious monitoring of one's own behavior, a self-imposed censorship born out of fear, can become an automatic reflex. Every email sent, every website visited, every search query typed, can be accompanied by a silent question: "Will this be misinterpreted? Will this draw unwanted attention? Will this be stored away and used against me later?" This constant internal vetting process can be exhausting, chipping away at mental peace and contributing to elevated stress levels.

The psychological impact can manifest in more profound ways, fostering a sense of paranoia. For individuals who have experienced genuine persecution or are particularly sensitive to perceived injustices, the knowledge of constant surveillance can amplify existing anxieties, creating a feedback loop where the surveillance itself becomes the perceived threat, even if no specific malevolent action has yet occurred. This state of hyper-vigilance can be debilitating, making it difficult to relax, to be spontaneous, or to engage in activities that might be considered unconventional or outside the mainstream. The erosion of a private mental space, where thoughts and feelings can be explored without external judgment, is a significant casualty. When individuals feel they are perpetually on display, the very act of self-discovery and personal growth can be stifled. This can lead to a retreat into carefully curated online personas, a façade of conformity designed to avoid any perceived transgression, however minor. The authentic self, with its inherent messiness and imperfections, becomes a liability in the eyes of an omnipresent observer.

This psychological burden can extend into social interactions, subtly altering how people connect with one another. Trust, a cornerstone of healthy social relationships, can be significantly eroded. When individuals suspect that their conversations, even with friends or family, might be intercepted or that their online social networks are being monitored for behavioral patterns, a natural hesitancy can emerge. The ease and openness that characterize genuine connection can be replaced by a cautious guardedness. People may self-censor their opinions, their jokes, or their expressions of vulnerability, fearing that these personal revelations could be used to categorize, judge, or even penalize them. This creates a social environment where genuine intimacy becomes a more precious and perhaps rarer commodity. The uninhibited sharing of thoughts and feelings, which is vital for building deep relationships and fostering empathy, is replaced by a more transactional and guarded form of communication. The fear of being misjudged or categorized based on fragmented pieces of data can lead to a reluctance to engage deeply, for fear of leaving an incriminating digital footprint.

Furthermore, the normalization of constant observation can lead to a gradual desensitization to privacy violations. When surveillance becomes an accepted, almost mundane, aspect of daily life – embedded in everything from our smartphones to our shopping experiences – the intrinsic value of privacy can begin to diminish in the collective consciousness. Individuals may become so accustomed to their data being collected and analyzed that they no longer question its appropriateness or its potential implications. This apathy, born out of prolonged exposure and a sense of powerlessness, is a dangerous societal trend. It creates a fertile ground for further encroachments on privacy, as the public may be less inclined to resist new forms of monitoring or to advocate for stronger privacy protections. The quiet acquiescence to pervasive surveillance can be more damaging than outright defiance, as it signals a tacit acceptance of a fundamental shift in the relationship between the individual and the state, or between the consumer and corporations.

The long-term effect of this desensitization and the underlying psychological stress can contribute to the formation of a more conformist society. When the fear of being noticed, being flagged, or being deemed an outlier is ever-present, individuals are incentivized to adhere to perceived norms and expectations. Deviating from the beaten path, whether in thought, expression, or behavior, becomes an act of greater personal risk. This can stifle creativity, innovation, and the vibrant diversity of ideas that are essential for a healthy, dynamic society. The pressure to blend in, to present a "clean" and unremarkable digital and physical presence, can lead to a homogenization of experience. Individuals may self-censor their artistic expression, their political leanings, or even their lifestyle choices, opting for the path of least resistance to avoid attracting the attention of surveillance apparatuses. This creates a society where originality is discouraged, where critical thinking is suppressed, and where the pursuit of individual autonomy is subtly but powerfully curtailed. The very essence of what it means to be a unique individual, capable of independent thought and action, is threatened when the omnipresent gaze of surveillance encourages uniformity and discourages the very exploration that leads to self-discovery and societal progress.

Consider the subtle ways this can manifest in everyday life. Imagine a young person exploring their identity, experimenting with different subcultures, or questioning established societal norms. In an environment of constant surveillance, the digital traces of these explorations – their searches for information, their interactions on social media, their musical preferences – could be collected and analyzed. The fear that these explorations might be interpreted as rebellious, deviant, or even dangerous could lead them to suppress these nascent aspects of their identity. Similarly, an artist might refrain from creating provocative or challenging work, fearing that it could be misconstrued or flagged by automated systems. A political activist might hesitate to engage in peaceful protest or to voice dissent online, concerned that their participation could be logged and used against them in the future. This isn't about preventing genuine criminal activity; it's about the stifling effect on legitimate personal and civic expression that falls outside a narrowly defined acceptable spectrum.

The economic implications also feed into this psychological and social dynamic. The increasing reliance on data for targeted advertising, credit scoring, and even employment opportunities means that an individual's digital footprint can have tangible consequences on their material well-being. The fear of a negative mark on one's digital record, accrued through perceived missteps or unconventional associations, can further incentivize conformity. People may avoid attending certain events, associating with certain individuals, or expressing certain opinions, not out of moral conviction, but out of a calculated fear of how these actions might be interpreted by data-driven algorithms and the entities that control them. This creates a system where individuals are constantly managing their digital persona, a performance of acceptable behavior that can be profoundly alienating. The very act of living, of making choices and experiencing life, becomes an exercise in risk management, with the ever-present specter of surveillance dictating the parameters of acceptable behavior.

The erosion of social trust extends beyond personal relationships to broader community engagement. When people believe that their neighbors, colleagues, or even public institutions might be under surveillance, or that their own actions are being monitored in relation to their community involvement, a sense of detachment can emerge. The willingness to participate in community initiatives, to engage in civic discourse, or to volunteer for public service can be diminished if individuals fear that their involvement will be scrutinized or that their associations will be tracked. This can weaken the social capital that binds communities together, making them more fragmented and less resilient. The collective action that is often necessary to address societal challenges becomes harder to mobilize when individuals are hesitant to openly associate or collaborate, fearing that such connections might draw unwelcome attention.

Moreover, the constant awareness of being observed can lead to a phenomenon known as "performance anxiety" on a societal scale. Just as an actor might struggle to perform naturally when aware of an audience, individuals in a surveillance society may find themselves performing their lives rather than truly living them. Every interaction can become a potential data point, every action a potential public record. This can lead to a draining of energy and a sense of inauthenticity. The spontaneity of human interaction, the ability to be present in the moment without the filter of constant self-monitoring, is severely compromised. This performance of normalcy, of acceptable behavior, can lead to a deep psychological disconnect between the public persona and the private self.

The psychological toll of living under such conditions is not uniform. Certain groups may be disproportionately affected. Marginalized communities, those with histories of political dissent, or individuals belonging to minority groups who are already subject to heightened scrutiny may experience these effects more acutely. For them, the knowledge of surveillance can reinforce existing feelings of alienation, distrust, and vulnerability. The potential for surveillance to be used as a tool of social control, to monitor and suppress dissent, or to enforce existing power structures, adds another layer of psychological burden. The very systems designed to ensure safety or security can become perceived instruments of oppression, amplifying existing societal inequalities and fueling a climate of fear within targeted populations.

The long-term consequences for a society characterized by pervasive surveillance are profound. It risks creating a population that is less inclined to question authority, less willing to engage in critical discourse, and more susceptible to manipulation. When the channels of communication are subtly policed by the fear of observation, the public sphere can become sanitized, devoid of the challenging ideas and diverse perspectives that are vital for democratic health. The psychological impact of constant monitoring can lead to a populace that is anxious, guarded, and ultimately less free, even if no overt acts of repression are directly witnessed. The subtle, internalized pressure to conform, coupled with the erosion of trust and the desensitization to privacy invasions, can create a society that is more orderly and predictable, but at the cost of its dynamism, its creativity, and its fundamental commitment to individual liberty. The human cost is not always measured in visible chains, but in the invisible shackles of a mind that is constantly self-policing, a spirit that is hesitant to explore, and a society that is subtly steered towards conformity by the ever-present, unseen gaze. The quiet anxieties and the suppressed desires that fester beneath the surface of a surveilled society represent a profound, yet often unacknowledged, human cost.

5: THE LIMITS OF OVERSIGHT AND ACCOUNTABILITY

The architecture of American democracy is built upon a foundational principle of checks and balances, designed to ensure that no single branch of government accumulates excessive power. Within this framework, congressional oversight of the executive branch, particularly its intelligence apparatus, is intended to serve as a critical safeguard. However, the practical implementation of this oversight often falls short of its theoretical promise, creating significant deficiencies that can allow surveillance programs to expand beyond their intended scope and without adequate public accountability. These shortcomings are not merely abstract systemic failures; they translate into tangible impacts on civil liberties and the balance of power in a democratic society.

One of the most significant structural impediments to effective congressional oversight is the inherent information asymmetry between the legislative and executive branches, especially concerning national security and intelligence matters. Intelligence agencies, by their very nature, operate under a veil of secrecy. They possess specialized knowledge, classified information, and technical expertise that is largely inaccessible to the vast majority of members of Congress and their staff. This information gap is not accidental; it is a consequence of the sensitive nature of intelligence work, but it also creates a profound power imbalance. Committees tasked with oversight, even those with dedicated staff and resources, often receive information curated and presented by the very agencies they are meant to scrutinize. This curated flow of information can obscure potential problems, downplay risks, or present a sanitized version of reality, making it exceedingly difficult for lawmakers to independently verify claims or uncover hidden activities. The reliance on agency briefings and declassified reports means that oversight is often reactive rather than proactive, and dependent on the willingness of the agencies to disclose potentially problematic information.

Furthermore, the pervasive use of classification protocols exacerbates this information asymmetry. While classification is necessary to protect genuinely sensitive sources and methods, its application can become overly broad, encompassing information that has little or no bearing on national security but instead serves to shield programs from public and congressional scrutiny. Members of Congress, even those with high-level security clearances, are often denied access to the full picture, or they are provided with information on a need-to-know basis, which can lead to a fragmented understanding of complex operations. The sheer volume of classified documents and the complexity of intelligence programs mean that even dedicated oversight committees can struggle to process and analyze the information adequately. This can lead to a situation where members are forced to rely on the interpretations and assurances of agency officials, rather than their own independent assessments. The incentives are often aligned for agencies to err on the side of classification, and for members of Congress, facing time constraints and the potential political fallout of mishandling classified information, to accept this as a necessary reality.

The partisan nature of American politics also casts a long shadow over the effectiveness of congressional oversight. While oversight committees are intended to be bipartisan bodies dedicated to the national interest, they are not immune to the divisive forces that characterize broader congressional dynamics. When one party controls the presidency and the other controls one or both houses of Congress, oversight can become a political weapon, used to embarrass the administration or to advance a particular policy agenda. Conversely, when the same party controls both the executive and legislative branches, the incentive for robust oversight can diminish, as lawmakers may be reluctant to challenge an administration from their own party, fearing it could undermine their collective political strength or alienate key constituencies. This can lead to periods where oversight is perfunctory, characterized by a lack of critical questioning and a willingness to accept agency assurances at face value. The desire for partisan advantage or unity can override the fundamental responsibility to hold the executive branch accountable, allowing potential abuses to go unchecked.

The specific mechanisms of oversight, primarily the intelligence committees in both the House of Representatives and the Senate, have faced persistent criticism regarding their efficacy. These committees, by design, are meant to be the primary conduits for congressional oversight of the intelligence community. However, their structure and operational realities present inherent challenges. Committee membership is often a secondary assignment for many lawmakers, who may have more pressing committee responsibilities or who may not possess a deep understanding of intelligence matters. This can lead to a lack of specialized expertise and a reliance on the intelligence community to educate them on complex issues. Furthermore, the committees themselves are often susceptible to the same partisan pressures that affect the broader Congress. Debates over declassification of reports, the scope of investigations, and public statements about intelligence activities can become deeply divided along party lines, hindering the ability to present a united and forceful message to the executive branch.

Instances where oversight has been demonstrably bypassed or proven inadequate are numerous. The expansion of surveillance programs following the September 11, 2001, attacks provides a stark example. While there was a perceived need for enhanced security measures, the subsequent growth of data collection programs, such as those conducted by the National Security Agency (NSA), often occurred with limited and, at times, insufficient congressional knowledge and debate. Legislation that expanded surveillance powers, such as amendments to the Foreign Intelligence Surveillance Act (FISA), was sometimes passed with broad support but with a lack of deep understanding by many members of the full implications of the technologies and data collection methods being authorized. The revelation of programs like PRISM, which collected vast amounts of data from internet companies, exposed a significant gap between what Congress had authorized and the reality of executive branch activities. While some members were briefed on aspects of these programs, the extent and nature of the data collection were not widely understood or debated by the full body of Congress, and the oversight that did exist was often reactive, triggered by leaks or external revelations rather than proactive inquiry.

The concept of "oversight theatre" has also been invoked to describe situations where congressional committees engage in public hearings and issue reports that create the appearance of robust oversight, but fail to enact meaningful change or impose significant constraints on executive branch activities. This can be driven by a desire to project an image of accountability to constituents, or by the difficulty of overcoming the institutional inertia and the inherent advantages of the executive branch in controlling information. When intelligence agencies can effectively argue that certain information cannot be disclosed for national security reasons, it becomes difficult for Congress to hold them accountable for actions that may have violated privacy or civil liberties. The classification system can become a shield, not just for sources and methods, but for potentially unlawful or unethical behavior.

Moreover, the revolving door between intelligence agencies and lobbying firms or defense contractors can create subtle but powerful influences on congressional decision-making. Former intelligence officials often possess intimate knowledge of agency operations and the legislative process, making them valuable assets for those seeking to shape policy or secure funding. This can create an environment where members of Congress are persuaded by arguments that may be influenced by vested interests rather than a dispassionate assessment of the public good or the necessity of particular surveillance activities. The deep personal and professional connections between individuals in these spheres can inadvertently, or sometimes deliberately, lead to a relaxation of oversight standards.

The structure of committee assignments further complicates effective oversight. Members of the intelligence committees often serve alongside members of other powerful committees, such as appropriations or armed services. This can create competing priorities, as members may be more inclined to focus on securing funding for programs or advancing their own legislative agendas than on conducting the meticulous, often tedious, work of intelligence oversight. The expertise required to effectively oversee complex technical surveillance programs is significant, and it is often difficult to develop and maintain that expertise when committee assignments are transient or when members have limited time to dedicate to these specialized issues.

The capacity of congressional staff also plays a crucial role. While oversight committees are supported by professional staff, their resources and access can be limited compared to the vast resources available to intelligence agencies. Staffers may face challenges in developing deep technical expertise, navigating complex legal frameworks, or acquiring the necessary security clearances to access all relevant information. Furthermore, the turnover of staff can impede the development of institutional memory and expertise, making it difficult to build a consistent and rigorous oversight practice over time.

The historical context of how oversight has evolved is also important. Following periods of significant public outcry and revelations of abuses, such as those uncovered by the Church Committee in the 1970s, Congress did enact reforms aimed at strengthening its oversight capabilities. The creation of specialized intelligence committees was a direct result of these efforts. However, in the decades since, particularly in the wake of increasing global threats and technological advancements, there has been a tendency for executive branch power to expand, often with legislative acquiescence, in the name of national security. The balance between security and liberty, a perpetual tension in democratic societies, often tips towards security during times of heightened perceived threat, and congressional oversight mechanisms can be swept up in this dynamic.

The classification of congressional oversight activities themselves can also be a problem. While the substance of intelligence operations is often classified, the processes and findings of oversight can also be shrouded in secrecy. This makes it difficult for the public and even for non-intelligence-focused members of Congress to assess the effectiveness of the oversight system. When oversight findings are classified, or when recommendations for reform are kept confidential, the very notion of public accountability is undermined. The public is largely reliant on leaks or the occasional, heavily redacted report to understand the extent to which their elected representatives are effectively monitoring the intelligence community.

The problem is further compounded by the fact that individual members of Congress, even those on oversight committees, may not fully grasp the implications of the technologies being deployed or the data being collected. The rapid pace of technological change, particularly in areas like artificial intelligence, machine learning, and mass data analytics, outstrips the capacity of even the most diligent lawmakers to stay fully informed. This knowledge gap makes it difficult to ask the right questions, to identify potential risks, and to enact appropriate safeguards. The intelligence community, with its dedicated technical experts, naturally possesses a deeper understanding of these evolving capabilities.

Ultimately, the deficiencies in congressional oversight of intelligence agencies are a complex interplay of structural, political, and practical factors. The information asymmetry, fueled by classification and the inherent secrecy of intelligence work, creates a fundamental imbalance. Partisan divisions can politicize oversight, rendering it a tool for partisan warfare rather than a genuine check on executive power. The specific structures of oversight committees, while designed to provide focus, are not immune to the broader dynamics of congressional life, including competing priorities and the demands of political expediency. The result is a system where the ability of Congress to effectively hold the intelligence community accountable for its surveillance activities is often compromised, leaving significant room for programs to expand and operate with less scrutiny than the principles of democratic governance demand. This creates a persistent challenge for safeguarding civil liberties in an era of increasingly sophisticated and pervasive intelligence operations.

The architecture of American democracy is built upon a foundational principle of checks and balances, designed to ensure that no single branch of government accumulates excessive power. Within this framework, congressional oversight of the executive branch, particularly its intelligence apparatus, is intended to serve as a critical safeguard. However, the practical implementation of this oversight often falls short of its theoretical promise, creating significant deficiencies that can allow surveillance programs to expand beyond their intended scope and without adequate public accountability. These shortcomings are not merely abstract systemic failures; they translate into tangible impacts on civil liberties and the balance of power in a democratic society.

One of the most significant structural impediments to effective congressional oversight is the inherent information asymmetry between the legislative and executive branches, especially concerning national security and intelligence matters. Intelligence agencies, by their very nature, operate under a veil of secrecy. They possess specialized knowledge, classified information, and technical expertise that is largely inaccessible to the vast majority of members of Congress and their staff. This information gap is not accidental; it is a consequence of the sensitive nature of intelligence work, but it also creates a profound power imbalance. Committees tasked with oversight, even those with dedicated staff and resources, often receive information curated and presented by the very agencies they are meant to scrutinize. This curated flow of information can obscure potential problems, downplay risks, or present a sanitized version of reality, making it exceedingly difficult for lawmakers to independently verify claims or uncover hidden activities. The reliance on agency briefings and declassified reports means that oversight is often reactive rather than proactive, and dependent on the willingness of the agencies to disclose potentially problematic information.

Furthermore, the pervasive use of classification protocols exacerbates this information asymmetry. While classification is necessary to protect genuinely sensitive sources and methods, its application can become overly broad, encompassing information that has little or no bearing on national security but instead serves to shield programs from public and congressional scrutiny. Members of Congress, even those with high-level security clearances, are often denied access to the full picture, or they are provided with information on a need-to-know basis, which can lead to a fragmented understanding of complex operations. The sheer volume of classified documents and the complexity of intelligence programs mean that even dedicated oversight committees can struggle to process and analyze the information adequately. This can lead to a situation where members are forced to rely on the interpretations and assurances of agency officials, rather than their own independent assessments. The incentives are often aligned for agencies to err on the side of classification, and for members of Congress, facing time constraints and the potential political fallout of mishandling classified information, to accept this as a necessary reality.

The partisan nature of American politics also casts a long shadow over the effectiveness of congressional oversight. While oversight committees are intended to be bipartisan bodies dedicated to the national interest, they are not immune to the divisive forces that characterize broader congressional dynamics. When one party controls the presidency and the other controls one or both houses of Congress, oversight can become a political weapon, used to embarrass the administration or to advance a particular policy agenda. Conversely, when the same party controls both the executive and legislative branches, the incentive for robust oversight can diminish, as lawmakers may be reluctant to challenge an administration from their own party, fearing it could undermine their collective political strength or alienate key constituencies. This can lead to periods where oversight is perfunctory, characterized by a lack of critical questioning and a willingness to accept agency assurances at face value. The desire for partisan advantage or unity can override the fundamental responsibility to hold the executive branch accountable, allowing potential abuses to go unchecked.

The specific mechanisms of oversight, primarily the intelligence committees in both the House of Representatives and the Senate, have faced persistent criticism regarding their efficacy. These committees, by design, are meant to be the primary conduits for congressional oversight of the intelligence community. However, their structure and operational realities present inherent challenges. Committee membership is often a secondary assignment for many lawmakers, who may have more pressing committee responsibilities or who may not possess a deep understanding of intelligence matters. This can lead to a lack of specialized expertise and a reliance on the intelligence community to educate them on complex issues. Furthermore, the committees themselves are often susceptible to the same partisan pressures that affect the broader Congress. Debates over declassification of reports, the scope of investigations, and public statements about intelligence activities can become deeply divided along party lines, hindering the ability to present a united and forceful message to the executive branch.

Instances where oversight has been demonstrably bypassed or proven inadequate are numerous. The expansion of surveillance programs following the September 11, 2001, attacks provides a stark example. While there was a perceived need for enhanced security measures, the subsequent growth of data collection programs, such as those conducted by the National Security Agency (NSA), often occurred with limited and, at times, insufficient congressional knowledge and debate. Legislation that expanded surveillance powers, such as amendments to the Foreign Intelligence Surveillance Act (FISA), was sometimes passed with broad support but with a lack of deep understanding by many members of the full implications of the technologies and data collection methods being authorized. The revelation of programs like PRISM, which collected vast amounts of data from internet companies, exposed a significant gap between what Congress had authorized and the reality of executive branch activities. While some members were briefed on aspects of these programs, the extent and nature of the data collection were not widely understood or debated by the full body of Congress, and the oversight that did exist was often reactive, triggered by leaks or external revelations rather than proactive inquiry.

The concept of "oversight theatre" has also been invoked to describe situations where congressional committees engage in public hearings and issue reports that create the appearance of robust oversight, but fail to enact meaningful change or impose significant constraints on executive branch activities. This can be driven by a desire to project an image of accountability to constituents, or by the difficulty of overcoming the institutional inertia and the inherent advantages of the executive branch in controlling information. When intelligence agencies can effectively argue that certain information cannot be disclosed for national security reasons, it becomes difficult for Congress to hold them accountable for actions that may have violated privacy or civil liberties. The classification system can become a shield, not just for sources and methods, but for potentially unlawful or unethical behavior.

Moreover, the revolving door between intelligence agencies and lobbying firms or defense contractors can create subtle but powerful influences on congressional decision-making. Former intelligence officials often possess intimate knowledge of agency operations and the legislative process, making them valuable assets for those seeking to shape policy or secure funding. This can create an environment where members of Congress are persuaded by arguments that may be influenced by vested interests rather than a dispassionate assessment of the public good or the necessity of particular surveillance activities. The deep personal and professional connections between individuals in these spheres can inadvertently, or sometimes deliberately, lead to a relaxation of oversight standards.

The structure of committee assignments further complicates effective oversight. Members of the intelligence committees often serve alongside members of other powerful committees, such as appropriations or armed services. This can create competing priorities, as members may be more inclined to focus on securing funding for programs or advancing their own legislative agendas than on conducting the meticulous, often tedious, work of intelligence oversight. The expertise required to effectively oversee complex technical surveillance programs is significant, and it is often difficult to develop and maintain that expertise when committee assignments are transient or when members have limited time to dedicate to these specialized issues.

The capacity of congressional staff also plays a crucial role. While oversight committees are supported by professional staff, their resources and access can be limited compared to the vast resources available to intelligence agencies. Staffers may face challenges in developing deep technical expertise, navigating complex legal frameworks, or acquiring the necessary security clearances to access all relevant information. Furthermore, the turnover of staff can impede the development of institutional memory and expertise, making it difficult to build a consistent and rigorous oversight practice over time.

The historical context of how oversight has evolved is also important. Following periods of significant public outcry and revelations of abuses, such as those uncovered by the Church Committee in the 1970s, Congress did enact reforms aimed at strengthening its oversight capabilities. The creation of specialized intelligence committees was a direct result of these efforts. However, in the decades since, particularly in the wake of increasing global threats and technological advancements, there has been a tendency for executive branch power to expand, often with legislative acquiescence, in the name of national security. The balance between security and liberty, a perpetual tension in democratic societies, often tips towards security during times of heightened perceived threat, and congressional oversight mechanisms can be swept up in this dynamic.

The classification of congressional oversight activities themselves can also be a problem. While the substance of intelligence operations is often classified, the processes and findings of oversight can also be shrouded in secrecy. This makes it difficult for the public and even for non-intelligence-focused members of Congress to assess the effectiveness of the oversight system. When oversight findings are classified, or when recommendations for reform are kept confidential, the very notion of public accountability is undermined. The public is largely reliant on leaks or the occasional, heavily redacted report to understand the extent to which their elected representatives are effectively monitoring the intelligence community.

The problem is further compounded by the fact that

individual members of Congress, even those on oversight committees, may not fully grasp the implications of the technologies being deployed or the data being collected. The rapid pace of technological change, particularly in areas like artificial intelligence, machine learning, and mass data analytics, outstrips the capacity of even the most diligent lawmakers to stay fully informed. This knowledge gap makes it difficult to ask the right questions, to identify potential risks, and to enact appropriate safeguards. The intelligence community, with its dedicated technical experts, naturally possesses a deeper understanding of these evolving capabilities.

Ultimately, the deficiencies in congressional oversight of intelligence agencies are a complex interplay of structural, political, and practical factors. The information asymmetry, fueled by classification and the inherent secrecy of intelligence work, creates a fundamental imbalance. Partisan divisions can politicize oversight, rendering it a tool for partisan warfare rather than a genuine check on executive power. The specific structures of oversight committees, while designed to provide focus, are not immune to the broader dynamics of congressional life, including competing priorities and the demands of political expediency. The result is a system where the ability of Congress to effectively hold the intelligence community accountable for its surveillance activities is often compromised, leaving significant room for programs to expand and operate with less scrutiny than the principles of democratic governance demand. This creates a persistent challenge for safeguarding civil liberties in an era of increasingly sophisticated and pervasive intelligence operations.

Beyond the halls of Congress, the judiciary plays a critical, albeit often unseen, role in overseeing government surveillance activities. At the forefront of this judicial oversight, particularly concerning foreign intelligence, is the Foreign Intelligence Surveillance Court (FISC), often referred to as the "FISA Court." Established by the Foreign Intelligence Surveillance Act of 1978, the FISC was intended to provide a judicial check on the executive branch's powers to conduct surveillance for national security purposes. Its unique structure and operational procedures, however, have generated considerable debate about its effectiveness as a bulwark for civil liberties and constitutional rights.

The FISC operates under a veil of profound secrecy. Its proceedings are almost entirely ex parte, meaning that typically only the government presents evidence and arguments. The targets of surveillance, who are often unaware that they are being investigated, have no opportunity to be present, represented by counsel, or challenge the government's assertions. This asymmetry is justified by the government on the grounds that notifying targets would compromise ongoing investigations and reveal sensitive intelligence sources and methods. However, this practice also means that the FISC judges, appointed by the Chief Justice of the United States and drawn from federal district and appellate courts, primarily hear one side of the story. Their task is to review applications from the executive branch, primarily the Department of Justice and intelligence agencies, for warrants to conduct electronic surveillance and physical searches against individuals suspected of being agents of foreign powers or involved in international terrorism.

The nature of the information presented to the FISC is crucial. Agencies submit detailed applications, often voluminous, outlining the factual basis for their request, the legal justification, and the specific targets and methods of surveillance. These applications are reviewed by the FISC judges, who must determine if there is probable cause to believe that the target is an agent of a foreign power and that the proposed surveillance is necessary and appropriate. While the statutory standard for obtaining FISA warrants is often described as less stringent than the probable cause required for domestic criminal investigations (focusing on "foreign intelligence information" rather than criminal activity), the ex parte nature of the proceeding places an immense burden of trust on the government to present accurate, complete, and unvarnished information.

Critics have long raised concerns about the FISC's perceived deference to the executive branch. Given the specialized nature of intelligence gathering and the classification of most information, judges are heavily reliant on the government's representations. This reliance can inadvertently create an environment where the FISC acts more as a rubber stamp than an independent arbiter. For instance, in the wake of the September 11th attacks, the FISC's role came under intense scrutiny. Revelations, particularly those stemming from the Snowden leaks in 2013, exposed the vast scale of data collection programs, such as the NSA's bulk collection of telephone metadata under Section 215 of the USA PATRIOT Act. It became apparent that the FISC had, for years, authorized these programs based on government applications that, in the view of many critics and some courts, did not fully disclose the scope of the collection or the legal interpretations underpinning it.

One significant controversy revolved around the government's assertions regarding the minimization procedures required under FISA. These procedures are meant to ensure that information collected on U.S. persons incidentally swept up in foreign intelligence surveillance is properly handled and minimized to protect their privacy. In the case of
"In re: All Circumstances", a declassified opinion from the FISC in 2011, it was revealed that the NSA had been knowingly

violating its own minimization procedures, collecting vast amounts of domestic data in violation of its own rules and the court's orders. The FISC, in this instance, acknowledged the violations but permitted the NSA to continue its programs, requiring only that the agency fix its procedures. Critics argued that this response was insufficient, demonstrating a lack of robust enforcement and accountability from the court, particularly given the unprecedented scale of the data collection.

The issuance of Section 702 of the FISA Amendments Act of 2008 also brought new dimensions to the FISC's oversight responsibilities. Section 702 allows the government to conduct warrantless surveillance of non-U.S. persons located outside the United States, but it inevitably sweeps up communications of U.S. persons who communicate with those targets. The FISC reviews the government's foreign intelligence surveillance program (FISP) certifications for Section 702, approving the targeting procedures. However, the nature of the "ancillary collection" of U.S. person data has become a major point of contention. In 2018, the FISC issued a highly critical opinion, later declassified, finding that the FBI had conducted thousands of warrantless backdoor searches of U.S. persons' data collected under Section 702 without adequate justification. This finding highlighted a significant breakdown in oversight and a repeated pattern of the FBI accessing data in ways that arguably exceeded the court's authorization and statutory limitations. The FISC's opinion described the FBI's practices as "a significant departure from the standard of accuracy and completeness required," and noted that the agency had "failed to adequately address the statutory and constitutional implications" of its actions.

The limited transparency of FISC proceedings makes it challenging to assess the full scope of its impact. While some opinions and orders are declassified, often years after their issuance and with significant redactions, the vast majority of the FISC's work remains secret. This opacity makes it difficult for the public, civil liberties advocates, and even many lawmakers to fully understand how the court is exercising its oversight function and whether it is effectively balancing national security needs with constitutional protections. The ability of the FISC to provide meaningful judicial review is hampered not only by the ex parte nature of its proceedings but also by the inherent advantages of the government in controlling the information presented.

Dissenting opinions within the FISC, though rare and rarely made public, offer glimpses into internal disagreements about the government's requests. When a judge does dissent, it suggests that even within the closed FISC system, there are instances where the executive branch's justifications are questioned. However, the lack of a formal adversarial process means that these internal dissents do not have the same impact as a public court ruling, where both sides can vigorously argue their cases.

The judicial branch's role in overseeing surveillance extends beyond the FISC, though the FISC is the primary specialized court. In cases where FISA warrants are challenged or where the legality of surveillance programs is litigated in Article III courts (the regular federal courts), judges have sometimes taken a more critical stance. For instance, in

In re: United States, a 2013 ruling by the Second Circuit Court of Appeals, the court found that the NSA's bulk collection of telephone metadata under Section 215 of the PATRIOT Act violated the statute itself, deeming the program's scope too broad and not sufficiently tied to specific, articulable facts. While this ruling was later rendered moot by statutory changes and the program's termination, it represented a significant judicial check on executive surveillance practices, originating from a public court rather than the FISC.

Similarly, in *Humanitarian Law Project v. Clinton,* the Ninth Circuit Court of Appeals examined the constitutionality of provisions of FISA that prohibited providing material support to designated foreign terrorist organizations. While not directly about surveillance, such cases illustrate how the broader judiciary engages with national security statutes and their impact on fundamental rights. The tension between national security and individual liberties is a recurring theme, and the judiciary's interpretation of laws like FISA shapes the boundaries of government power.

However, the FISC's unique position as the gatekeeper for much of this intelligence surveillance means that its decisions have a disproportionate impact. The court's authority to issue anticipatory warrants, its interpretation of "agent of a foreign power," and its approval of novel surveillance technologies all shape the landscape of government intelligence gathering. The fact that the FISC's decisions are largely insulated from public scrutiny and adversarial testing creates a significant accountability gap. Unlike the regular courts, where public opinion, media scrutiny, and the advocacy of multiple parties can inform judicial reasoning, the FISC operates in a near-total information vacuum from the public perspective.

The reforms enacted by Congress over the years, particularly in response to revelations about past abuses, have aimed to strengthen the FISC's oversight capacity. The USA FREEDOM Act of 2015, for example, was a direct response to the NSA's bulk telephone metadata collection program. It ended that specific program, replacing it with a system where telecommunications companies retain the data and the government must obtain specific court orders from the FISC to access it, with specific numbers tied to terrorism investigations. This reform sought to re-center the FISC's role in authorizing data access, moving away from the broad, retrospective authorization of bulk collection.

Despite such reforms, questions persist about the FISC's ability to provide a truly independent check. The appointment process, where judges are selected by the Chief Justice, and their continued service on the FISC are discretionary. This could create subtle pressures to maintain cordial relations with the executive branch. Furthermore, the sheer volume of applications—often thousands per year—necessitates a swift review process, which might not always allow for the deep, deliberative consideration that constitutional rights often demand. The specialized knowledge required to understand complex digital surveillance techniques and cryptographic methods further complicates the task of independent judicial review.

The debate over judicial review of surveillance programs, particularly through the FISC, highlights a fundamental challenge in balancing national security with the protection of civil liberties. While the FISC was designed to provide a crucial judicial safeguard, its structure and the inherent secrecy of its operations raise legitimate concerns about whether it can effectively fulfill that role. The reliance on ex parte presentations, the potential for deference to executive branch expertise, and the limited transparency of its proceedings all contribute to a system where the judiciary's oversight function, though present, may not always provide the robust check that constitutional principles require. Landmark cases and declassified FISC opinions have, at times, revealed instances where the court has pushed back or expressed significant concerns, but these are often viewed against a broader backdrop of routine approvals, leaving a persistent question about whether the court's power is adequately exercised to safeguard fundamental rights in an era of pervasive digital surveillance. The effectiveness of the FISC, therefore, remains a critical point of contention in the ongoing struggle to define the boundaries of government power in the name of national security.

In the intricate machinery of government oversight, where information is meticulously controlled and accountability can often be obscured by layers of classification and bureaucratic process, whistleblowers emerge as a vital, albeit often embattled, force. They are the individuals who, by choice or by circumstance, find themselves privy to information that suggests significant transgressions of public trust or, more critically, the erosion of fundamental rights. Their decision to speak out, to pierce the veil of secrecy, is not undertaken lightly, for it often comes at a profound personal cost. The story of modern government surveillance, particularly in the United States, is inextricably linked to the actions of these courageous individuals who dared to expose what was being done in the shadows.

Perhaps no figure better embodies this modern reality than Edward Snowden. A former contractor for the National Security Agency (NSA), Snowden, in 2013, leaked a trove of classified documents that irrevocably altered the public's understanding of the extent of government surveillance capabilities. The documents revealed the vast scale of digital data collection undertaken by intelligence agencies, including programs that gathered telephone metadata from millions of Americans, intercepted communications from foreign leaders, and accessed user data from major technology companies through initiatives like PRISM. These revelations were not abstract; they represented a tangible intrusion into the private lives of ordinary citizens and demonstrated a reach of government power that many had not imagined. Snowden's actions, undertaken at immense personal risk, forced a global conversation about privacy, security, and the balance of power between the state and the individual. He operated under the belief that the public had a right to know how its government was using its resources and what technologies were being employed in its name. His willingness to forfeit his freedom, his career, and his ability to return to his home country underscores the profound moral weight of his decision.

Snowden's actions, and those of other whistleblowers before and after him, highlight a critical vulnerability in the system of checks and balances. When legislative and judicial oversight mechanisms prove insufficient to uncover or address potential abuses, individuals within the system often become the last line of defense. However, the legal landscape for whistleblowers, particularly those within the intelligence community, is fraught with peril. While there are laws and protections in place designed to encourage reporting of waste, fraud, and abuse within government, these protections are often inadequate or selectively applied, especially when national security is invoked. For intelligence community employees, the espionage statutes and other laws can be used to prosecute those who leak classified information, regardless of their intent or the public benefit derived from the disclosure. This creates a chilling effect, where the very individuals who might possess knowledge of illegal or unethical activities are deterred from coming forward by the threat of severe legal repercussions, including lengthy prison sentences.

The government's response to whistleblowers like Snowden has often been swift and severe. Rather than engaging with the substance of the revelations, the emphasis has frequently been on identifying the source of the leak and prosecuting that individual. This approach, critics argue, serves to punish dissent and discourage future whistleblowing, thereby protecting the government's secrecy apparatus rather than addressing the underlying issues raised. The prosecution of Snowden, under the Espionage Act of 1917, exemplifies this dynamic. He was charged with violations of this century-old law, a statute originally intended for wartime espionage, but which has increasingly been used to target leakers of classified information. The severity of these charges, and the possibility of spending decades in prison, sends a clear message about the risks involved in challenging government secrecy. Other individuals who have come forward with information about government overreach have faced similar fates, including David Petraeus, who was prosecuted for mishandling classified information, and Chelsea Manning, who leaked classified military documents detailing wartime conduct in Iraq and Afghanistan, and who faced court-martial and imprisonment.

The public debate ignited by Snowden's revelations was profound and far-reaching. It forced a reckoning with the practical implications of digital surveillance in an interconnected world. Citizens, policymakers, and technology companies alike were compelled to confront questions about the scope of government data collection, the adequacy of legal frameworks governing surveillance, and the fundamental right to privacy in the digital age. This public discourse led to legislative reforms, such as the USA FREEDOM Act of 2015, which aimed to curb the NSA's bulk collection of telephone metadata by requiring the government to obtain specific court orders for data held by telecommunications companies. It also spurred increased scrutiny of Section 702 of the Foreign Intelligence Surveillance Act, which permits the surveillance of non-U.S. persons abroad but can sweep up communications of Americans. The revelations also prompted widespread discussion about encryption, the role of technology companies in protecting user data, and the need for greater transparency in government surveillance activities.

However, the question remains whether these reforms have gone far enough to address the systemic issues exposed by whistleblowers. The legal protections for whistleblowers themselves remain a contentious issue. While some legislative efforts have sought to strengthen these protections, particularly for those reporting within specific agencies or programs, the broader landscape remains challenging. The national security context often provides a powerful justification for limiting transparency and prosecuting leaks, creating an inherent tension between the public's right to know and the government's perceived need for secrecy.

The challenge for future whistleblowers is immense. The government's capacity to detect and investigate unauthorized disclosures has grown exponentially with advancements in technology. Identifying leakers has become a sophisticated endeavor, often involving extensive forensic analysis of digital footprints and communications. This increased capacity for detection, coupled with the severity of the penalties, can act as a powerful deterrent. Moreover, the narrative surrounding whistleblowers can be manipulated by the government to portray them not as public servants exposing wrongdoing, but as traitors or criminals who have endangered national security. This framing can undermine public sympathy and support, making it more difficult for whistleblowers to find allies or protection.

Furthermore, the concept of "whistleblowing" itself can be politically charged. What one person considers a courageous act of public service, another might view as a dangerous act of disloyalty or a violation of sworn oaths. The distinction between leaking information that serves a clear public interest and leaking information that could genuinely harm national security or endanger lives is often blurred in the public and legal discourse. This ambiguity allows for the classification system and national security justifications to be used as a broad brush to silence legitimate criticism or to conceal governmental overreach.

The public's role in this ecosystem is also critical. Without an informed and engaged citizenry, the revelations of whistleblowers can be easily dismissed or distorted. The media plays a crucial part in responsibly reporting on classified information, balancing the public's right to know with the potential risks of disclosure. Investigative journalism, which often partners with whistleblowers, serves as a vital conduit for bringing such information to light, but it too operates under constraints, including legal threats and the difficulty of verifying and contextualizing highly sensitive material. The public's ability to understand the complex technical and legal issues involved in government surveillance is also essential for fostering informed debate and demanding appropriate accountability.

The legacy of Edward Snowden and others like him is that they have irrevocably shifted the conversation around surveillance and privacy. They have demonstrated that even within highly secure intelligence agencies, individual conscience can compel extraordinary action. However, their experiences also underscore the ongoing struggle to create a system that can effectively hold government accountable for its actions, particularly in the realm of national security, without unduly suppressing the flow of vital information that the public needs to understand and to govern itself. The absence of robust protections for whistleblowers, and the government's tendency to prioritize punishment over introspection following major leaks, suggests that the tension between secrecy and transparency, between security and accountability, remains a defining challenge for democratic societies. Without mechanisms that allow for the safe and effective disclosure of wrongdoing, the potential for unchecked power and the erosion of fundamental rights persists, leaving future oversight reliant on the courage and sacrifice of individuals willing to defy the established order.

The sheer scale of resources dedicated to intelligence activities, particularly within the United States, is almost incomprehensible to the average citizen, and indeed, often to those tasked with overseeing it. This is largely due to the existence of what is commonly known as the "black budget." This term refers to the portion of the national budget allocated to classified intelligence programs and operations, the specifics of which are kept from public view, and often even from large segments of elected officials. While the precise figure fluctuates annually, it consistently runs into the tens of billions of dollars, funding a vast array of agencies, initiatives, and technological developments that operate well beyond the typical scrutiny applied to other government expenditures. The very nature of the "black budget" is designed to shield sensitive operations from adversaries, but this shield inadvertently, or perhaps intentionally, also obscures the activities from domestic oversight, creating a significant accountability deficit.

The secrecy surrounding the black budget is not merely a matter of hiding operational details or technological capabilities. It extends to the very allocation of funds and the justification for these expenditures. When billions of dollars are dispensed within a framework of extreme classification, the traditional mechanisms of accountability – public debate, legislative review, and even detailed auditing – are severely compromised. Congress, while theoretically holding the purse strings, operates with significant limitations. Committees responsible for intelligence oversight, such as the House Permanent Select Committee on Intelligence and the Senate Select Committee on Intelligence, receive classified briefings and have access to some level of detail. However, the sheer volume of classified information, coupled with the need for absolute discretion, means that even these committees cannot fully scrutinize every program or expenditure. The culture of secrecy within the intelligence community, reinforced by national security imperatives, often means that information is provided on a "need-to-know" basis, leaving many with legitimate oversight responsibilities in the dark.

This opacity creates a fertile ground for unchecked growth in surveillance capabilities. Without clear and public understanding of what is being funded, how it is being used, and what its efficacy and implications are, agencies can expand their reach and develop new technologies with minimal resistance. The black budget, in essence, becomes a slush fund for innovation in surveillance, driven by internal priorities and perceived threats, rather than by explicit public mandate or rigorous, open debate. When the true costs and capabilities of these programs are hidden, it becomes impossible for lawmakers or the public to make informed decisions about whether these activities are truly necessary, proportionate, or consistent with democratic values and civil liberties.

Consider, for instance, the development and deployment of sophisticated data mining and analytical tools. These technologies, often born out of classified research and development funded through the black budget, can process and analyze vast quantities of information – much of it personal data – at unprecedented speeds. While proponents argue these tools are essential for national security, the lack of transparency means that the public has no way of knowing the extent to which their communications, online activities, or even their physical movements are being monitored and analyzed. The ethical implications of such widespread data collection and analysis, the potential for bias in algorithms, and the safeguards against misuse are all obscured by the classification veil.

The historical context of the black budget further illustrates the problem. Following the creation of the modern intelligence apparatus after World War II, particularly with the National Security Act of 1947, there was a recognized need for specialized funding and operations that could not be subjected to public scrutiny. However, over decades, the scope and ambition of intelligence activities have grown exponentially, outpacing the development of robust and effective oversight mechanisms. The evolution of technology has played a significant role; the digital revolution has made mass surveillance technically feasible in ways that were unimaginable even a few decades ago. The black budget has been the engine for acquiring and developing these cutting-edge technologies, often at the expense of transparency.

Moreover, the complex web of intelligence agencies and their overlapping jurisdictions further complicates oversight. The United States intelligence community is not a monolithic entity but a collection of 17 agencies, many of which operate under different departmental umbrellas and have their own specific mandates and funding streams within the broader black budget. This fragmentation, while sometimes serving the purpose of specialization, also creates gaps and blind spots for oversight bodies. When information is siloed within different agencies, or when programs are deliberately structured to avoid centralized oversight, the ability to gain a comprehensive understanding of the overall intelligence enterprise becomes exceedingly difficult. This is particularly true for programs that involve interagency cooperation or the sharing of data and capabilities, where the ultimate source of funding and authority can become deliberately obscured.

The reliance on contractors also adds another layer of complexity and opacity. A significant portion of the black budget is channeled through private defense and technology contractors. These entities, while subject to government oversight to some degree, operate with their own corporate structures and profit motives. The exact nature of their work, the data they collect, and the technologies they develop or maintain are often shielded by proprietary concerns and the classification requirements of their government contracts. This raises questions not only about the cost-effectiveness of outsourcing intelligence functions but also about the accountability for the actions of private actors operating under the broad umbrella of national security. When private companies are involved in sensitive surveillance activities, the lines of responsibility can become blurred, making it even harder to pinpoint who is accountable if something goes wrong or if rights are violated.

The legislative branch's role, while constitutionally mandated, is often hampered by the very nature of classified information. Members of Congress may have security clearances, but the sheer volume of classified material, often presented in dense, technical reports, requires significant time and expertise to review. Furthermore, the political dynamics within Congress can also play a role. Agencies, knowing that their funding is subject to legislative approval, can strategically manage the information they provide, highlighting successes and downplaying failures, or framing issues in a way that garners political support for continued funding. The pressure to appear tough on national security, coupled with the difficulty of publicly debating classified programs, can lead to a passive acceptance of agency requests for resources, even when those resources fund activities that might otherwise be deemed excessive or inappropriate.

The classification system itself, a necessary tool for protecting genuine national security secrets, becomes a weapon against accountability when applied too broadly or arbitrarily. The "black budget" thrives in an environment where almost any aspect of intelligence operations can be deemed classified, thereby preventing public discourse, independent review, or even meaningful congressional oversight. This creates a "black hole" where funds disappear into a complex, opaque system, fueling activities that, if brought into the light, might be subject to intense public criticism or legal challenge. Without the ability to know what is being done, for what purpose, and at what cost, the foundational principle of democratic governance – that the government is accountable to the people – is fundamentally undermined. The "black budget" is not just a line item in a government ledger; it is a symbol and an enabler of a system where oversight and accountability are perpetually struggling to keep pace with the expansion of state power and secrecy. The challenge, therefore, is not simply to increase the budget for oversight, but to fundamentally reform the way intelligence activities are funded, managed, and subjected to scrutiny, ensuring that the pursuit of security does not come at the permanent expense of democratic accountability.

The erosion of accountability stemming from the black budget is not a theoretical concern; it has tangible consequences for the functioning of a democratic society. When the vast majority of intelligence spending and operations are shielded from public view, citizens are left with an incomplete picture of their government's activities and priorities. This lack of transparency can foster a sense of detachment and distrust, as the public suspects that significant actions are being taken on its behalf without its knowledge or consent. Moreover, the absence of robust external review means that potential abuses of power, overreach in surveillance, or inefficient or wasteful spending can go undetected and unaddressed for extended periods.

The argument is often made that the very nature of intelligence work—dealing with foreign adversaries, sensitive sources and methods, and highly classified information—necessitates a degree of secrecy in budgeting and operations. Proponents of this view contend that revealing the specifics of the black budget would provide invaluable intelligence to those seeking to harm national security interests. While there is a legitimate basis for this concern, the current level of opacity extends far beyond what is strictly necessary for operational security. The challenge lies in drawing a clear line between information that genuinely requires protection and information that is classified primarily to avoid political embarrassment, stifle dissent, or shield inefficient or ineffective programs from scrutiny.

Furthermore, the budgetary process itself, for classified programs, often bypasses the robust debate and scrutiny that publicly funded initiatives undergo. While congressional committees review classified budget requests, the information presented is often highly sanitized and framed to elicit approval. The absence of a public record means that alternative approaches, cost-benefit analyses that consider societal impacts, or even simple checks on the necessity of certain programs are difficult to conduct. This can lead to the perpetuation of programs that are no longer relevant, effective, or proportionate to the threats they are intended to address. The "black budget" can, in effect, shield bureaucratic inertia and institutional self-interest from the corrective forces of accountability.

The evolution of technology, particularly in the realm of digital surveillance, has amplified the challenges associated with the black budget. The development of sophisticated capabilities for data collection, analysis, and prediction—often funded through these classified channels—has outpaced the legal and ethical frameworks designed to govern them. Without clear public understanding and debate about the deployment and implications of these technologies, agencies can acquire and utilize them with minimal checks and balances. This has led to situations where surveillance capabilities are deployed broadly, impacting large segments of the population, without explicit authorization or informed consent from the governed. The black budget, by funding these technologies in secrecy, allows for their proliferation and integration into government operations before the public or its representatives have had a meaningful opportunity to weigh the trade-offs between security and liberty.

The impact of this opacity on oversight is profound. When oversight bodies lack access to fundamental information about what is being funded, who is benefiting, and what the intended and actual outcomes are, their ability to provide meaningful oversight is severely curtailed. They are essentially asked to approve and monitor activities based on incomplete or highly filtered information. This dynamic can lead to a situation where oversight becomes a rubber-stamping exercise, reinforcing the power of the executive branch and the intelligence agencies, rather than serving as a genuine check on their activities. The very purpose of congressional oversight – to ensure that government operates legally, ethically, and effectively in accordance with the will of the people – is compromised when a significant portion of its activities are conducted in the dark, funded by a budget that itself is largely invisible.

Moreover, the concept of accountability extends beyond just fiscal responsibility. It also encompasses accountability for the ethical implications of surveillance activities, the adherence to legal frameworks, and the protection of civil liberties. When these aspects of intelligence operations are obscured by the black budget, it becomes nearly impossible to hold agencies accountable for any potential transgressions. The lack of transparency prevents the public, the media, and even the courts from fully understanding the scope and nature of surveillance programs, making it exceedingly difficult to challenge their legality or their impact on fundamental rights. This creates a situation where intelligence agencies can operate with a significant degree of impunity, shielded by the classification of their budgets and operations.

The cycle of secrecy perpetuates itself. As more programs and technologies are developed and funded through the black budget, the need for secrecy to protect these investments and operations grows, further entrenching the lack of transparency. This can lead to a self-reinforcing loop where secrecy breeds more secrecy, and accountability becomes increasingly elusive. The challenge for those seeking to strengthen oversight and accountability is to find ways to inject sunlight into this opaque system without compromising legitimate national security interests. This is a delicate balancing act, but one that is essential for the health of a democratic society. Without a fundamental understanding of how intelligence resources are being used, and without robust mechanisms for holding those who wield these powers accountable, the potential for unchecked expansion of government power and the erosion of individual liberties remains a significant and present danger. The "black budget" serves as the financial engine of this opacity, a critical choke point where the flow of information essential for accountability is deliberately dammed.

The pervasive and often impenetrable veil of secrecy surrounding intelligence operations and their funding mechanisms creates a profound deficit in public knowledge. This inherent lack of transparency is not a byproduct of occasional necessary discretion; rather, it has become a systemic feature of the modern surveillance state, fundamentally hindering any genuine pursuit of democratic accountability. The mechanisms by which this opacity is maintained are multifaceted, ranging from the explicit operational secrecy demanded by the very nature of intelligence gathering to the deliberate application of classification practices and the strategic exploitation of legal loopholes. Together, these elements conspire to keep the vast majority of citizens not only unaware of the specific technologies and methods employed in data collection and analysis but also ignorant of the sheer scale and scope of these activities.

At the core of this issue is the operational necessity that intelligence agencies often cite for their clandestine nature. The argument is consistently made that to effectively gather information on adversaries, protect sources and methods, and conduct covert operations, a degree of secrecy is indispensable. While this premise holds undeniable truth in specific, narrowly defined circumstances, its application has, over decades, expanded far beyond its original intent. What began as a shield for truly sensitive operations has morphed into a broad justification for withholding information about activities that have direct and significant implications for the daily lives of ordinary citizens. The collection and analysis of vast troves of personal data, for instance, are presented as operational necessities, yet the public is rarely informed about the specifics of what data is being gathered, from whom, and for what precise analytical purposes. This creates a chasm between the perceived need for secrecy and the public's right to know how its government is operating, particularly when those operations involve its most private information.

Classification practices, while intended to protect genuine national security secrets, have become a powerful tool in maintaining this deficit of public knowledge. The sheer volume of information classified by intelligence agencies far exceeds what could reasonably be considered vital to national security. Reports suggest that the vast majority of classified documents never see the light of day and are unlikely to contain information that, if revealed, would genuinely imperil national security. Instead, classification is often employed to safeguard programs from public scrutiny, insulate agencies from criticism, or protect against political embarrassment. When the details of data collection programs, the methodologies of algorithmic analysis, or the scale of metadata harvesting are routinely classified, the public is rendered incapable of understanding the extent to which their digital lives are being surveilled. This is not merely an academic concern; it directly impacts the ability of individuals and civil society to assess whether these pervasive surveillance activities are proportionate, necessary, or even lawful.

Legal loopholes further solidify this wall of secrecy. For instance, the interpretation and application of legislation like the Foreign Intelligence Surveillance Act (FISA) have, in practice, often expanded the reach of surveillance beyond what was originally envisioned or publicly understood. Provisions related to "incidental collection" or the use of Section 702 of FISA to collect information on foreign nationals that may incidentally sweep up data from U.S. citizens have been sources of significant controversy and public concern. However, the specifics of how these provisions are implemented, the volume of data collected, and the mechanisms for reviewing and safeguarding the information of U.S. persons are often shrouded in secrecy, making it exceedingly difficult for the public and even for lawmakers to fully grasp the implications. The lack of transparency here is not a passive occurrence; it is an active consequence of how laws are written, interpreted, and executed in the name of national security, often with limited public input or understanding.

This deficit of public knowledge acts as a formidable barrier to democratic accountability. In a democratic society, the government's legitimacy and its operations are ideally informed by the consent of the governed. This consent, however, can only be meaningful if the governed are sufficiently informed about what their government is doing. When the extent of surveillance, the nature of data analysis, and the implications for civil liberties are hidden from public view, informed debate becomes virtually impossible. Citizens cannot meaningfully participate in discussions about privacy, security, and the appropriate balance between the two if they do not know the facts of the situation. This lack of informed consent means that powerful surveillance capabilities can be developed and deployed without the public having the opportunity to weigh in, challenge their necessity, or demand safeguards. The result is a chilling effect on civic engagement and a weakening of the democratic process itself.

The consequences of this opacity extend beyond the realm of informed public debate. It also significantly impedes meaningful reform. Without a clear understanding of the problems—the specific types of data being collected, the technologies used, the potential for misuse, and the actual efficacy of these programs—it becomes exceedingly difficult to advocate for or implement effective reforms. How can one demand limits on a particular form of data collection if its existence and nature are classified? How can one advocate for stronger privacy protections if the scope of data analysis remains unknown? This creates a situation where potential abuses can persist unchecked, and where well-intentioned efforts to improve oversight can be hampered by a fundamental lack of accessible information. The cycle is self-perpetuating: secrecy prevents knowledge, lack of knowledge prevents accountability, and lack of accountability allows secrecy to flourish.

The role of investigative journalism and public advocacy becomes, therefore, not merely supplementary but absolutely essential to the functioning of a democratic society in the face of pervasive state secrecy. It is through the dedicated efforts of journalists and advocacy groups that cracks in the wall of silence are often revealed. Whistleblowers, often risking severe professional and legal repercussions, have historically been instrumental in bringing to light the true extent of government surveillance programs. Investigative reporting then takes this information, contextualizes it, verifies it, and presents it to the public in a comprehensible form, thereby injecting much-needed transparency into otherwise opaque operations. Organizations dedicated to civil liberties and privacy rights play a crucial role in analyzing the information that does become public, informing the public about their rights, and advocating for legislative and policy changes to curb overreach and enhance accountability.

The impact of such efforts cannot be overstated. Revelations about expansive surveillance programs, often made possible by the courage of whistleblowers and the persistence of investigative journalists, have, at various times, galvanized public opinion, sparked legislative action, and prompted judicial review. These moments of transparency, however infrequent, serve as critical reminders that even in the face of immense state power and secrecy, the pursuit of accountability is possible. They underscore the vital role of a free press and an active civil society in holding power to account, especially when that power operates in the shadows. Without these crucial checks, the deficit in public knowledge would likely become an unbridgeable chasm, leaving citizens entirely at the mercy of whatever operations the state chooses to conduct in their name, unseen and unquestioned.

The challenge, therefore, is to move beyond reactive transparency—the kind that often emerges only after significant breaches or leaks—towards a more proactive and systemic approach. This involves re-examining the classification system to ensure it is used judiciously and not as a tool to stifle legitimate public inquiry. It requires a commitment to making the broad outlines of surveillance activities, the legal authorities under which they operate, and the general nature of the data collected accessible to the public. Furthermore, strengthening the oversight capabilities of legislative bodies and empowering independent review mechanisms are crucial steps. These measures, however, can only be truly effective if they are informed by a bedrock of public knowledge, accessible to the very citizens whose privacy and liberties are at stake. The fight for transparency in surveillance is, in essence, a fight for the very foundations of democratic governance, a continuous effort to ensure that the pursuit of security does not inadvertently dismantle the liberties it is meant to protect. The lack of public knowledge about surveillance is not merely an inconvenience; it is an existential threat to the principles of a free and open society, a fertile ground where unaccountable power can take root and flourish, shielded from the essential light of democratic scrutiny. Without this essential knowledge, the citizenry is rendered a passive observer, incapable of fulfilling its fundamental role in governing itself.

6: LEGAL BATTLES AND PRIVACY RIGHTS

The advent of the digital age has irrevocably altered the landscape of personal privacy, presenting novel and complex challenges to established legal frameworks designed to protect citizens from unreasonable searches and seizures. At the heart of this ongoing legal evolution lies a fundamental question: what constitutes a 'search' in a world where information is increasingly intangible, dispersed, and constantly collected? The Fourth Amendment of the United States Constitution, a cornerstone of individual liberty, guarantees the right of the people to be secure in their persons, houses, papers, and effects, against unreasonable searches and seizures. However, the abstract nature of digital data, residing not in physical 'houses' or 'papers' but in vast server farms, cloud storage, and transient electronic signals, strains the applicability of the Amendment's literal protections. The legal battles that have emerged in response reflect a continuous, often contentious, effort to bridge the gap between the Fourth Amendment's enduring principles and the realities of 21st-century technology.

Historically, the Supreme Court's interpretation of the Fourth Amendment was largely tethered to physical intrusion. The seminal case of *Olmstead v. United States* (1928), for example, established a "trespass doctrine," holding that a search only occurred if there was a physical invasion of a constitutionally protected area. In *Olmstead*, the Court ruled that wiretapping telephone lines, which did not involve physical entry into the defendants' homes or offices, did not constitute a search. This physical trespass requirement, however, proved increasingly inadequate as technology advanced, allowing for government intrusion without any tangible encroachment. The understanding of what constitutes a "legally protected interest" in privacy began to shift, moving away from property-based rights towards a more subjective, privacy-focused analysis.

This paradigm shift was most notably marked by the Supreme Court's decision in *Katz v. United States* (1967). In *Katz*, the Court abandoned the strict trespass doctrine, famously stating, "The Fourth Amendment protects people, not places." Justice Harlan's concurring opinion introduced the now-iconic two-pronged test for determining whether a person's Fourth Amendment rights have been violated: first, whether the individual has exhibited an actual (subjective) expectation of privacy, and second, whether that expectation is one that society is prepared to recognize as "reasonable" (objective). This "reasonable expectation of privacy" test became the bedrock for analyzing Fourth Amendment claims in the digital age, even though it was articulated in a pre-digital era. The challenge has been to apply this subjective-objective standard to the ephemeral and interconnected nature of digital information.

The initial application of the *Katz* standard to digital communications proved challenging. For instance, in *United States v. Jones* (2012), the Supreme Court confronted the issue of GPS tracking devices attached to a suspect's vehicle. While a majority of the Court ultimately found that the physical attachment of the device constituted a search under the Fourth Amendment, a significant plurality, including Justice Sotomayor, focused on the privacy implications of prolonged, continuous government monitoring of a person's movements. This movement-tracking, they argued, revealed intimate details of a person's life, constituting an intrusion into their "private affairs" even without a physical trespass in the traditional sense. The Court acknowledged that the aggregation of granular location data over an extended period could reveal a great deal about an individual's habits, associations, and even their constitutionally protected activities, underscoring the need for a broader understanding of what constitutes a search in the digital realm.

A related and highly debated area concerns the privacy of data held by third parties, often referred to as the "third-party doctrine." Under this doctrine, established in cases like *United States v. Miller* (1976) concerning bank records and *Smith v. Maryland* (1979) concerning telephone company pen register (call detail) records, individuals generally have no reasonable expectation of privacy in information they voluntarily share with third parties. The reasoning was that by disclosing information to a bank or a phone company, individuals are essentially entrusting that information to those entities, and thus their expectation of privacy diminishes. This doctrine has been critically important in the digital age, as much of our online activity involves interacting with third-party service providers – email providers, social media platforms, cloud storage services, and internet service providers (ISPs).

The application of the third-party doctrine to modern digital services has led to significant concern among privacy advocates. In *Riley v. California* (2014), the Supreme Court unanimously held that police generally cannot search the digital contents of a cell phone seized from an individual who has been arrested without a warrant. The Court reasoned that cell phones are vastly different from the physical items, like wallets or briefcases, that could be searched incident to arrest. A modern smartphone contains an "immense storage capacity," capable of holding not just phone numbers and call logs, but also photographs, emails, text messages, browsing history, location data, and a vast array of personal information. To search a cell phone without a warrant, the Court concluded, would be to allow the government to access the "privacies of life" that the Fourth Amendment was designed to protect. This decision, while focused on digital content, implicitly questioned the extent to which the third-party doctrine could or should apply to the vast amounts of data contained within personal electronic devices.

The debate over metadata further illustrates the challenges in defining a 'search.' Metadata, often described as "data about data," can include information about communications rather than their content – for example, who you called, when you called them, for how long, and from where. In *United States v. Jones*, while the Court was divided on the primary reasoning for finding a search, several justices suggested that the collection of cell-site location information (CSLI) – the data that cell towers collect to route calls and texts – might constitute a search if done prospectively and for an extended period. Justice Sotomayor, in her concurrence, argued that the government's collection of CSLI for 28 days constituted a "profoundly intimate window into [Jones's] life, revealing the attendances of his political and religious beliefs, the span of his dawns and the timing of his meals, his medical appointments and his associations." This suggested that even if not directly accessing content, the aggregation of metadata could reveal deeply personal information, thereby implicating a reasonable expectation of privacy.

The Supreme Court revisited the issue of CSLI in *Carpenter v. United States* (2018). In a landmark decision, the Court held that the government generally needs a warrant to access historical CSLI held by a mobile phone provider. The Court explicitly rejected the government's argument that the third-party doctrine applied to CSLI, stating that "an individual maintains a legitimate expectation of privacy in the whole of his physical movements." The justices reasoned that obtaining CSLI for even a short period could reveal a wealth of information about a person's life, far exceeding the scope of information revealed by a simple pen register. By aggregating data from multiple cell towers over extended periods, the government could reconstruct a person's movements and thereby gain access to "the privacies of life." The *Carpenter* decision signaled a significant re-evaluation of the third-party doctrine's applicability in the digital age, particularly concerning the vast quantities of location data generated by modern mobile devices. The Court emphasized that the nature of the information gathered and the extent of government surveillance were crucial factors in determining whether a search had occurred.

The implications of *Carpenter* extend beyond just cell-site location data. It has raised questions about the privacy of other types of digital information shared with third parties, such as browsing history, search queries, and even the content of emails or messages stored in the cloud. If the government seeks to access such data, particularly in aggregate or over extended periods, *Carpenter* suggests that a warrant may be required, even if that data is technically held by a third-party service provider. This evolving understanding acknowledges that the ease with which vast amounts of personal data can be collected and analyzed in the digital age necessitates a more robust application of Fourth Amendment protections. The Court's willingness to depart from a rigid application of the third-party doctrine in *Carpenter* reflects a growing judicial awareness that technological advancements can profoundly alter the nature of privacy expectations.

Moreover, the concept of "electronic surveillance" itself has broadened considerably. It now encompasses not just wiretaps of phone conversations but also the monitoring of emails, text messages, social media activity, and internet browsing. The Electronic Communications Privacy Act (ECPA) of 1986, and its subsequent amendments, attempts to regulate government access to electronic communications. However, ECPA's framework, particularly its distinction between accessing "stored" versus "in-transit" communications, has been a source of significant legal debate. Under ECPA, government access to stored emails, for instance, might require a warrant, while access to emails in transit could be obtained through less stringent legal processes. This distinction has become increasingly blurred as cloud computing and asynchronous communication methods become prevalent.

The ongoing legal interpretation of what constitutes a 'search' is a dynamic process, constantly adapting to new technologies and surveillance techniques. For example, the use of sophisticated data analytics, artificial intelligence, and facial recognition technology raises further questions about privacy. When the government uses algorithms to sift through massive datasets, identifying patterns, flagging individuals, or predicting behavior, is that a search? If these systems analyze publicly available information, or information obtained through less intrusive means, do they still implicate Fourth Amendment rights? The legal system is grappling with how to apply the 'reasonable expectation of privacy' standard to situations where data is not being directly accessed from an individual but rather aggregated and analyzed in ways that can reveal intimate details.

The challenge lies in ensuring that Fourth Amendment protections keep pace with technological innovation. As governments and private entities develop increasingly sophisticated methods for collecting, analyzing, and storing personal data, the traditional boundaries of privacy become more porous. Court decisions like *Riley* and *Carpenter* represent important steps in adapting constitutional law to this new reality. They acknowledge that the digital footprint individuals leave behind is as private, if not more so, than their physical presence. The ongoing legal battles over digital privacy underscore the critical need for continuous judicial and legislative attention to ensure that the Fourth Amendment remains a meaningful shield for individual liberties in the digital age, safeguarding against unreasonable intrusions into the increasingly vast and complex digital lives of citizens. The definition of 'search' is no longer simply about physical intrusion; it is about the nature of the information accessed, the methods of access, and the potential for those methods to reveal the intimate details of a person's life, regardless of whether that information is stored in a physical file cabinet or on a remote server. This ongoing redefinition is crucial for preserving fundamental rights in an era of ubiquitous surveillance.

The digital age has ushered in an era where the fundamental tenets of the Fourth Amendment—the right to be secure against unreasonable searches and seizures —are facing unprecedented challenges. While past legal interpretations often revolved around physical intrusion, the contemporary landscape is dominated by the collection and analysis of vast digital datasets, electronic communications, and the deployment of sophisticated surveillance technologies. This dynamic environment compels a re-examination of what constitutes a "search" and whether existing constitutional protections can adequately safeguard individual privacy in the face of pervasive government data acquisition. The core of this ongoing debate lies in the tension between the government's asserted need for information, particularly in the name of national security, and the citizenry's expectation of privacy in their increasingly digitized lives.

At the heart of many government surveillance justifications is the continued reliance on, or reinterpretation of, established legal doctrines, most notably the "third-party doctrine." This doctrine, as established in cases like *United States v. Miller* and *Smith v. Maryland*, posits that individuals relinquish their reasonable expectation of privacy when they voluntarily disclose information to third parties. In the digital realm, this translates to information held by internet service providers, telecommunications companies, social media platforms, and cloud storage providers. The government often argues that since data is entrusted to these entities, it is no longer protected by the Fourth Amendment, allowing for its acquisition through subpoenas or other less stringent legal processes than a warrant, especially when the data is deemed to be "in transit" or when the quantity involved is considered manageable. However, privacy advocates argue that this doctrine was conceived in an era with fundamentally different technological capabilities and that its application to the massive, often involuntary, collection of data today is a misapplication that strips citizens of meaningful privacy. They contend that the sheer volume and sensitivity of information held by third parties—our call logs, browsing histories, email content, location data, and social connections —collectively paint an intimate portrait of an individual's life, a portrait that society reasonably expects to remain private.

The government's rationale frequently leans heavily on national security imperatives. Following events like the September 11th terrorist attacks, legislative and executive actions expanded government surveillance powers, often under the umbrella of counterterrorism. Programs aimed at collecting vast amounts of metadata, such as call detail records or internet browsing habits, are defended as essential tools for identifying potential threats before they materialize. The argument is that these broad data sweeps are necessary to connect the dots and prevent attacks, and that the incidental collection of data belonging to innocent individuals is a necessary cost of ensuring public safety. This approach often frames the issue as a trade-off between privacy and security, suggesting that an overly strict interpretation of the Fourth Amendment could hinder vital intelligence gathering. The legal theories employed to bypass traditional warrant requirements have included arguments that certain types of data collection do not constitute a "search" under the Fourth Amendment because they do not involve physical intrusion or because the information has been voluntarily disclosed to a third party. For instance, before the *Carpenter* decision, the government successfully argued that cell-site location information (CSLI) was discoverable through less than a warrant, often citing *Smith v. Maryland* and the notion that phone companies, as third parties, held this data.

Civil libertarians and privacy advocates counter these arguments by emphasizing the chilling effect that mass surveillance can have on freedom of speech, association, and even thought. They argue that knowing one's communications and movements are constantly being monitored can deter individuals from expressing dissenting opinions, engaging in lawful protests, or associating with certain groups, thereby undermining core democratic principles. The aggregation of seemingly innocuous data points, they contend, can reveal an individual's most private habits, beliefs, and relationships, making the collection of metadata as intrusive, if not more so, than a physical search of one's home. The *Carpenter v. United States* decision, in which the Supreme Court ruled that the government generally needs a warrant to access historical CSLI, marked a significant pushback against the government's expansive use of the third-party doctrine. The Court recognized that the nature of CSLI —providing a detailed, chronological record of a person's movements over extended periods—was inherently private and that its acquisition constituted a search implicating the Fourth Amendment. This ruling underscored the idea that the *method* of collection and the *nature* of the data are critical to the analysis, suggesting that traditional doctrines may not be sufficient in the digital age.

The application of these principles to other forms of digital data continues to be a battleground. For example, the collection of internet search histories, browsing data, and location information stored by applications on smartphones raises similar questions. If individuals use private browsing modes or take steps to anonymize their online activity, do they not retain a reasonable expectation of privacy in that data, even if it is technically held by a third-party service provider? Privacy advocates argue that such actions demonstrate a subjective expectation of privacy, and that society recognizes this expectation as objectively reasonable given the sensitive nature of the information. The government's counterarguments often revert to the third-party doctrine or argue that such data is not "private" because it is transmitted across public networks or stored on servers accessible by multiple parties.

Furthermore, the development and deployment of advanced surveillance technologies, such as facial recognition software, predictive policing algorithms, and sophisticated data analytics tools, present novel challenges. When these technologies analyze vast datasets, often compiled from publicly available information or data acquired through less intrusive means, do they constitute a search? The debate here centers on whether the aggregation and algorithmic analysis of information, even if individually public, can create a privacy interest in the aggregate. For instance, if a program uses facial recognition to scan public surveillance footage and cross-reference it with other data, creating a detailed profile of an individual's movements and associations, does this constitute an unreasonable search? The Fourth Amendment jurisprudence has historically focused on specific instances of intrusion, not necessarily on the systematic analysis of broadly accessible information that, when combined, reveals deeply personal insights.

The legal system is attempting to grapple with these issues through a variety of means.

Some courts have narrowly interpreted the third-party doctrine in light of *Carpenter*, suggesting that it should not apply to the continuous, comprehensive collection of data that reveals intimate details about an individual's life. Others have focused on the concept of "reasonable expectation of privacy," arguing that in the digital age, individuals have a heightened expectation of privacy in the data they generate, even when shared with third parties. The government, meanwhile, continues to advocate for broader access to digital information, often citing national security and law enforcement needs, and has sought to expand exceptions to warrant requirements. This includes arguments for "exigent circumstances" in the digital realm, or claims that certain data is not protected because it is "readily available to the public."

The ongoing legal battles over the Fourth Amendment's reach in the digital age highlight the critical need for legislative action and judicial adaptation. While landmark decisions like *Riley v. California* and *Carpenter v. United States* have provided crucial protections for digital privacy, the rapid evolution of technology means that the legal landscape remains in flux. The very definition of what constitutes a "search" and "seizure" is being continuously tested and redefined. Privacy advocates argue that a proactive approach is necessary, one that either updates existing laws like the Electronic Communications Privacy Act (ECPA) to reflect modern technological realities or creates new legal frameworks that explicitly protect digital privacy. They contend that without such measures, the Fourth Amendment risks becoming an anachronism, unable to provide meaningful safeguards against the pervasive surveillance capabilities of the modern state. The government's argument for expansive surveillance powers, while often framed in terms of security, must be balanced against the fundamental rights guaranteed by the Constitution, ensuring that the pursuit of safety does not inadvertently erode the very liberties it is meant to protect. The core question remains: can the bedrock principles of the Fourth Amendment, designed for a world of physical objects and tangible spaces, effectively govern the intangible realm of digital information and pervasive data collection, or will these principles be irrevocably diluted in the crosshairs of modern surveillance? The answer to this question will shape the future of privacy rights in the 21st century.

The Constitution, particularly the Fourth Amendment, has long served as the bedrock of privacy protection in the United States. However, the advent of the digital age and the evolving nature of government surveillance have compelled the Supreme Court to revisit and reinterpret these fundamental protections. This ongoing legal evolution is best understood through an examination of key Supreme Court cases that have incrementally, and sometimes dramatically, shaped the landscape of privacy rights in an increasingly interconnected world. These rulings are not merely historical footnotes; they are the living expressions of constitutional principles grappling with technological realities, setting precedents that continue to influence how government intrusion is assessed and what individuals can reasonably expect in terms of privacy.

One of the foundational cases in understanding Fourth Amendment protections, even predating the digital era, is *Katz v. United States* (1967). While the case itself concerned a physical search—the electronic bugging of a public telephone booth—its significance lies in its revolutionary redefinition of what constitutes a "search" under the Fourth Amendment. Prior to *Katz*, Fourth Amendment protections were largely tied to physical trespass onto property, a concept known as the "physical penetration" or "trespassory" test. The government could wiretap a phone booth, for instance, without a warrant, as long as they did not physically intrude upon the booth itself. Justice Potter Stewart, writing for the majority, famously shifted this focus, declaring that "the Fourth Amendment protects people, not places." This pivotal statement introduced the "reasonable expectation of privacy" test. The Court held that if an individual exhibits an actual (subjective) expectation of privacy and that expectation is one that society is prepared to recognize as "reasonable" (objective), then the Fourth Amendment applies, regardless of whether there was a physical trespass. In *Katz*, the Court found that a person using a public telephone booth had a subjective expectation of privacy and that society reasonably recognizes such an expectation, thus making the warrantless bugging a Fourth Amendment violation. This principle of "reasonable expectation of privacy" became the cornerstone for analyzing Fourth Amendment claims in all contexts, including the digital realm, setting the stage for future debates about what activities and data fall within constitutionally protected spheres. The legacy of *Katz* is profound, as it moved the analysis from the physical nature of the intrusion to the privacy interests of the individual, a conceptual leap that would prove crucial as technology began to mediate and record human activities in ways previously unimaginable. It established that the government cannot simply disregard privacy interests by choosing a location that, while public, is used for private communication, a concept that resonates strongly today with our reliance on public Wi-Fi networks and cloud-based services where our data is stored and transmitted.

Decades later, the Supreme Court confronted the implications of new technologies in *United States v. Jones* (2012). This case involved the warrantless installation of a GPS tracking device on a suspect's vehicle, which was then used to monitor the vehicle's movements for 28 days. The government argued that since the vehicle was in public places and the GPS device was small, there was no physical trespass and therefore no Fourth Amendment search. However, the Court, in a unanimous decision, found the warrantless GPS tracking to be an unconstitutional search. The plurality opinion, authored by Justice Scalia, revived the trespassory test from pre-*Katz* jurisprudence, emphasizing that the physical attachment of the device to the Joneses' property constituted a search. Scalia reasoned that "the government's installation of a GPS device on the respondent's vehicle, and its subsequent use of that device to monitor the vehicle's movements, constituted a 'search.'" This was not merely about the aggregate of data, but the physical intrusion to install the device, which was a clear governmental intrusion onto private property.

However, the concurring opinions in *Jones* offered a more forward-looking interpretation that would prove highly influential. Justice Sotomayor, in her concurrence, articulated a powerful concern about the cumulative effect of technology on privacy, even without physical trespass. She argued that the aggregation of GPS data over extended periods allowed the government to gain an "intimate window into [the Joneses'] lives," revealing not just where they went, but their habits, routines, and associations. She questioned whether the *Katz* reasonable expectation of privacy test was sufficient in the face of such technologically enabled pervasive surveillance, suggesting that the mere retention and aggregation of data, even if collected from public spaces, could constitute a search. Justice Alito, also concurring, focused on the "aggregate volume and temporal length of the data" collected, agreeing that prolonged tracking could intrude upon privacy expectations. He proposed a "longer-term, more general conspiracy to monitor the public citizen's movements" as a potential threshold for Fourth Amendment protection, even if the individual data points were not considered private. The *Jones* decision, therefore, was significant not only for its revival of property-based analysis but also for highlighting the emerging challenge of "dataveillance"—the systematic collection and analysis of data about individuals—and how it might necessitate new interpretations of Fourth Amendment protections, even if the justices did not fully coalesce on a single rationale. The plurality's reliance on trespass, while seemingly a step backward from *Katz*, was seen by some as a pragmatic way to address the technological intrusion without upending the established privacy test entirely. Yet, the concurring opinions laid the intellectual groundwork for future decisions that would more directly confront the privacy implications of digital data aggregation.

The tension between the physical trespass approach and the reasonable expectation of privacy, amplified by the data aggregation concerns raised in *Jones*, came to a head in *Carpenter v. United States* (2018). This case directly addressed the government's ability to obtain cell-site location information (CSLI) from wireless carriers without a warrant. CSLI is data that shows which cell towers a mobile phone connected to at various times, effectively creating a historical record of a person's movements. The government had argued that CSLI was akin to information voluntarily disclosed to third parties, such as phone records, and therefore not protected by the Fourth Amendment under the third-party doctrine established in cases like *Smith v. Maryland*.

The Supreme Court, in a 5-4 decision, delivered a landmark ruling that significantly curbed the government's ability to access historical CSLI. Chief Justice Roberts, writing for the majority, explicitly rejected the government's attempt to extend the third-party doctrine to CSLI. He reasoned that CSLI is fundamentally different from the type of information at issue in *Smith v. Maryland*. Unlike simply revealing the numbers dialed or the duration of calls, CSLI provides a "detailed chronicle of a person's past movements," essentially a diary of one's physical presence throughout each day. Roberts wrote, "A person does not by his voluntary commingling of his address with that of his telephone company surrender all Fourth Amendment protection of that address." The Court emphasized that the acquisition of a comprehensive record of a person's movements over an extended period constitutes a "search" under the Fourth Amendment, and that such a search generally requires a warrant supported by probable cause.

Crucially, the *Carpenter* Court declined to definitively overrule the third-party doctrine, stating that it was "not persuaded that... [the] third-party doctrine has any applicability to cell-site location information." This careful delimitation suggests that the doctrine might still apply to other forms of data, or to less extensive collections of CSLI. However, the ruling strongly implied that the doctrine is not infinitely elastic and that the nature of the data and the scope of its collection are critical factors in the Fourth Amendment analysis. The Court also considered the privacy implications of aggregated data, acknowledging that CSLI paints "an]- exhaustive record of the holder's whereabouts." This acknowledgment of the privacy harm inherent in data aggregation, even without a physical trespass, represents a significant development in Fourth Amendment jurisprudence, aligning with the concerns raised in the concurring opinions of *Jones*.

The *Carpenter* decision was a watershed moment, signaling a judicial willingness to adapt Fourth Amendment principles to the realities of modern technology. It recognized that the ubiquitous nature of smartphones and the data they generate create new vulnerabilities for privacy, and that traditional legal frameworks, particularly the third-party doctrine, may not adequately protect individuals from pervasive government surveillance. By requiring warrants for historical CSLI, the Court provided a crucial safeguard for a type of data that offers deep insights into an individual's life. This ruling has had immediate and ongoing implications, prompting law enforcement agencies to seek warrants more frequently for such information and prompting debates about what other forms of digital data collection might similarly require Fourth Amendment protections. The decision also raised questions about prospective CSLI, or real-time tracking, which the Court did not directly address, leaving further legal battles on the horizon. Moreover, the emphasis on the "comprehensive" and "chronological" nature of the data collected suggests that other forms of aggregated digital information, such as location data from apps, browsing history, or social media activity, might also warrant similar protections.

Beyond these foundational cases, other legal developments and lower court decisions continue to refine the understanding of digital privacy rights. For instance, the *Riley v. California* (2014) decision, decided shortly before *Carpenter*, held that police generally need a warrant to search the digital contents of a cell phone seized from an individual arrested. The Court recognized that the vast amount of private information stored on modern smartphones—emails, texts, photos, location history, browsing data—makes them qualitatively different from physical items like wallets or address books, which were historically subject to search incident to arrest. The *Riley* Court's reasoning emphasized that a digital search of a phone could reveal far more intimate details of a person's life than a search of any physical item, and that the concerns about officer safety or evidence destruction that justify searches incident to arrest did not extend to the digital contents of a phone. This ruling firmly established that the Fourth Amendment's protections extend to the data stored on digital devices, a critical affirmation in an era where phones are digital extensions of ourselves.

The ongoing legal battles underscore the dynamic nature of constitutional law in the face of technological advancement. The Supreme Court's rulings in cases like *Katz*, *Jones*, *Riley*, and *Carpenter* illustrate a clear trend towards recognizing and protecting privacy interests in the digital realm, even as the government continues to assert its needs for surveillance. These decisions, while providing critical guidance, also open new avenues for legal debate. For example, the question of how much data aggregation is too much, or what constitutes a "voluntary" disclosure in the age of pervasive data collection by digital platforms, remains unsettled. The government's arguments often rely on interpretations of the third-party doctrine or claims that data is "publicly available," but the Supreme Court's recent jurisprudence suggests a growing skepticism towards such broad claims when they lead to comprehensive surveillance.

Privacy advocates continue to push for clearer legal standards that acknowledge the unique privacy challenges of the digital age. They argue that existing statutes, such as the Electronic Communications Privacy Act (ECPA), may be outdated and require updating to reflect current technologies and surveillance practices. The debate over how to balance national security and law enforcement needs with individual privacy rights is a constant and evolving one. The Supreme Court, through its decisions, is actively engaged in this balancing act, seeking to ensure that the Fourth Amendment remains a robust protector of liberty in the 21st century. The legacy of these cases is not a final answer, but a continuous process of re-evaluation, ensuring that the fundamental right to privacy is not eroded by the relentless march of technological innovation. Each ruling serves as a marker, defining the boundaries of permissible government intrusion and reinforcing the enduring principle that even in the digital ether, individuals retain a sphere of protected privacy. The judiciary's role in interpreting and adapting constitutional rights to new technological realities is paramount, and the path forged by these landmark decisions continues to guide the evolving understanding of our digital privacy rights.

The intricate web of national security legislation and evolving surveillance technologies has inevitably led to significant legal challenges, testing the boundaries of constitutional rights in an era defined by data and interconnectedness. At the forefront of these challenges are statutes like Section 702 of the Foreign Intelligence Surveillance Act (FISA) and key provisions of the Patriot Act. These laws, enacted in the wake of national security crises, have been lauded by proponents for their effectiveness in preventing terrorism and other threats, but have simultaneously drawn sharp criticism from civil liberties advocates who argue they grant the government excessive power, often at the expense of fundamental privacy and free speech rights.

One of the most prominent legal battlegrounds has been Section 702 of FISA, which authorizes the government to conduct warrantless surveillance of foreign nationals located outside the United States. While ostensibly targeting non-U.S. persons, the statute's implementation has raised profound concerns about the incidental collection of data pertaining to American citizens. The Foreign Intelligence Surveillance Court (FISC), a secret court that oversees surveillance requests, operates largely outside public scrutiny. This opacity, coupled with the sheer volume of data collected, has fueled arguments that Section 702 facilitates a form of backdoor surveillance, where the communications of U.S. persons can be swept up and accessed without a warrant.

Legal challenges against Section 702 have often been spearheaded by organizations such as the American Civil Liberties Union (ACLU), the Electronic Frontier Foundation (EFF), and the Brennan Center for Justice. Their arguments typically hinge on alleged violations of the Fourth Amendment's protection against unreasonable searches and seizures. Plaintiffs contend that the querying of the vast databases collected under Section 702, which can include the communications of Americans communicating with foreign targets, constitutes a search requiring individualized suspicion and a warrant. They point to instances where U.S. persons' data has been accessed, arguing that the government's broad justification for these queries, often citing national security, is insufficient to overcome constitutional protections.

Moreover, challenges have also been mounted on First Amendment grounds. Critics argue that the knowledge that one's communications might be monitored, even if incidentally, can have a chilling effect on free speech and association. Individuals may self-censor their communications, refrain from contacting certain foreign nationals, or avoid discussing sensitive political topics for fear of being caught in the surveillance net. This argument posits that the broad scope of Section 702 surveillance, and the potential for its misuse, undermines the robust public discourse essential for a functioning democracy.

The procedural hurdles in litigating these challenges are substantial. Many cases have been dismissed on grounds of standing, with courts finding that plaintiffs could not adequately demonstrate they had been directly harmed by the surveillance programs. The highly classified nature of these programs means that plaintiffs often cannot know with certainty if their communications have been intercepted, making it difficult to prove concrete injury. This has led to a reliance on affidavits from government officials that are often vague or uncorroborated, and to legal arguments that must navigate a complex landscape of national security exemptions and secrecy.

Despite these difficulties, some legal actions have achieved partial successes or have significantly shaped the public discourse and legislative debate. For example, in *In re National Security Agency Telecommunications Records Litigation*, various plaintiffs challenged the legality of the NSA's bulk collection of U.S. telephone metadata, alleging violations of the Fourth Amendment and statutory protections like the Stored Communications Act. While the government initially invoked broad secrecy, the case eventually led to admissions about the scope of the program and contributed to reforms, such as the USA FREEDOM Act of 2015, which ended the government's bulk collection of U.S. phone records and shifted to a system where telecommunication companies hold the data and require specific court orders to release it.

The Patriot Act, enacted in the wake of the September 11th attacks, also became a focal point for legal challenges. Sections like 215, which allowed the government to obtain "any tangible things" relevant to terrorism investigations, were particularly controversial. This provision was famously used to authorize the NSA's telephony metadata collection program. Numerous lawsuits were filed against Section 215, with plaintiffs arguing that it violated the Fourth Amendment, the First Amendment, and the principle of statutory limitation. These challenges culminated in significant rulings, including a landmark decision by the Second Circuit Court of Appeals in *ACLU v. Clapper* (later reversed on standing grounds by the Supreme Court), which found that Section 215 did not authorize the NSA to collect the telephony metadata of all Americans. The Supreme Court's eventual decision in *Clapper v. Amnesty International USA* focused on the difficulty of proving standing, but the underlying legal arguments against the broad interpretation of Section 215 continued to resonate.

The legislative landscape itself has been a site of intense debate and legal interpretation. Following revelations by Edward Snowden in 2013 about the extent of government surveillance, public outcry and legal pressure led to significant reforms. The USA FREEDOM Act, for instance, was a direct response to the controversies surrounding Section 215 of the Patriot Act. It aimed to rein in bulk data collection and strengthen oversight mechanisms. Similarly, debates surrounding Section 702 have led to periodic reauthorizations by Congress, often with amendments designed to increase transparency and accountability, though many advocates argue these reforms do not go far enough.

The legal challenges against these statutes highlight a persistent tension between the government's asserted need for broad surveillance powers to protect national security and the fundamental rights of citizens. The classified nature of many surveillance programs creates a significant imbalance, making it incredibly difficult for individuals and advocacy groups to mount effective legal defenses. The courts, in turn, have grappled with how to apply established constitutional principles, such as the Fourth Amendment's requirement for probable cause and particularity, to the novel forms of data collection and analysis that characterize modern intelligence gathering.

One of the core difficulties in challenging these statutes lies in the interpretation of "foreign intelligence" versus "domestic law enforcement." While FISA is designed to govern foreign intelligence collection, the lines can blur when intelligence gathered under FISA is subsequently used for domestic law enforcement purposes. Critics argue that this can be a way to circumvent domestic legal requirements, such as the warrant requirement. The government often maintains that such data usage is permissible under its national security mandate, but this interpretation is frequently contested in legal forums.

Furthermore, the role of the Foreign Intelligence Surveillance Court (FISC) has been a subject of intense scrutiny. As a court that operates in secret and rarely denies government requests, its effectiveness as a check on executive power is questioned. Lawsuits have attempted to open up the FISC's proceedings or to challenge the legal interpretations it has adopted. However, these efforts have generally been met with resistance, citing the need for secrecy in matters of national security. The Supreme Court's decision in *Clapper v. Amnesty International USA* further complicated matters by making it harder for plaintiffs to establish standing to challenge the FISC's rulings.

The ongoing litigation and advocacy surrounding FISA, the Patriot Act, and other surveillance statutes underscore the dynamic and often contentious relationship between government power, national security imperatives, and individual liberties in the digital age. While legal victories for privacy advocates have been hard-won and often incremental, the persistent challenges have undeniably contributed to greater public awareness and legislative reform, signaling a continuous effort to adapt constitutional protections to the realities of modern surveillance. The debate remains active, with advocates continually seeking to clarify the scope of government surveillance and to ensure that technological advancements do not erode fundamental privacy rights and freedoms. The difficulty in bringing successful legal challenges against highly classified government programs operating under broad national security justifications remains a significant barrier, yet the persistence of these legal battles demonstrates a deep-seated commitment to upholding constitutional principles even in the face of evolving threats and advanced surveillance capabilities.

The persistent tension between the imperative of national security and the fundamental right to privacy forms the bedrock of many legal battles in the post-9/11 era. On one side stand privacy advocates and a coalition of civil liberties organizations, who fervently argue that robust privacy protections are not merely a personal preference, but a cornerstone of a free and democratic society. They contend that unfettered government surveillance, even if ostensibly aimed at preventing threats, erodes the very freedoms that national security efforts are meant to preserve. Their philosophy is rooted in the belief that a society where citizens fear constant monitoring is a society that stifles dissent, limits free expression, and ultimately undermines democratic accountability.

These advocates, often operating through influential organizations like the American Civil Liberties Union (ACLU), the Electronic Frontier Foundation (EFF), the Brennan Center for Justice, and numerous smaller, grassroots groups, champion a legal and ethical framework that prioritizes individual autonomy and the presumption of innocence. They draw heavily on the Fourth Amendment of the U.S. Constitution, which guards against unreasonable searches and seizures, arguing that modern surveillance technologies, with their capacity for mass data collection and analysis, inherently violate this principle when implemented without specific, individualized suspicion and judicial oversight in the form of warrants. They posit that the ability of government agencies to collect vast troves of data – communications metadata, internet browsing history, location data, and even the content of communications – on millions of people, regardless of any suspected wrongdoing, constitutes an unreasonable intrusion into private lives.

The arguments put forth by privacy advocates are

multifaceted. Firstly, they highlight the "chilling effect" that widespread surveillance can have on free speech and association. When individuals know or suspect that their communications and activities are being monitored, they may self-censor, hesitate to express controversial opinions, avoid contacting certain people or organizations, or refrain from engaging in political activism for fear of being flagged or scrutinized. This fear, they argue, can lead to a less vibrant public sphere, where critical discourse is muted and intellectual exploration is curtailed. For example, a journalist investigating government misconduct might be hesitant to communicate with sources if they believe their correspondence is being routinely intercepted. Similarly, an activist organizing a peaceful protest could be deterred from using online platforms if they fear their communications will be used to identify and potentially suppress their movement.

Secondly, privacy advocates emphasize the potential for abuse and mission creep. While surveillance programs are often justified by national security concerns, history has shown that the tools developed for one purpose can easily be repurposed for others. They worry that information collected for counter-terrorism could be used to target political opponents, monitor activists, or even discriminate against certain communities. The lack of transparency surrounding many of these programs makes it difficult to identify and correct such abuses. Without clear limits and robust oversight, these powerful tools, they argue, represent a significant threat to the balance of power between the state and the individual. The ability to access and analyze personal data can provide an unprecedented level of insight into citizens' lives, creating a power imbalance that is antithetical to a democratic society.

Furthermore, privacy advocates challenge the notion that mass surveillance is an effective or necessary means of ensuring security. They often point to studies and expert opinions suggesting that targeted surveillance, based on specific intelligence and judicial authorization, is more effective than broad data sweeps. They argue that the sheer volume of data collected often overwhelms intelligence analysts, making it harder to identify genuine threats amidst the noise. Moreover, they contend that focusing solely on technological solutions can distract from addressing the root causes of terrorism and other security threats, such as socioeconomic factors, political grievances, and ideological extremism.

On the opposing side of this complex debate stand national security officials, intelligence agencies, and policymakers who advocate for the necessity of sophisticated surveillance tools. Their arguments are grounded in the perceived realities of modern threats, which they characterize as increasingly transnational, technologically sophisticated, and capable of inflicting mass casualties. From their perspective, the primary duty of the government is to protect its citizens from harm, and in an interconnected world where threats can materialize rapidly and originate from anywhere, effective intelligence gathering is paramount.

National security proponents argue that traditional methods of intelligence gathering are no longer sufficient to counter contemporary threats. They point to the speed at which individuals can communicate, plan, and organize online, often using encrypted channels and sophisticated anonymization techniques, as evidence of the need for advanced capabilities. They contend that surveillance is not about spying on innocent citizens, but about identifying and disrupting plots before they can be executed. Without the ability to monitor communications and activities of suspected individuals and groups, they argue, intelligence agencies would be operating blind, leaving the nation vulnerable to attacks.

These officials often invoke the principle of "necessity," arguing that while privacy is important, it cannot come at the expense of public safety. They emphasize that surveillance is conducted under legal frameworks and oversight mechanisms, even if those mechanisms operate in secrecy for national security reasons. They might point to the Foreign Intelligence Surveillance Act (FISA) and its amendments, or to presidential directives, as evidence of a structured approach to surveillance, designed to balance security needs with legal constraints. They often assert that the vast majority of collected data is never reviewed by analysts, and that access to information pertaining to U.S. persons is subject to strict legal limitations and oversight.

The arguments from the national security perspective also often highlight the difficulty in predicting where the next threat will emerge. They argue that intelligence gathering must be proactive rather than reactive. This requires the ability to collect and analyze data broadly, identifying patterns and connections that might not be apparent through targeted methods alone. For instance, understanding the communication networks of a terrorist organization might involve monitoring the communications of many individuals, some of whom may not be directly involved in plotting an attack but are nevertheless part of the broader network. The argument is that a comprehensive understanding of these networks is essential for identifying key players and disrupting their activities.

The debate, therefore, often boils down to a fundamental disagreement about risk assessment and the acceptable trade-offs between security and liberty. National security officials operate under the assumption that the potential for catastrophic harm is ever-present, and that preemptive measures, even if they intrude upon privacy, are a necessary cost of doing business in a dangerous world. Privacy advocates, conversely, tend to focus on the gradual erosion of civil liberties that can occur through pervasive surveillance, viewing the potential for harm from government overreach as a more insidious, though equally real, threat to societal well-being.

The quest for equilibrium between these competing interests is fraught with challenges. The very nature of intelligence work, involving clandestine operations and sensitive information, makes transparency and public accountability inherently difficult. Critics often argue that the secrecy surrounding surveillance programs allows governments to operate with insufficient checks and balances, shielding potentially overreaching or abusive practices from public scrutiny. This opacity fosters distrust and makes it harder for the public to engage in informed debate about the appropriate scope of government power.

Moreover, the rapid evolution of technology constantly shifts the landscape. As new surveillance tools and data analysis techniques are developed, the potential for intrusion into privacy grows, often outpacing legal and ethical frameworks. Encryption technologies, artificial intelligence-driven analytics, facial recognition, and the ubiquitous nature of data collection through smartphones and the internet all present new challenges to established notions of privacy. This necessitates continuous adaptation and re-evaluation of legal boundaries, a process that is often slow and contentious.

The legal battles, as explored in previous sections, are a direct manifestation of this societal struggle. Lawsuits filed by privacy advocates, while often facing steep hurdles such as proving standing and navigating classified information, serve a crucial purpose in challenging government interpretations of law, forcing greater transparency, and influencing legislative reform. The outcomes of these cases, whether they result in outright victories, partial concessions, or dismissals, contribute to the ongoing public and legal discourse on the proper balance between national security and individual privacy. They keep the dialogue alive, ensuring that the potential for governmental overreach remains a subject of scrutiny and that the constitutional rights of citizens are not quietly set aside in the name of security. Ultimately, this persistent tension and the legal challenges it engenders are indicative of a healthy, albeit often difficult, democratic process striving to define the boundaries of state power in an increasingly complex and surveilled world. The challenge lies in ensuring that the pursuit of security does not inadvertently undermine the foundational principles of freedom that national security is intended to protect.

7: CORPORATE COMPLICITY AND PROFITEERING

The digital age has birthed an intricate, largely invisible ecosystem operating beneath the surface of everyday life: the data broker industry. These are not the tech giants whose names are household words, but a sprawling network of companies whose primary, and often sole, purpose is to collect, aggregate, analyze, and ultimately, sell vast quantities of personal information about individuals. Their business model is built on the premise that data is the new oil, and they are the prospectors, refiners, and distributors. They function as the unseen architects of personalized advertising, sophisticated marketing campaigns, and increasingly, as a critical, albeit often clandestine, source of intelligence for governmental and law enforcement agencies. Understanding this ecosystem is crucial to grasping the full scope of surveillance in the modern era, as it represents a powerful, profit-driven engine that fuels many of the information-gathering practices that elude traditional public scrutiny.

The sheer breadth of data collected by these entities is staggering. It begins with the seemingly innocuous: public records, readily available information that paints a foundational portrait of an individual. This includes property records, voter registration information, marriage and divorce filings, court records, and business licenses. These are the accessible building blocks, often scraped and digitized with remarkable efficiency. But the data acquisition extends far beyond these public repositories, delving deep into the private sphere through myriad channels. Online activity forms a significant trove. Every website visited, every search query entered, every social media interaction, every app downloaded and used – all generate data points that are meticulously collected, often through sophisticated tracking mechanisms embedded across the internet. Cookies, web beacons, and third-party trackers are not merely tools for targeted advertising; they are the digital tendrils reaching out to capture the nuances of our online behavior.

Furthermore, purchasing histories represent another rich vein for data brokers. Point-of-sale data from retailers, loyalty program information, online purchase histories from e-commerce platforms, and even transaction data from financial institutions (often anonymized or aggregated, but the aggregation itself can be de-anonymized) are all grist for the data broker mill. These transactions reveal consumption patterns, brand preferences, lifestyle choices, and even sensitive personal information such as health-related purchases or financial commitments. The data can be granular enough to infer life events, such as a recent move, a pregnancy, or a change in employment status. This information is then purchased by brokers from data aggregators, data management platforms (DMPs), and directly from data-rich companies, often under complex contractual agreements that obscure the ultimate use of the data.

The aggregation process is where the true power, and the inherent danger, of the data broker ecosystem emerges. Individual data points, while informative, are often incomplete. Data brokers excel at the art of stitching together these disparate pieces to create comprehensive, detailed profiles of individuals. They employ sophisticated algorithms and data science techniques to link data from various sources, building a mosaic of a person's life. This profile might include demographic information (age, gender, income, education level), geographic location and movement patterns, consumer behavior, online interests, political leanings, religious affiliations, family status, and even inferred psychological traits. These profiles are then segmented and categorized, creating valuable audiences for marketers and, critically, for entities seeking to understand and influence populations.

The business model is straightforward: collect data, enrich it, package it, and sell it. The customers are diverse. Primarily, they are marketers and advertisers seeking to reach specific demographics with tailored messages, thereby increasing the efficiency and effectiveness of their campaigns. Companies utilize this data to understand their customer base, identify potential new customers, and personalize their outreach. Beyond marketing, however, the utility of this data extends into areas with far more significant implications for civil liberties. Financial institutions use it for risk assessment and fraud detection. Insurance companies employ it to underwrite policies and set premiums. Real estate companies use it to identify potential buyers and sellers. And, as will be further explored, government agencies and law enforcement entities have become increasingly significant, often covert, consumers of this commercially gathered data.

What makes the data broker ecosystem particularly concerning from a civil liberties perspective is its operation largely outside the purview of traditional privacy regulations that govern government surveillance. While government agencies are typically bound by legal frameworks like the Fourth Amendment, requiring warrants or specific legal justifications for accessing personal information, data brokers operate in a commercial space where data collection and sale are often viewed as legitimate business activities. Laws like the general data protection regulation (GDPR) in Europe and, to a lesser extent, state-level privacy laws in the United States, are beginning to impose some constraints, but the industry remains vast and largely self-regulated, particularly in its cross-border operations and the opaque nature of its data flows.

This creates what can be termed a "shadow surveillance network." Government agencies, seeking to circumvent the legal hurdles and public scrutiny associated with direct data acquisition, can simply purchase the information that data brokers have already amassed. This might involve buying access to vast databases of consumer information, purchasing detailed location data triangulated from mobile devices, or acquiring data analytics services that can process and interpret information for specific intelligence needs. This purchasing of data effectively allows government entities to acquire personal information on a massive scale without the need for traditional warrants or the direct application of surveillance laws. The data is already collected, processed, and curated by private entities, and it can be acquired through a simple transaction, bypassing the checks and balances that are meant to safeguard individual privacy from state intrusion.

For instance, consider the proliferation of mobile location data. Many applications on smartphones collect precise location data, often with user consent buried deep within lengthy terms of service agreements. This data is then frequently sold to data aggregators, who in turn sell it to data brokers. These brokers can then aggregate and package this location data, creating detailed movement histories for millions of individuals. Government agencies, such as immigration and customs enforcement (ICE) or federal law enforcement agencies, have been documented to purchase this commercially available location data, using it to track individuals without obtaining warrants or even informing them of the surveillance. This method sidesteps the legal requirements for probable cause and judicial oversight that would be necessary if the government were to directly request such location data from mobile carriers. The ease of acquisition and the lack of direct governmental intrusion in the initial collection phase create a plausible deniability and a significant gap in accountability.

Similarly, data brokers amass extensive information about individuals' online activities, browsing histories, and app usage. This information, often deemed sensitive and protected under various legal interpretations, can be purchased by government entities for a variety of purposes, ranging from counter-terrorism investigations to assessing social media sentiment, or even conducting background checks on individuals without their knowledge or consent. The commercial nature of the data collection means that it is not subject to the same legal protections as data directly collected by the government. This creates a tiered system of privacy, where individuals have fewer protections against information that has been intermediated through private commercial entities.

The opacity of the data broker ecosystem is a significant contributing factor to its ability to operate in the shadows. The complex web of data collection, sharing, and reselling makes it incredibly difficult to trace the provenance of data or to understand precisely who has access to what information about whom. Companies that collect data often sell it to other companies, which then further aggregate, refine, and resell it. This multi-layered transaction process means that by the time data reaches a government agency, it may have passed through multiple intermediaries, each adding its own layer of commercial value and obscuring its origins. This lack of transparency makes it challenging for individuals to know if their data has been compromised, to correct inaccuracies, or to opt out of data collection and sale.

Moreover, the lack of robust federal privacy legislation in the United States exacerbates the problem. Unlike many other developed nations, the U.S. lacks a comprehensive federal law that grants individuals broad rights over their personal data. Instead, privacy protections are often sector-specific, meaning that different types of data are subject to different, often weaker, regulations. This patchwork of laws creates significant loopholes that the data broker industry, and by extension, government agencies that utilize their services, can exploit. The result is a landscape where personal information can be treated as a commodity, bought and sold with minimal oversight, and then leveraged for surveillance purposes without the legal safeguards typically associated with government intelligence gathering.

The ethical implications of this commercialized surveillance are profound. When personal data, including highly sensitive information, is treated as a product to be bought and sold, it strips individuals of their agency and their right to control their own digital identities. The profit motive of data brokers can incentivize ever more intrusive data collection methods, as the value of data increases with its granularity and breadth. This commercial imperative can clash directly with the fundamental right to privacy, creating a system where personal information is exploited for financial gain, and then repurposed for state intelligence gathering, often without the knowledge or consent of the individuals concerned.

Furthermore, the data held by brokers can be used for more than just surveillance; it can be used for manipulation. Detailed profiles can reveal vulnerabilities, preferences, and biases that can be exploited in political campaigns, social engineering, or even to subtly influence public opinion. When combined with sophisticated analytical tools, this data allows for microtargeting of individuals with tailored messages, which can be used to persuade, dissuade, or even polarize populations. The potential for misuse in democratic processes, by both domestic and foreign actors, is a significant and often underestimated threat.

The data broker ecosystem is not a monolithic entity but a complex and dynamic industry. It includes companies that specialize in data aggregation, data enrichment, data analytics, and data marketing. Some brokers focus on specific types of data, such as consumer credit information, healthcare data, or mobile device location data, while others aim to build comprehensive profiles by integrating data from across all domains. The industry is characterized by a high degree of consolidation, with larger players acquiring smaller ones to expand their data holdings and analytical capabilities. This consolidation further concentrates power and data within a few dominant entities, making oversight and accountability even more challenging.

The legal arguments surrounding the data broker industry often revolve around the definition of "personally identifiable information" and the scope of consent. Proponents of the industry argue that much of the data they collect is either publicly available or anonymized, and that users consent to data collection through clickwrap agreements and terms of service. Critics, however, point out that these agreements are often unread and incomprehensible, and that the "anonymization" processes are frequently imperfect, allowing for de-anonymization through cross-referencing with other datasets. Moreover, they argue that the sheer breadth of data collection and the creation of comprehensive profiles go far beyond what a reasonable person would understand to be implied consent for commercial use, let alone for potential government surveillance.

The increasing reliance of government agencies on data brokers represents a fundamental shift in how surveillance can be conducted. It allows for the procurement of capabilities that might otherwise be illegal or require significant legal justification. By outsourcing data collection and processing to the private sector, agencies can amass intelligence on individuals and populations without the direct accountability and transparency that would accompany traditional, government-led surveillance operations. This privatized surveillance model creates a critical vulnerability in the protection of civil liberties, as it leverages the profit-driven motives of the commercial sector to circumvent the legal and ethical boundaries designed to protect citizens from state intrusion. The data broker ecosystem, therefore, is not merely an ancillary part of the digital economy; it is a foundational element of a pervasive, often invisible, surveillance infrastructure that has profound implications for privacy, autonomy, and democracy. Its operations highlight a critical gap in regulatory oversight, where the commercialization of personal data has inadvertently paved the way for extensive, low-scrutiny intelligence gathering by government entities, operating under a veil of commercial necessity and technological advancement.

The digital landscape, increasingly dominated by a handful of colossal technology firms, has inadvertently or perhaps deliberately transformed these private enterprises into indispensable partners of the state in its quest for surveillance. Companies like Google, Meta (formerly Facebook), Apple, and major telecommunications providers are not merely platforms for communication and information sharing; they are custodians of an immense, deeply personal, and granular dataset pertaining to billions of individuals worldwide. Their infrastructures, designed for immense data processing and user engagement, have become conduits through which government agencies can access capabilities that would be exceedingly difficult, if not impossible, to replicate through traditional means. This symbiotic relationship, often shrouded in legal mandates and complex operational agreements, is a critical, yet frequently understated, pillar of modern surveillance architecture.

The mechanisms through which these technology behemoths cooperate with government surveillance vary, ranging from overt compliance with legal demands to more subtle forms of data provision facilitated by their core business models. At the most direct level, these companies are subject to legal process, including warrants, subpoenas, and, more controversially, National Security Letters (NSLs) issued by the FBI. NSLs, a powerful administrative subpoena, allow the government to compel the production of certain types of sensitive customer records, such as subscriber information and billing records, from telecommunications common carriers and other entities. Crucially, NSLs can be issued without judicial oversight and often come with gag orders, preventing the recipient company from disclosing that it received such a request. For tech giants, this means they can be legally obligated to hand over vast troves of user data – ranging from search histories and email content to location data and social connections – under a veil of secrecy. The sheer volume of users and data processed by these companies means that compliance with even a single NSL can provide intelligence on an individual or a group with significant breadth and depth.

Beyond these formal legal requisitions, the relationship becomes more intricate with programs like PRISM, famously revealed by Edward Snowden. PRISM provided the National Security Agency (NSA) with direct access to the servers of major U.S. technology companies. This access allowed the NSA to collect data in various categories, including emails, instant messages, videos, file transfers, and social networking details, often in near real-time. While companies involved have historically maintained that they only provide data when legally compelled and under strict oversight, the existence of such programs implies a level of cooperation that goes beyond passive compliance. The technical architectures of these companies, built to seamlessly transfer and process user data for their own purposes, were leveraged to facilitate government access. This raises profound questions about the extent to which the infrastructure of private communication and data storage has been integrated into national security apparatuses.

The ethical tightrope walked by these technology companies is precarious and often contradictory. On one hand, they publicly champion user privacy and data protection, investing heavily in encryption technologies and privacy policies designed to reassure their user base. They often argue that they push back against government data requests when they deem them overly broad or lacking proper legal basis. For example, companies have engaged in legal battles to challenge the scope of certain government demands for data, particularly when those demands extend beyond U.S. borders or seek information in bulk without specific targeting. These efforts are often framed as defending user rights and maintaining public trust. However, the reality of their operational engagement with intelligence agencies often involves a complex negotiation between legal obligations, business continuity, and public relations.

The public perception of these companies' complicity in the surveillance state is a multifaceted issue. For many users, the realization that the platforms they rely on for daily life are also conduits for government data access can be a profound shock, leading to a sense of betrayal and a desire for greater transparency. Social media posts, private messages, and location histories, once perceived as private interactions within a digital ecosystem managed by private companies, are now understood to be potentially accessible to state actors. This awareness can foster a chilling effect, discouraging open expression and potentially leading to self-censorship. The constant tension between the companies' stated commitment to privacy and their documented cooperation with government surveillance creates a significant trust deficit.

Furthermore, the business models of many of these tech giants are intrinsically linked to data collection and analysis. While some data is used for targeted advertising, other forms of data analysis can serve broader governmental intelligence needs, creating a potential conflict of interest. When a company's core business involves extracting and leveraging personal data, the impulse to cooperate with entities that can offer significant financial incentives or legal assurances is understandable, even if it strains their stated privacy commitments. The argument that such cooperation is necessary to avoid more intrusive, less controlled government methods (e.g., direct hacking of systems) is often made, but it doesn't fully assuage concerns about the privatization of surveillance capabilities.

The telecommunications sector, often overlooked in discussions focused on internet giants, plays an equally critical role. Mobile carriers, internet service providers (ISPs), and other network operators are the gatekeepers of raw communication data. They possess records of call logs, text messages, internet browsing activity, and, crucially, precise location data derived from cell tower triangulation or GPS. These companies are often directly subject to statutes like the Communications Assistance for Law Enforcement Act (CALEA) in the U.S., which mandates that they build capabilities to facilitate lawful interception of communications. While CALEA was initially designed to support wiretaps, its application has expanded with the evolution of communication technologies, encompassing internet traffic and mobile data. The networks these companies manage are the arteries through which digital information flows, making them prime targets and willing or unwilling collaborators for government surveillance.

The nature of these collaborations often involves establishing secure, direct channels for data transfer. For example, the U.S. government may establish dedicated lines or secure portals for intelligence agencies to request and receive specific data sets from telecommunications providers or tech companies. These arrangements are typically governed by strict protocols and legal frameworks, but the sheer volume and sensitivity of the data being transferred are immense. The companies argue that these processes are designed to protect the integrity of the data and the privacy of their users to the greatest extent possible under legal constraints, but the fundamental reality is that they are facilitating government access to private information.

Apple, for instance, has famously positioned itself as a staunch defender of user privacy, particularly with its emphasis on end-to-end encryption for iMessage and FaceTime, and its efforts to limit the data it collects and retains on users. However, even Apple has been compelled to provide metadata, such as iCloud backups, contact lists, and location history, when presented with valid legal orders. The company has also faced intense pressure from law enforcement to create backdoors or unlock encrypted devices, a battle that highlights the fundamental tension between strong encryption, which protects users from malicious actors and overreaching governments, and the demands of law enforcement agencies for unfettered access. While Apple has largely resisted the creation of backdoors, its provision of other forms of data demonstrates its role as a data partner, albeit one that selectively resists certain types of access.

Google's vast data ecosystem, encompassing search, email, maps, and mobile operating systems, makes it an unparalleled source of personal information. The company regularly publishes transparency reports detailing the number of government requests for user data it receives. These reports reveal a consistent pattern of a high volume of requests, with Google complying with a significant percentage of them, often on a per-record basis. The company's sophisticated data analytics capabilities, designed to personalize user experiences and optimize services, are also inherently valuable for intelligence analysis. Thus, Google's role as a data partner extends beyond simply providing raw data; it can also offer analytical insights derived from that data, further enhancing governmental surveillance capabilities.

Meta, with its ownership of Facebook, Instagram, and WhatsApp, controls social graphs and communication patterns for billions of people. The intimate details shared on these platforms – relationships, personal opinions, daily activities, and private conversations – constitute a rich tapestry for surveillance. While WhatsApp is end-to-end encrypted, metadata concerning who communicated with whom, when, and for how long, remains accessible and is often subject to legal disclosure. Facebook and Instagram data, on the other hand, provides a wealth of information about user interests, political leanings, social networks, and personal life events, all of which can be highly valuable for intelligence agencies seeking to understand individuals and groups.

The argument of "voluntary cooperation" versus "legal mandate" is often a fine line for these companies. While they may technically be compelled by law, the proactive development of systems to facilitate such access, or the willingness to interpret legal requests in a manner that maximizes data disclosure, can be seen as a form of active partnership. The companies are not merely passive repositories; they are active participants in the infrastructure that enables surveillance. The financial incentives are also not insignificant. While direct payments for data access are not always publicly disclosed, the economic value of government contracts, the cost savings for agencies in not having to conduct their own extensive data collection, and the potential for future business opportunities all contribute to a complex web of interdependence.

The ethical dilemma is stark: these companies derive immense profit and market power from user data, which they promise to protect. Yet, they are also integral to systems that allow governments to monitor citizens, often without the knowledge or explicit consent of those individuals for this secondary purpose. This creates a chilling precedent where fundamental rights to privacy and freedom of expression can be eroded not by overt authoritarian control, but by the seemingly mundane, yet profound, collaboration between private industry and state security apparatuses. The very technologies designed to connect and empower individuals are, in this context, repurposed to monitor and control them, making the tech giants unwitting or complicit partners in the expansion of the surveillance state. The public trust placed in these platforms is leveraged, in part, to build and maintain a pervasive surveillance infrastructure, raising critical questions about accountability, transparency, and the future of privacy in the digital age.

The burgeoning landscape of modern surveillance is not solely an endeavor of government agencies and their internal capabilities. A parallel, and increasingly dominant, force shaping how intelligence is gathered, analyzed, and acted upon is the pervasive involvement of private contractors. This outsourcing of critical functions represents a fundamental shift, transforming the very nature of state surveillance from a strictly public enterprise to one deeply intertwined with private enterprise and its profit-driven imperatives. Governments, facing escalating demands for information and complex technological challenges, have found it expedient, and often more efficient in the short term, to contract with external companies. These entities possess specialized expertise, cutting-edge technologies, and a flexible workforce that government bureaucracies can struggle to match. The spectrum of contracted services is vast, encompassing everything from the mundane yet essential—background checks, physical security, and logistical support—to the highly specialized and sensitive domains of data analytics, cybersecurity operations, signal intelligence, and even the direct operation of surveillance platforms.

This reliance on private contractors for intelligence and surveillance functions introduces a complex web of ethical and practical considerations. Foremost among these is the question of accountability. When a government agency outsources a core surveillance capability, where does the ultimate responsibility lie? If a contractor's actions lead to privacy violations, data breaches, or intelligence failures, tracing the chain of command and assigning culpability becomes significantly more challenging than within a traditional governmental structure. The layers of contractual agreements, subcontracting, and the geographic dispersion of these private entities can obscure lines of responsibility, making it difficult for oversight bodies and the public to scrutinize operations effectively. The profit motive, inherent in any private enterprise, also raises distinct concerns. While government employees are theoretically motivated by public service, contractors are driven by profitability. This can create an incentive structure where the maximization of contracts, the expansion of service offerings, and the efficient (and potentially cost-cutting) delivery of services might subtly or overtly influence the scope and methods of surveillance. The pressure to demonstrate value and secure future contracts could, in theory, lead to the overzealousness in data collection or a less rigorous approach to privacy protections if those are perceived as hindering efficiency or increasing costs.

The growth of the private contracting sector in surveillance is not a new phenomenon, but its scale and depth have dramatically expanded in the post-9/11 era. Following the September 11th attacks, there was an unprecedented surge in government spending on national security, much of which flowed into the private sector. Companies that had previously specialized in commercial IT services, defense contracting, or data management rapidly pivoted to offer services tailored to the burgeoning intelligence community. These companies often recruited former government intelligence officers and military personnel, creating a revolving door that further cemented the symbiosis between the public and private sectors. This influx of talent and capital allowed private contractors to develop sophisticated capabilities that directly mirrored and, in some cases, surpassed those of government agencies. They became adept at managing massive datasets, developing advanced algorithms for pattern analysis, and operating complex electronic surveillance systems.

Consider the realm of data analysis. Governments possess vast amounts of raw data collected through various means —signals intelligence, open-source intelligence, financial records, communication metadata, and more. The sheer volume and complexity of this data often exceed the analytical capacity of in-house government teams. This is where private contractors excel. Companies employ legions of data scientists, linguists, cultural experts, and intelligence analysts, many of whom possess specialized skills honed through academic research or prior intelligence work. They develop and deploy sophisticated software tools capable of sifting through terabytes of information, identifying patterns, anomalies, and connections that might otherwise remain hidden. These tools can include sophisticated machine learning algorithms for predictive analysis, natural language processing for sentiment analysis and topic extraction from text, and advanced network analysis software to map relationships between individuals and organizations. The contracting companies not only provide the analytical personnel but also develop and maintain the proprietary software and hardware infrastructure necessary to perform these tasks. This outsourcing allows government agencies to scale their analytical capabilities up or down as needed, without the long-term commitment and bureaucratic overhead associated with expanding their permanent workforce.

Cybersecurity and network defense represent another critical area where private contractors are indispensable. In an increasingly interconnected world, the infrastructure of government agencies and critical national infrastructure are constant targets for cyberattacks from state adversaries, terrorist groups, and criminal organizations. Private cybersecurity firms provide a range of services, including penetration testing, threat intelligence, incident response, and the development of secure network architectures. They are often at the forefront of identifying new vulnerabilities and developing countermeasures. However, the same expertise that protects government networks can also be applied to offensive cyber operations, or the development of surveillance tools that exploit network vulnerabilities. The lines can blur, and the proprietary nature of the tools and techniques developed by these companies can limit transparency about their exact capabilities and applications.

The operation of surveillance technologies themselves is increasingly outsourced. This can range from the management of sensor networks and intelligence-gathering platforms to the provision of specialized aerial surveillance capabilities through drones or manned aircraft. Companies may be contracted to operate sophisticated signal intercept equipment, process intercepted communications, or manage the vast databases that store this information. In some cases, private contractors may even be responsible for the deployment and maintenance of surveillance equipment in sensitive or covert operational environments. This delegation of operational control to private entities raises significant questions about oversight and the potential for mission creep. When private employees are operating under a contract, their incentives might differ from those of military or intelligence personnel. While they are bound by contractual obligations and potentially by government directives, the underlying motivation of profit can influence decision-making in subtle ways, particularly when faced with competing priorities or resource constraints.

The legal and ethical frameworks governing intelligence activities were largely designed with public servants in mind. Adapting these frameworks to encompass private contractors is an ongoing challenge. Existing laws and regulations concerning the handling of classified information, the use of surveillance technologies, and the protection of privacy were not necessarily built with the profit-driven motivations and employment structures of private industry in mind. For example, the background investigation and security clearance processes for contractor personnel, while rigorous, may differ in depth and scope from those applied to government employees. Similarly, the contractual clauses that govern their behavior regarding privacy and data handling can be complex and may not always provide the same level of explicit protection as statutory requirements for government agencies.

The concept of "privatized surveillance" extends beyond simply hiring contractors to perform specific tasks. It can also involve the development and sale of surveillance technologies and services directly to governments. Companies are actively innovating in areas such as facial recognition, gait analysis, social media monitoring, and advanced data mining, creating sophisticated tools that governments can purchase and deploy. This market-driven approach to surveillance technology development means that innovation is often driven by what is commercially viable and what governments are willing to buy, rather than necessarily by what is ethically justifiable or what best serves democratic principles. The competitive nature of this market can also lead to a race to the bottom, where companies may cut corners on privacy safeguards or ethical considerations to gain a competitive edge.

The lack of transparency surrounding contractor involvement is a significant impediment to public oversight. While government agencies are increasingly publishing transparency reports detailing certain aspects of their surveillance activities, the specifics of contractor roles, the scope of their operations, and the value of contracts awarded are often kept confidential, citing national security or proprietary business interests. This opacity makes it difficult for lawmakers, civil society organizations, and the public to fully understand the extent to which private entities are involved in surveillance, the types of data being collected, and the potential risks to civil liberties. Without adequate transparency, the potential for unchecked power and abuse by private actors operating under the guise of national security is considerably heightened.

Furthermore, the intellectual property rights associated with the surveillance technologies and analytical methodologies developed by contractors can create additional layers of complexity. When a government agency pays a private company to develop a sophisticated algorithm or a unique surveillance platform, that company may retain ownership of the intellectual property. This can limit the government's ability to share or independently audit the technology, and it can also mean that the same technology, or variations of it, can be sold to other governments, potentially including those with questionable human rights records. This commercialization of surveillance capabilities means that the tools and techniques of monitoring are not confined to a single national security apparatus but can proliferate across the global market.

The implications of this privatization for the nature of the surveillance state are profound. It moves the capabilities and operations of surveillance further away from direct public control and democratic accountability. While government agencies are subject to a range of oversight mechanisms, including congressional committees, Inspectors General, and judicial review (albeit often limited in national security contexts), the oversight of private contractors is often indirect and mediated through contractual relationships. This can create a situation where the most sensitive and impactful aspects of surveillance are conducted by entities that are less directly answerable to the public or to elected representatives. The profit motive, while a legitimate driver in the private sector, is fundamentally misaligned with the principles of public service and the protection of civil liberties that should underpin state surveillance activities.

The "revolving door" phenomenon, where individuals move between government intelligence agencies and private contracting firms, further complicates the issue. Former intelligence officials often possess intimate knowledge of government needs, operational procedures, and classified technologies, making them highly valuable assets to private companies. Conversely, individuals who have worked for contractors may later join government agencies, bringing with them their industry experience and potentially their company's perspectives and interests. This constant flow of personnel can lead to a blurring of the lines between public and private interests, potentially influencing procurement decisions, policy development, and the overall direction of surveillance programs. It raises concerns about whether decisions are being made in the best interest of national security and public welfare, or in ways that benefit particular private companies.

The challenge for governments is to strike a balance: leveraging the expertise and innovation of the private sector while ensuring robust oversight, accountability, and adherence to legal and ethical standards. This requires clear contractual terms that explicitly address privacy protections, data handling protocols, and prohibitions against certain types of surveillance activities. It also necessitates strong independent oversight mechanisms capable of scrutinizing contractor performance and ensuring compliance with regulations. Transparency is paramount; the public has a right to know how their governments are conducting surveillance and the extent to which private entities are involved. Without these safeguards, the privatization of surveillance risks creating a shadow intelligence apparatus, operating with reduced transparency and accountability, and driven by profit motives that may not always align with the public good or fundamental human rights. The outsourcing of surveillance is not merely a logistical decision; it is a fundamental reshaping of the relationship between the state, its citizens, and the private sector, with profound implications for democracy and civil liberties.

The digital economy, at its core, is a data economy. For a vast array of private companies, particularly those operating in the online space, the collection, analysis, and exploitation of user data is not a secondary activity but the very engine of their business models. This fundamental economic imperative drives an insatiable appetite for information, a drive that has profound implications for the nature and scale of surveillance, both private and public. The prevailing paradigm is one where user data is the raw material, transformed through various proprietary processes into valuable commodities, primarily targeted advertising and personalized services. However, this commodification of personal information creates an unprecedented reservoir of data, readily accessible and increasingly detailed, that inevitably becomes a crucial resource for government intelligence agencies.

Consider the ubiquitous social media platforms. Their revenue streams are overwhelmingly generated through advertising. To effectively serve these advertisements, these companies must understand their users with granular precision. This necessitates the collection of an astonishing breadth of data points: not just what users post, but also who they interact with, what they click on, what they 'like,' their stated interests, demographic information (often inferred if not explicitly provided), location data, device information, browsing history, and even the duration of their attention on specific content. Each interaction, each piece of shared information, is logged, analyzed, and used to build an intricate profile of the individual. This profile is then leveraged to categorize users into specific demographic, psychographic, and behavioral segments, which are then sold to advertisers looking to reach particular audiences with tailored messages. The more data a platform possesses, the more accurate its profiling becomes, and the more valuable its advertising inventory is. This creates a powerful economic incentive to maximize data collection, pushing the boundaries of what is technically feasible and, at times, what is ethically permissible or transparent to the user.

Beyond social media, the digital ecosystem is replete with

services that operate on similar data-centric principles. E-commerce platforms track not only purchases but also browsing habits, wish lists, abandoned carts, and even how long users spend examining product pages. Search engines meticulously record every query, building detailed histories that reveal evolving interests, needs, and concerns. Streaming services monitor viewing habits, taste preferences, and even the specific moments users pause or rewind content. Navigation apps collect vast amounts of real-time location data and route histories. Even seemingly innocuous services, like weather apps or online games, often collect extensive data on user behavior and device information. In each instance, the underlying economic logic remains consistent: gather as much data as possible, as it enhances the ability to personalize services, improve algorithms, and, crucially, monetize user attention and behavior, primarily through advertising.

This relentless pursuit of data is not merely about improving user experience or delivering relevant ads. It is a strategic business imperative. Companies invest heavily in sophisticated data analytics infrastructure, employing data scientists, machine learning engineers, and AI specialists to extract maximum value from the terabytes and petabytes of information they accumulate. The ability to identify subtle correlations, predict future behavior, and understand consumer trends is a significant competitive advantage. This data becomes a company's most valuable asset, a proprietary intelligence network built from the aggregated digital footprints of millions, if not billions, of individuals. The sheer scale of this data accumulation is staggering; it represents an unprecedented aggregation of information about human behavior, preferences, relationships, and activities, collected with a depth and breadth that was unimaginable just a few decades ago.

The critical juncture where this private data collection

intersects with government surveillance lies in the accessibility and utility of this amassed information. For intelligence agencies, the challenge has historically been the acquisition of relevant data, often requiring complex legal processes, technical intrusions, or extensive human intelligence operations. However, the pervasive data collection by private companies dramatically lowers these barriers. Instead of initiating a difficult and potentially overt intelligence-gathering operation from scratch, agencies can often tap into existing, vast repositories of data already collected by the private sector. This transforms the economics of intelligence gathering, making it significantly more cost-effective and efficient.

One of the primary mechanisms through which this data flows from the private sector to government agencies is through direct acquisition or access agreements. While specific details are often shrouded in secrecy, it is well-documented that various government entities, including law enforcement and intelligence agencies, have engaged with data brokers and technology companies to obtain vast datasets. These datasets can include commercially available information, such as aggregated location data, consumer purchasing histories, or publicly available social media posts. However, the scope can extend to more sensitive information, depending on the agreements and legal frameworks in place. For instance, bulk data purchases, particularly concerning mobile device location data, have become a significant avenue for intelligence agencies to track individuals' movements and associations without necessarily needing to issue individual warrants for each data point. Companies that aggregate this location data from myriad apps and services, often anonymized or pseudonymized, can then sell it to government entities, effectively creating a surveillance tool from otherwise ordinary commercial activity.

Another significant pathway is through data requests or legal demands, such as subpoenas or national security letters (NSLs). While these are legal processes, the sheer volume of requests made by governments highlights the reliance on private data. For example, companies that store user communications, browsing histories, or other personal data are frequently asked by law enforcement and intelligence agencies to provide specific information related to investigations. The ability of companies to respond to such requests is directly correlated to the amount of data they have collected and the sophistication of their data management systems. A company that collects more data is inherently a more valuable source of information for a government investigation. This creates a feedback loop: the incentive for companies to collect more data is amplified by the knowledge that this data is highly sought after by government entities, and the ability of governments to conduct surveillance is enhanced by the data-hoarding practices of private companies.

Furthermore, the nature of proprietary data analysis tools developed by private companies also plays a role. Companies specializing in big data analytics, artificial intelligence, and machine learning often develop sophisticated algorithms capable of identifying patterns, connections, and anomalies within massive datasets. These tools are designed for commercial purposes, such as optimizing marketing campaigns or identifying fraud. However, the underlying analytical capabilities can be readily adapted or directly applied to intelligence analysis. When a government agency contracts with these companies for services, or acquires their software, they are essentially gaining access to advanced analytical power that would be prohibitively expensive and time-consuming to develop in-house. This outsourcing of analytical capability, driven by the economic incentives of the private sector to create and sell these tools, directly facilitates government surveillance by providing powerful methods for processing and interpreting vast quantities of data.

The economic model of "freemium" services—offering basic services for free with the expectation of upselling or generating revenue through data—also contributes significantly. Users may not directly pay for many online services, but they pay with their data. This data is then monetized by the companies, and its availability to governments becomes a downstream consequence. The economic incentive is to attract as many users as possible to generate the largest possible datasets, creating a powerful engine for both commercial exploitation and potential government access. This model effectively subsidizes government surveillance by making the collection and aggregation of personal information a core component of private sector profitability.

The competitive landscape further intensifies this drive for data acquisition. In a crowded digital marketplace, companies are constantly seeking ways to differentiate themselves and gain a competitive edge. Enhanced personalization, more accurate recommendations, and more effective targeted advertising are all achieved through superior data analytics. This leads to a constant arms race in data collection and processing capabilities. Companies that are more aggressive in their data gathering practices and more adept at analyzing it are more likely to succeed commercially. This competitive pressure directly fuels the expansion of data collection, ensuring that even companies that might otherwise be more conservative in their data practices feel compelled to keep pace with rivals.

The implications of this economic symbiosis between private data collection and government surveillance are far-reaching. It blurs the lines between commercial interests and national security objectives. It creates a situation where the commercial imperative to collect data can inadvertently, or sometimes deliberately, serve the surveillance needs of the state. The data collected for targeted advertising can become the raw material for identifying potential threats, tracking individuals of interest, or mapping social networks for intelligence purposes. This commodification of personal information essentially transforms the daily digital activities of ordinary citizens into a resource that can be readily accessed and utilized by government intelligence apparatuses, often with less public scrutiny than traditional forms of state surveillance.

Moreover, the opaque nature of many of these data-sharing agreements and the proprietary algorithms used for analysis further shield these activities from public view. When data is purchased from a third-party broker, or when analysis is conducted by a contractor using proprietary software, the trail of accountability becomes more diffuse. This lack of transparency makes it difficult for citizens to understand the extent to which their personal data, initially shared for commercial purposes, is being utilized by government agencies. The economic incentives driving private data collection are thus directly enabling and amplifying government surveillance capabilities, creating a powerful, albeit often unseen, engine of information gathering that operates at the intersection of commerce and state security. The question then becomes not just how governments surveil, but how the fundamental economic drivers of the private sector have inadvertently, yet powerfully, reshaped the landscape of surveillance for everyone.

The digital age has ushered in a new economic paradigm, one that fundamentally reshapes the relationship between individuals, corporations, and the state. At its heart lies a model that has been critically termed "surveillance capitalism." This is not merely a euphemism for aggressive data collection; it describes a distinct economic system where human experience, translated into behavioral data, is claimed, traded, and sold for profit. The core logic is deceptively simple: collect as much raw behavioral data as possible from individuals, analyze it to predict future behavior, and then use these predictions to influence and modify that behavior for lucrative outcomes. The ultimate commodity is not the data itself, but the certainty it provides about future actions, a certainty that advertisers, for instance, are willing to pay a premium for.

The architects of this model, many of whom operate at the vanguard of the internet economy, have masterfully turned the very act of being online into a perpetual data-gathering exercise. Every click, every search, every keystroke, every message sent or received, every product viewed, every location visited, and every interaction with content is meticulously logged. Initially, this data served the purported purpose of improving user experience—personalizing search results, recommending products, or curating news feeds. However, as the sophistication of analytical tools, particularly artificial intelligence and machine learning, advanced, the true economic potential of this data became apparent. It was not just about understanding users; it was about predicting and, crucially, shaping their future behavior.

This predictive capability is where the profitability of surveillance capitalism truly flourishes. Companies build sophisticated predictive models that can forecast what a user is likely to do next: what they might buy, what they might click on, where they might go, or even how they might feel. These predictions are then bundled and sold as behavioral futures. Advertisers, for example, can purchase access to these predictions to target specific audiences with unprecedented precision. Instead of broadcasting a message to a general demographic, they can target individuals who are statistically predicted to be most receptive to a particular product or service at a specific moment. This targeted advertising, powered by vast reserves of behavioral data, offers a far higher return on investment than traditional mass marketing, creating an insatiable demand for more and better data.

However, the cycle of surveillance capitalism extends beyond mere prediction; it actively seeks to modify behavior. By understanding individual vulnerabilities, preferences, and decision-making processes, companies can subtly nudge users towards desired actions. This can manifest in various ways: presenting certain options more prominently, framing choices in specific ways, or delivering personalized content designed to evoke particular emotional responses. Think of the algorithms that curate your social media feed, always learning what keeps you engaged, what makes you click, what keeps you scrolling. The objective is not necessarily to serve your interests but to maximize your engagement, thereby gathering more data and increasing the opportunities for monetization. This continuous feedback loop of data collection, prediction, and behavioral modification forms the bedrock of the surveillance capitalist enterprise.

The economic engine of surveillance capitalism operates on a foundation of what Shoshana Zuboff, who coined the term, describes as "extraction." This refers to the systematic appropriation of human experience as free raw material for profit. Services that appear to be free to the end-user—social media, search engines, email, mapping applications—are, in reality, complex systems for extracting personal data. Users are not customers in the traditional sense; they are the source of the raw material. The actual customers are the businesses that pay for access to the predictions and behavioral modifications derived from this data. This asymmetrical relationship, where users provide valuable data without direct financial compensation, and often without full awareness of the extent of data extraction, is fundamental to the model's success.

The infrastructure built to facilitate this data extraction is immensely powerful and pervasive. Companies invest billions in servers, data centers, sophisticated analytical software, and legions of data scientists and engineers. This creates a digital nervous system that maps and monitors human activity on a global scale. The aggregation of data from millions, even billions, of individuals results in datasets of unparalleled richness and depth. These datasets are not simply static records; they are dynamic pools of information that are constantly being analyzed, refined, and leveraged to generate further insights and profits. The sheer scale and complexity of these operations mean that the companies at the forefront of surveillance capitalism possess a level of insight into human behavior that far surpasses that of any government agency, at least in the initial stages of data acquisition.

This leads us to the critical nexus between surveillance capitalism and government surveillance. The economic incentives that drive private companies to collect and analyze vast amounts of personal data have, perhaps unintentionally at first but increasingly by design, created an infrastructure that is perfectly suited for state surveillance. The data repositories, analytical tools, and predictive models developed for commercial purposes are inherently valuable to intelligence agencies and law enforcement. They represent a pre-existing, highly efficient, and incredibly detailed surveillance apparatus that governments can, and do, tap into.

One of the primary ways this synergy manifests is through the direct purchase or acquisition of data. Data brokers, companies that specialize in aggregating and selling personal information, often obtain their raw materials from the very platforms and services that embody surveillance capitalism.

They then repackage and sell this data to government entities. For instance, location data, harvested from mobile applications that track users' movements, can be bought by intelligence agencies to map the comings and goings of individuals without the need for traditional warrants. This data, initially collected to provide a personalized experience or targeted advertising, is transformed into a tool for tracking and monitoring, effectively privatizing the acquisition of intelligence. The commercial drive to monetize location data directly fuels government surveillance capabilities, often circumventing privacy protections that might otherwise be in place.

Beyond direct acquisition, government agencies can also leverage the analytical capabilities developed by surveillance capitalists. Companies that excel at identifying patterns, predicting behavior, and profiling individuals for commercial purposes possess sophisticated algorithms and analytical frameworks. These are not merely tools for selling more products; they are powerful engines for identifying trends, anomalies, and potential threats within large populations. When governments contract with these companies for services, or acquire the software they develop, they gain access to these advanced analytical capabilities. This means that insights derived from billions of data points, originally collected for profit, can be applied to national security objectives, allowing agencies to sift through immense volumes of information with unprecedented efficiency. The proprietary nature of these analytical tools, however, often creates a black box, obscuring how data is being interpreted and what conclusions are being drawn.

Furthermore, the legal frameworks governing data access often facilitate this convergence. While laws may require warrants for direct government access to data held by service providers, the involvement of third-party data brokers or contractors can create loopholes. Data purchased from a commercial entity may be considered legally distinct from data directly obtained from a user's service provider, potentially reducing the need for judicial oversight. This creates a tiered system of surveillance, where data collected for commercial ends can be more easily accessed by the state than data collected directly by the state through traditional, albeit often intrusive, methods. The profit motive behind data aggregation thus inadvertently, and sometimes deliberately, lowers the barrier to entry for government surveillance.

The concept of "zero-day" vulnerabilities, often discussed in cybersecurity, can be seen as an analogy here. In the context of surveillance, the "zero-day" is the personal data itself. Companies have already extracted and aggregated it, creating a readily available intelligence asset. Governments can then exploit this pre-existing infrastructure, much like a hacker exploits a software vulnerability. The privacy investment made by individuals in using these digital services is essentially exploited for state intelligence purposes, all facilitated by the commercial imperative of surveillance capitalism.

The normalization of data collection for commercial purposes also plays a significant role in conditioning the public to accept broader surveillance. When users are accustomed to their online activities being monitored for targeted advertising, the idea of their data being used for other purposes, including government intelligence, may seem less of a departure. This gradual erosion of privacy expectations, driven by the ubiquitous nature of commercial surveillance, makes it easier for governments to expand their own surveillance activities without encountering significant public resistance. The commercial infrastructure for pervasive monitoring thus serves as a de facto foundation for state monitoring.

Moreover, the economic model of surveillance capitalism incentivizes the continuous expansion of data collection. As companies seek new revenue streams and competitive advantages, they are driven to collect ever more granular data about individuals and to find new ways to analyze and monetize it. This creates a constantly expanding data footprint for each individual, a richer and more detailed tapestry of their lives, which in turn becomes a more valuable resource for any entity seeking to monitor or understand them. This relentless growth of data collection means that the potential for surveillance, both commercial and governmental, is perpetually increasing.

The sheer ubiquity of digital devices and online services means that very few aspects of modern life remain outside the purview of data collection. From the smart devices in our homes that monitor our conversations and habits, to the wearable technology that tracks our physical activity and even our physiological responses, to the online platforms that catalog our social interactions and intellectual pursuits, nearly every human activity is being translated into data. This data is then processed, analyzed, and often aggregated by companies operating under the surveillance capitalist model. The byproduct of this commercial activity is a comprehensive, highly detailed, and constantly updated dossier on virtually every individual, a resource of incalculable value for intelligence gathering.

The economic logic of surveillance capitalism is inherently intertwined with the practice of mass surveillance. The profit motive drives the creation of systems that collect and analyze personal data on an unprecedented scale. The insights generated from this data are then leveraged not only to predict and modify individual behavior for commercial gain but also to provide intelligence and insights to government agencies. This symbiotic relationship blurs the lines between the private sector and the state, creating an environment where the infrastructure and practices developed for corporate profit simultaneously enable and amplify governmental surveillance. The question then shifts from whether we are being surveilled, to how deeply, by whom, and for whose ultimate benefit, in a system where the engine of commerce has become the primary engine of observation.

8: PUBLIC PERCEPTION AND AWARENESS

The pervasive myth of online anonymity is one of the most significant and persistent public misconceptions of our digital era. Many users, even those who are otherwise digitally savvy, operate under the comforting, yet ultimately false, belief that their online activities are cloaked in a veil of privacy, shielded from prying eyes. This perceived anonymity is a crucial psychological lubricant for the engine of surveillance capitalism, allowing the vast collection and monetization of personal data to proceed with minimal friction. However, the reality is starkly different: virtually every interaction in the digital sphere leaves an indelible trace, a breadcrumb that can be collected, aggregated, and, with remarkable ease, attributed to an individual. The intricate web of tracking technologies and data brokers ensures that the illusion of privacy is precisely that – an illusion.

At the most fundamental level, every time a device connects to the internet, it does so through an Internet Protocol (IP) address. This address, unique to a particular network connection at a specific point in time, acts like a digital mailing address. While IP addresses can fluctuate, particularly with dynamic IP assignments common for home internet connections, they can still provide a significant geographical marker and, when combined with other data points, can be used to identify or at least narrow down the identity of a user. Furthermore, many users, especially those accessing the internet via mobile devices or public Wi-Fi, might assume this renders them anonymous. However, mobile devices themselves are treasure troves of data. Beyond IP addresses, they transmit cell tower information, Wi-Fi network identifiers, and GPS coordinates, all of which paint a detailed picture of a user's physical location. This location data, often collected by apps under broad privacy policies that users rarely scrutinize, is a potent identifier. Even when an IP address changes, the persistent identifiers within a device – such as cookies, device IDs, or even more sophisticated browser fingerprinting techniques – can link disparate online activities back to a single user.

Websites themselves are primary nodes in this data collection network. When you visit a website, your browser communicates with the website's servers. This communication inherently includes information about your browser type, operating system, screen resolution, and the referring URL (the website you came from). This basic information, known as browser metadata, helps websites optimize their content for different devices and browsers, but it also contributes to your digital fingerprint. More significantly, most websites employ tracking technologies, the most common of which are "cookies." These small text files are stored on your device by the websites you visit. First-party cookies are set by the website you are directly interacting with and are often used to remember your login status, shopping cart contents, or language preferences, enhancing usability. However, the real ubiquity of tracking comes from third-party cookies. These are set by domains other than the one you are visiting, typically by advertisers or analytics companies that have placed their code on the website. When you visit a website with third-party cookies enabled, those cookies can communicate back to the third-party's servers, allowing them to track your browsing activity across multiple websites. If you visit Website A, then Website B, and both use cookies from the same advertising network, that network can link your activities on both sites, building a profile of your interests.

The proliferation of online advertising networks and analytics providers has created a complex ecosystem where multiple entities are constantly trying to track users. Every page load, every click on an advertisement, every product viewed on an e-commerce site, and every article read on a news portal can be logged by these third parties. This happens in near real-time, often through complex "real-time bidding" auctions where advertisers bid to display ads to you as a webpage loads. Your digital profile, built from cookies and other identifiers, is used to determine how much they are willing to pay to show you an ad, making your activity directly commodified. The sheer volume of data points collected from these interactions – what you search for, what you click on, what you watch, what you buy, and even how long you linger on a particular piece of content – is staggering. This data is then used to create granular user profiles, categorizing individuals based on demographics, interests, purchase intent, and even predicted future behavior.

Social media platforms are arguably the most potent data-gathering machines. Beyond the explicit information users provide – their posts, photos, connections, likes, and comments – these platforms meticulously track engagement metrics. They monitor how long you watch a video, which comments you read, which profiles you visit, and which advertisements you interact with. The algorithms that curate your feed are not designed for your passive consumption; they are sophisticated engines that learn what keeps you engaged, what triggers a reaction, and what can be most effectively monetized. Every scroll, every pause, every share is data that refines these algorithms and strengthens your user profile. Furthermore, social media often extends its tracking beyond its own walls. Through "like" buttons, share widgets, and embedded content on other websites, social media companies can monitor your activity even when you are not actively using their platforms. This pervasive tracking builds a comprehensive picture of your social graph, your opinions, and your daily routines.

The act of online purchasing is another critical juncture for data collection. When you buy something online, you provide a wealth of sensitive information: your name, address, payment details, and purchase history. E-commerce platforms, and the payment processors and shipping companies they work with, retain this data. This purchase history is invaluable, revealing your consumption habits, your brand preferences, and your economic status. Beyond the direct transaction, browsing behavior on e-commerce sites – what you look at, what you add to your cart but don't buy, what you compare – is also tracked. This data can be used to personalize future recommendations, but it also feeds into broader profiles that advertisers and data brokers can leverage. Even seemingly innocuous online activities, like using a search engine, are deeply tracked. Search queries reveal intentions, curiosities, and needs, offering a window into an individual's immediate interests and concerns. Search engine companies aggregate this data to personalize results and target advertising, effectively transforming your queries into a marketable commodity.

Location services, often enabled on smartphones and other devices, further erode any sense of online anonymity. When location services are turned on, your device constantly communicates its position. This data can be used for navigation, to find nearby businesses, or to receive location-specific alerts. However, this same data is also collected and sold by app developers and location data aggregators. This allows for the creation of detailed movement histories, mapping where you go, when you go there, and how long you stay. This granularity of location data can reveal deeply personal aspects of your life, such as your home address, workplace, regular haunts, and even sensitive visits to healthcare providers or places of worship. The aggregation of this location data, often anonymized in theory but often re-identifiable when combined with other data points, forms a critical layer of surveillance.

The technical mechanisms for linking these disparate data points are sophisticated and ever-evolving. Beyond cookies, technologies like browser fingerprinting use a combination of browser settings, installed fonts, plugins, and other characteristics to create a unique identifier for your device, even if you clear your cookies or use private browsing modes. Mobile devices have unique identifiers such as the Identifier for Advertisers (IDFA) on iOS and the Advertising ID on Android, which, while designed to be reset or limited by users, can still be used to track activity across apps and websites. Furthermore, techniques like "cross-device tracking" attempt to link your activities across different devices – your phone, your laptop, your tablet – by matching patterns of behavior, IP addresses, Wi-Fi networks, and account logins. If you log into your Google account on your laptop, and that same Google account is logged into on your phone, your activities can be linked. Even without direct logins, sophisticated algorithms can infer that different devices likely belong to the same person based on their usage patterns and network proximity.

The role of data brokers is pivotal in solidifying this tracking network and transforming raw data into actionable intelligence for various entities, including advertisers, marketers, and, as previously discussed, government agencies. These companies specialize in acquiring, aggregating, and selling personal data. They collect information from a multitude of sources: public records, social media, online purchase histories, data breaches, and the data collected by apps and websites. They then clean, enrich, and segment this data, creating detailed profiles that can be purchased by businesses seeking to target specific audiences. This creates a secondary market for your personal information, where data that was once associated with your online activities is repackaged and sold, often without your knowledge or consent. These brokers essentially create comprehensive dossiers on millions of individuals, linking online behavior with offline identities and personal attributes.

The concept of "incognito" or "private" browsing modes, offered by most major web browsers, is a significant source of the illusion of anonymity. While these modes do prevent the browser from saving your browsing history, cookies, and site data locally on your device after the session ends, they do not make you invisible to the websites you visit or your internet service provider (ISP). Websites can still see your IP address, and if you log into an account on a website (like your email or social media), that website knows exactly who you are, regardless of whether you are in incognito mode. Your ISP can still monitor your internet traffic and see which websites you are connecting to. Furthermore, any tracking scripts embedded on websites can still collect information about your browsing habits during the incognito session, and if these scripts are linked to a persistent identifier (like a device fingerprint or a logged-in account), your activity can still be attributed to you. The promise of privacy in these modes is therefore largely superficial, offering a false sense of security.

Similarly, many users believe that using a Virtual Private Network (VPN) guarantees anonymity. While a VPN can mask your IP address and encrypt your internet traffic, making it harder for your ISP and websites to track your activity, it is not a foolproof shield. The VPN provider itself can see your original IP address and your online activity. If the VPN provider keeps logs of user activity, and these logs are compromised or subpoenaed, your data can be exposed. Moreover, even with a VPN, websites can still use cookies, browser fingerprinting, and other client-side tracking methods to identify you if you are logged into accounts or have previously visited the site without the VPN. The effectiveness of a VPN largely depends on the provider's privacy policy and the user's understanding of its limitations.

The constant stream of notifications from apps and websites, the personalized recommendations that seem eerily accurate, and the targeted advertisements that appear uncannily relevant – these are all tangible manifestations of the pervasive data collection and analysis described. They are not accidental byproducts of a user-friendly digital environment; they are the intended outcomes of a sophisticated surveillance apparatus built on the commodification of personal experience. The public's relative unawareness of the extent and technical sophistication of this tracking allows the system to perpetuate itself, as users are often desensitized or resigned to the idea that their data is being collected, without fully grasping the implications or the mechanisms at play. This gap between perceived privacy and the reality of constant digital surveillance is a critical vulnerability exploited by the architecture of the modern internet.

The way government surveillance programs are understood and discussed by the public is significantly influenced by the media. News organizations act as crucial intermediaries, translating complex technical details and often classified government operations into narratives that resonate with a broader audience. The media's role extends beyond mere reporting; it involves investigation, contextualization, and the framing of issues, all of which shape public perception and consequently, policy debates. The depth and breadth of public awareness regarding surveillance, therefore, are intricately tied to the media's capacity and willingness to delve into these often opaque activities.

Historically, the media's engagement with government surveillance has varied considerably. In times of national security crises, or perceived threats, reporting often adopts a more deferential tone, focusing on the necessity of surveillance for protecting citizens. However, periods of scandal or major revelations, such as those following the post-9/11 era and the revelations by Edward Snowden, have seen the media pivot towards more critical and investigative stances. These pivotal moments often trigger intense public discourse, forcing governments to account for their actions and prompting legislative reviews. The media's ability to secure and disseminate information during these times is paramount; it acts as a watchdog, holding power accountable by bringing clandestine activities into the light.

Investigative journalism plays a particularly vital role in illuminating government surveillance. When journalists undertake the arduous task of uncovering evidence of mass data collection, warrantless monitoring, or the misuse of surveillance technologies, they challenge the official narratives often presented by government agencies. The process of investigation can be fraught with peril, requiring sources within government who are willing to risk their careers and personal safety to leak information. Journalists must then meticulously verify this information, often facing legal challenges and immense pressure from powerful entities seeking to suppress the stories. The success of such investigations—and their subsequent impact on public discourse—hinges on the journalist's tenacity, ethical rigor, and the resources available to support in-depth, long-term reporting.

The extent to which the media investigates and exposes clandestine surveillance activities has a direct correlation with the level of public awareness and engagement. When major newspapers and broadcast networks dedicate significant resources to covering surveillance, conducting deep dives into programs like those revealed by Snowden, the public becomes more informed. These reports often go beyond simply stating that surveillance exists; they explain the technologies involved, the legal justifications (or lack thereof), the potential implications for civil liberties, and the scope of data being collected. The use of infographics, explainer videos, and detailed articles helps demystify the technical aspects, making the issue accessible to a wider audience. This comprehensive approach fosters a more nuanced understanding, moving beyond simplistic notions of security versus privacy to a more complex appreciation of the trade-offs and risks involved.

However, the media landscape is not monolithic, and the approach to covering surveillance can differ significantly between outlets. Some news organizations may prioritize the security angle, framing surveillance as an essential tool for preventing terrorism and crime. Others might lean more heavily on the civil liberties perspective, highlighting the potential for abuse, the erosion of privacy, and the chilling effect on free speech and association. This spectrum of coverage contributes to the complexity of public discourse. While a diversity of perspectives can be healthy, it can also lead to a polarized public debate, where individuals may only engage with media that confirms their pre-existing beliefs, hindering a comprehensive understanding of the issue.

A significant challenge for journalists covering government surveillance is the pervasive culture of classification and government secrecy. Many surveillance programs are developed and operated under classifications that restrict access to information, even for elected officials and oversight committees. This deliberate opacity makes it incredibly difficult for journalists to obtain concrete details about the scope, capabilities, and targets of these programs. When information is classified, journalists must rely on leaks from whistleblowers, which, while invaluable, can be incomplete or presented without full context. The government's narrative, often framed around national security imperatives, can therefore dominate the public discourse, especially when counter-narratives are hampered by a lack of verifiable information.

The legal framework surrounding surveillance also plays a critical role in how the media can report on it. Laws like the Espionage Act can be used to prosecute individuals who leak classified information, creating a chilling effect not only on potential whistleblowers but also on journalists who rely on them. The fear of government retribution, whether through legal action, surveillance of their own communications, or pressure on their sources, can constrain the scope of reporting. Furthermore, the sheer volume of data collected by governments and private entities means that even if specific programs are revealed, understanding the totality of surveillance can be an overwhelming task for any news organization. The aggregation of data from various sources —telecommunications metadata, internet browsing history, social media activity, location data—creates a surveillance infrastructure that is far more extensive than any single program.

The impact of media coverage on public opinion and policy debates is undeniable. When major surveillance revelations occur and are widely reported, public concern often rises, leading to increased calls for oversight, transparency, and legislative reform. For instance, the revelations about the National Security Agency's (NSA) bulk data collection programs in 2013, brought to light by Snowden and reported extensively by outlets like The Guardian and The Washington Post, sparked a global debate about privacy and government power. This media-driven discourse led to significant discussions in Congress, court challenges, and eventually, some legislative changes aimed at reforming surveillance practices. The media provided the public with the information necessary to understand the threat to privacy and empowered citizens to demand accountability from their governments.

Conversely, when media attention wanes or when government agencies effectively manage the narrative through public relations efforts or further classifications, public awareness can recede. The ongoing nature of surveillance, with new technologies and methods constantly emerging, means that the media must maintain a sustained focus to keep the public informed. The challenge lies in maintaining public interest in what can often be perceived as dry, technical, or abstract issues, especially when immediate, tangible threats are not apparent. The media's ability to connect the abstract concept of data collection to concrete impacts on individual lives—such as the potential for misuse of personal information, chilling effects on dissent, or the erosion of democratic freedoms—is crucial for sustaining public engagement.

The rise of digital media and social platforms has also transformed the way surveillance is reported and discussed. While these platforms can facilitate rapid dissemination of information and enable citizen journalism, they also present new challenges. The proliferation of misinformation and disinformation can muddy the waters, making it harder for the public to discern credible reporting from propaganda. Furthermore, the algorithms that govern social media feeds can create echo chambers, reinforcing existing beliefs and limiting exposure to diverse viewpoints. For journalists, social media can be a double-edged sword: a tool for reaching new audiences and gathering information, but also a potential vector for targeted harassment and the spread of unverified claims.

Moreover, the media's own data practices are increasingly coming under scrutiny. Many news organizations rely on advertising and analytics, which often involve collecting user data to understand audience behavior and personalize content. This can create a tension between the role of the media as a watchdog on surveillance and its own participation in data collection. Transparency about these practices is essential for maintaining credibility. When news outlets are themselves significant data collectors, their critique of government surveillance can be perceived as hypocritical by some segments of the public.

The effectiveness of media coverage is also dependent on the audience's media literacy. A public that is adept at critically evaluating information, identifying bias, and understanding the motivations behind different media reports is better equipped to engage with complex issues like government surveillance. Educational initiatives that promote media literacy are therefore indirectly supportive of informed public discourse on surveillance. Without these skills, individuals may be more susceptible to simplistic narratives or deliberate misinformation campaigns designed to either allay or inflame public opinion about surveillance activities.

The dynamic between government secrecy and media investigation is a perpetual tug-of-war that fundamentally shapes public perception. When governments are transparent, proactively share information about their surveillance programs, and establish robust independent oversight mechanisms, the media's investigative role becomes more about scrutinizing the effectiveness and fairness of these systems rather than unearthing hidden operations. However, in environments characterized by extensive classification and a reluctance to disclose operational details, the media's primary function shifts to uncovering what is being deliberately concealed. This often means relying on the bravery of whistleblowers and the investigative prowess of journalists to provide the public with essential insights into the surveillance state. The narrative that emerges from this interplay directly influences the public's trust in government, their understanding of their own rights, and ultimately, the trajectory of surveillance policy. The media's capacity to consistently and effectively challenge secrecy, verify information, and communicate the implications of government surveillance to the public remains a cornerstone of democratic accountability.

The landscape of public perception regarding government surveillance is not static; it is a dynamic field influenced by a multitude of factors, including the pervasive evolution of technology and the differing life experiences of various demographic groups. Among these, generational differences stand out as a particularly significant lens through which to understand how privacy is valued and how tolerance for surveillance manifests. This subsection delves into these potential divergences, exploring how individuals who have come of age in distinct technological and societal eras may hold contrasting views on data collection and governmental oversight.

Younger generations, often characterized as digital natives, have grown up immersed in an environment where the lines between the public and private spheres have been continually redrawn by the internet and ubiquitous digital devices. From their earliest memories, social media platforms, online sharing, and constant connectivity have been part of the fabric of daily life. This prolonged exposure to digital environments, where personal information is routinely shared, collected, and analyzed, could foster a different baseline expectation regarding privacy. For many within these cohorts, the concept of a purely private, offline existence might seem more abstract or even anachronistic. The normalized practice of creating digital footprints, whether through social media posts, online purchases, or location-based services, suggests a potentially higher degree of comfort with the pervasive collection of personal data. This is not necessarily an endorsement of surveillance but rather a reflection of a lived experience where data generation and sharing are intrinsic to social interaction and daily functioning.

Conversely, older generations, particularly those who experienced their formative years before the widespread adoption of the internet and digital technologies, may possess a fundamentally different conception of privacy. For them, privacy might be more closely aligned with older notions of physical privacy, secure personal spaces, and the expectation that personal communications and activities would not be routinely monitored or recorded by external entities. The advent of mass surveillance technologies, therefore, might represent a more jarring departure from their established norms and expectations. The shift from a predominantly analog world to a digital one, where personal information can be aggregated, analyzed, and potentially misused on an unprecedented scale, can evoke greater concern and skepticism. This perspective is often rooted in a deep-seated belief in the sanctity of personal autonomy and the potential for intrusions into one's private life that were unimaginable in earlier eras.

Several factors contribute to these potential generational divides. Digital literacy and familiarity with the underlying technologies play a crucial role. Younger individuals often possess a more intuitive understanding of how data is collected, processed, and utilized, even if that understanding is not always comprehensive. They are more likely to be adept at navigating privacy settings, employing anonymization tools, or simply accepting the data-sharing inherent in many digital services as a necessary trade-off for convenience and connectivity. Older generations, while not necessarily technologically illiterate, may have a steeper learning curve when it comes to understanding the nuances of digital privacy and the sophisticated methods of data collection employed by both governments and corporations. The abstract nature of data collection, often occurring in the background without explicit user action, can be particularly disorienting for those not accustomed to such constant digital interaction.

Exposure to surveillance technologies also shapes attitudes. For younger generations, the presence of CCTV cameras in public spaces, the data collection practices of social media companies, and the constant monitoring by online platforms are often simply part of the background noise of modern life. They may have grown up with these technologies, leading to a form of desensitization or an implicit acceptance of their presence. Older generations, on the other hand, may perceive these technologies as more intrusive and indicative of a surveillance society, recalling a time when such pervasive monitoring was not the norm. The experience of having one's communications or movements tracked, even for seemingly benign purposes, can feel like a significant overreach to those who have not internalized the data-centric nature of contemporary society.

Trust in institutions is another significant differentiator. Generational cohorts that have experienced periods of significant institutional distrust, perhaps due to political events, economic downturns, or perceived governmental overreach, may exhibit a more critical stance towards surveillance. Older generations, having witnessed more instances of governmental power being exercised in ways that impacted civil liberties, might harbor a more inherent skepticism towards claims of necessity and benign intent surrounding surveillance programs. Younger generations, while not immune to distrust, may also exhibit a greater degree of faith in technological solutions and the pronouncements of authority figures, particularly if they have not personally experienced or extensively studied historical instances of surveillance abuse. However, this is a generalization; a significant portion of younger individuals are acutely aware of privacy concerns and actively advocate for stronger protections.

It is crucial, however, to guard against oversimplification and broad generalizations about entire age cohorts. Within any generation, there exists a wide spectrum of opinions and experiences. Not all young people are unconcerned about privacy; many are highly engaged activists and advocates for digital rights. Similarly, not all older individuals are resistant to technological advancements or uncritical of government actions. Factors such as education level, socioeconomic status, personal experiences with data breaches or surveillance, political ideology, and exposure to advocacy groups can all significantly influence an individual's privacy attitudes, irrespective of their generational cohort. The nuances within each generation are as important, if not more so, than the broad strokes that differentiate them. For instance, a young person whose parents have been involved in civil liberties advocacy might develop a more pronounced concern for privacy than an older individual who prioritizes national security above all else.

Furthermore, the very definitions of "privacy" and "surveillance" may evolve differently across generations. For younger individuals, privacy might be increasingly understood in terms of controlling the narrative of their digital identity, managing who sees what online, and ensuring that their data is not used in ways that could lead to discrimination or reputational damage. For older generations, privacy might still be more closely tied to the protection of personal correspondence, family matters, and freedom from unwarranted physical or digital intrusion. The concept of "surveillance" itself can also be interpreted differently; while older generations might associate it with traditional forms of espionage or state monitoring, younger generations might also perceive the data collection practices of large technology companies as a form of surveillance, blurring the lines between state and corporate oversight.

The digital divide, while diminishing, still plays a role. Those who have less access to technology or lower levels of digital literacy, regardless of age, may be less aware of the extent to which their data is being collected and how it is being used. This can lead to a passive acceptance of surveillance practices due to a lack of understanding, rather than a deliberate choice or acceptance of risk. Conversely, those with higher digital literacy, regardless of age, are more likely to be informed about the potential implications of data collection and to adopt strategies for mitigating privacy risks.

The way information about surveillance is disseminated also impacts generational perceptions. Younger individuals are more likely to encounter news and discussions about surveillance through social media, online forums, and digital news aggregators. The framing and presentation of information in these digital spaces, often characterized by brevity and immediate engagement, can lead to different interpretations compared to more in-depth reporting found in traditional media. Older generations may still rely more on traditional news sources, television broadcasts, and print media, which can provide different contexts and levels of detail. This difference in information consumption can lead to varying degrees of awareness and understanding of the complexities surrounding surveillance.

The experience of being a target of surveillance, whether direct or indirect, can profoundly shape an individual's attitudes. For example, individuals from minority groups who may feel disproportionately targeted by law enforcement or national security surveillance might develop a heightened sensitivity to privacy concerns and a deep distrust of governmental data collection, regardless of their age. Similarly, individuals who have had their personal data compromised in a breach, or whose online activities have led to negative consequences, are likely to be more vigilant about privacy than those who have not had such experiences. These personal encounters can override broader generational trends.

Moreover, the evolving legal and ethical frameworks surrounding data privacy are crucial. As regulations like GDPR and CCPA are introduced and debated, public discourse on privacy intensifies. How these discussions are framed and how effectively younger and older generations engage with them can further shape their respective attitudes. Younger generations, being more digitally engaged, might be more exposed to the debates surrounding data rights and the implications of new technologies for privacy.

Ultimately, understanding generational differences in privacy attitudes requires a nuanced approach that acknowledges the significant impact of technology, life experience, and institutional trust, while simultaneously recognizing the vast diversity of individual perspectives that exist within any age cohort. It is not a simple dichotomy of digitally fluent, less concerned youth versus analog-reared, more concerned elders. Instead, it is a complex interplay of factors that shape how privacy is perceived, valued, and protected across different segments of the population. The challenge for policymakers, educators, and advocates alike is to bridge these potential divides, fostering a universal understanding of privacy rights and the importance of robust safeguards in an increasingly data-driven world, ensuring that the concerns of all generations are heard and addressed.

The steady hum of technological advancement and the subtle, yet persistent, creep of data collection into every facet of modern life can often lull the public into a state of passive acceptance. For years, the intricacies of government surveillance programs, their scope, and their potential impact on civil liberties remained largely shrouded in opacity, understood only by a select few within intelligence agencies, legislative oversight committees, and the specialized circles of cybersecurity. The average citizen, while perhaps harboring a vague unease, lacked concrete evidence and clear understanding of the extent to which their digital lives were being monitored. This equilibrium, however, was profoundly disrupted by high-profile revelations that irrevocably altered the public perception of government surveillance and ignited widespread debate.

The seismic impact of disclosures made by individuals like Edward Snowden in 2013 cannot be overstated. Prior to these revelations, the discourse surrounding national security and data collection was often framed by government assurances of necessity, targeted operations, and minimal intrusion into the lives of ordinary citizens. The narrative emphasized safeguarding against threats, with the underlying assumption that such measures were judiciously applied and subject to rigorous oversight. Public understanding, therefore, was largely shaped by official statements and a general faith in the institutions tasked with protecting national security. However, Snowden's leaks provided an unprecedented, and for many, chilling, glimpse behind the curtain. Documents revealed the vast scale of metadata collection, including the telephone records of millions of Americans, the pervasive monitoring of internet communications through programs like PRISM, and the backdoor access granted to intelligence agencies to tap into the infrastructure of major technology companies.

These revelations did more than just inform; they fundamentally shifted public consciousness. The sheer volume and reach of the surveillance programs exposed were staggering, far exceeding what most citizens believed was happening. The idea that a government could, without specific warrants for individual suspects, systematically collect and store the communication data of its entire population was a concept that resonated deeply and disturbingly. Polling data from the period immediately following the Snowden disclosures starkly illustrates this shift. Surveys conducted by reputable organizations showed a significant increase in the number of Americans who believed that government surveillance had gone too far. For instance, a Pew Research Center poll released in June 2013, shortly after the initial leaks, indicated that 56% of Americans believed that the government's counterterrorism programs infringed upon people's basic civil liberties. This represented a marked increase from earlier surveys where such concerns were less pronounced.

The disclosures also injected a new urgency into the public discourse surrounding privacy. Previously, privacy concerns were often viewed as abstract or relevant only to those with something to hide. Snowden's revelations demonstrated that the issue was far more pervasive and affected everyone. The argument that privacy was merely a concern for criminals or those with nefarious intentions was effectively countered by the evidence that innocent citizens' communications were being routinely vacuumed up and stored. This realization fostered a sense of shared vulnerability and galvanized public opinion against unchecked surveillance powers. The conversation moved from the theoretical to the concrete, prompting widespread debate not just among civil liberties advocates and policymakers, but also among everyday citizens sharing their concerns on social media, in coffee shops, and in their communities.

The political landscape regarding surveillance policy was irrevocably altered. Legislators who had previously supported or remained silent on expansive surveillance measures found themselves under increased public scrutiny. Constituents, armed with detailed information about programs like the NSA's bulk telephone metadata collection, began demanding accountability and reform. The disclosures led to a flurry of legislative activity, with various proposals aimed at reining in surveillance powers and enhancing oversight. While the effectiveness and scope of these reforms have been a subject of ongoing debate, the sheer fact that such legislative efforts were initiated and gained traction is a testament to the impact of the revelations on public awareness and political will. Debates in Congress, previously dominated by national security imperatives, now had to contend with the significant public demand for privacy protections.

The role of media and technology in amplifying these revelations and shaping public perception was crucial. News organizations, empowered by the leaked documents, dedicated extensive resources to investigating and reporting on the surveillance programs. The dissemination of this information across various platforms, including traditional news outlets, online publications, and social media, ensured that the revelations reached a broad audience. Furthermore, the very technologies that were the subject of the surveillance, such as the internet and social media, became powerful tools for public discussion and mobilization. Hashtags related to privacy and surveillance trended globally, and online forums became hubs for debating the implications of the leaks. This digital amplification effect ensured that the conversation was not confined to elite circles but became a widespread public phenomenon.

Beyond the immediate surge in awareness, the Snowden disclosures also fostered a more critical and informed public regarding the trade-offs between security and liberty. The narrative often presented by governments, that enhanced surveillance is an indispensable tool for security, was challenged by the reality of mass data collection that seemed to capture indiscriminable amounts of information. This prompted citizens to question whether the perceived security benefits truly justified the erosion of privacy and the potential for abuse. The emphasis shifted from simply accepting government assurances to demanding empirical evidence of efficacy and robust safeguards against misuse. The very definition of what constitutes a reasonable balance between national security and individual privacy began to be re-evaluated by the public at large.

The impact was not monolithic, however. While a significant portion of the public became more concerned about government surveillance, there remained segments of the population who prioritized national security and viewed the revelations as a threat to intelligence operations. Public opinion was, and continues to be, divided along these lines, often correlating with broader political ideologies and views on the role of government. Some viewed the leaks as a courageous act of whistleblowing, while others condemned them as treasonous acts that endangered national security. This division highlighted the complex and often irreconcilable nature of the security versus liberty debate. However, even among those who prioritized security, the revelations forced a more conscious consideration of the privacy implications, leading to a more nuanced understanding of the issue.

The long-term impact of these high-profile revelations

continues to unfold. The increased public awareness has led to sustained pressure on governments and technology companies to be more transparent about their data collection practices. It has also spurred the development and adoption of privacy-enhancing technologies and practices by individuals and organizations alike. Concepts that were once considered niche, such as end-to-end encryption and the use of virtual private networks (VPNs), have become more mainstream, driven by a public desire to reclaim a degree of control over their digital lives. The legacy of these disclosures is, therefore, not just a moment of heightened public concern, but a foundational shift in the public's understanding and engagement with issues of privacy and surveillance.

The aftermath of the Snowden revelations also saw a more critical examination of the legal frameworks governing surveillance. In the United States, for instance, Section 702 of the Foreign Intelligence Surveillance Act (FISA) and Section 215 of the Patriot Act, which facilitated some of the bulk data collection, became focal points of intense debate and reform efforts. The public outcry contributed to the passage of the USA Freedom Act in 2015, which, while not eliminating bulk data collection entirely, did introduce reforms aimed at increasing transparency and limiting the government's ability to collect phone metadata in bulk. This legislative action, directly influenced by the public discourse ignited by the leaks, represented a tangible, albeit debated, outcome of the increased public awareness.

Furthermore, the global nature of the revelations meant that the impact extended far beyond the United States. In countries around the world, citizens and policymakers alike began to scrutinize their own governments' surveillance capabilities and the legal frameworks that underpinned them. International bodies and civil society organizations amplified calls for stronger privacy protections and greater accountability for intelligence agencies. The discussion about data sovereignty, the right to privacy in the digital age, and the potential for extraterritorial surveillance became a prominent feature of international human rights discourse.

The enduring legacy of these high-profile revelations lies in their ability to demystify complex government operations and empower the public with knowledge. While the debate over the appropriate balance between security and privacy continues, the disclosures ensured that this debate is now conducted with a greater degree of public awareness and engagement. The casual acceptance of pervasive data collection has been replaced by a more informed skepticism and a demand for greater transparency and accountability. The efforts of individuals like Edward Snowden, however controversial, undeniably succeeded in bringing the intricate and often hidden world of government surveillance into the public square, forcing a reckoning with the implications for civil liberties in the digital age. This heightened awareness continues to shape policy, drive technological innovation in privacy protection, and empower individuals to advocate for their fundamental right to privacy. The public discourse initiated by these monumental revelations has irrevocably changed the way societies grapple with the ever-expanding capabilities of state surveillance, ensuring that the conversation about privacy remains a central and urgent concern.

The initial shockwaves from major surveillance revelations, while undeniably powerful, often give way to a subtler, yet equally significant, challenge: the insidious creep of public apathy and the gradual normalization of surveillance. This phenomenon is not born of a single cause but rather a confluence of psychological, societal, and technological factors that, over time, can erode active concern and create a widespread acceptance of pervasive monitoring as an inescapable feature of modern life. While many may have recoiled at the scale of data collection exposed in the wake of events like the Snowden leaks, the day-to-day reality of living in an increasingly monitored world often leads to a resigned acceptance, a quiet surrender to the status quo.

One of the most potent drivers of this apathy is the perceived lack of direct, personal impact. For individuals who do not believe they are engaged in activities that would attract the attention of authorities or who simply feel they have "nothing to hide," the abstract concept of mass surveillance can appear distant and irrelevant. This "nothing to hide" argument, though frequently deconstructed by privacy advocates, remains a deeply ingrained justification for passive acceptance. It operates on a flawed premise that privacy is solely a shield for wrongdoing, rather than a fundamental human right essential for autonomy, free expression, and the development of personal identity. When surveillance is framed implicitly or explicitly as a tool to catch criminals or terrorists, those who identify as law-abiding citizens may see themselves as outside the scope of its concern, thereby excusing themselves from active engagement with the issue. This selective blindness allows individuals to compartmentalize their awareness of surveillance, acknowledging its existence but disassociating it from their personal lived experience. The consequence is a public that, while perhaps vaguely uneasy, lacks the personal impetus to demand change or scrutinize the underlying systems that facilitate this pervasive monitoring.

Furthermore, the very design and deployment of surveillance technologies often contribute to their normalization by presenting them as conveniences or unavoidable necessities. From smart home devices that listen to household conversations to social media platforms that meticulously track user interactions, surveillance is frequently embedded within the fabric of daily life, often under the guise of enhancing user experience, providing personalized services, or ensuring safety. The ubiquitous presence of CCTV cameras in public spaces, once a topic of heated debate, has largely faded into the background, becoming as commonplace as streetlights. Similarly, the constant collection of location data by smartphones, essential for navigation and other services, is often accepted without significant pushback. This constant exposure, coupled with the integration of these technologies into essential aspects of modern living, blurs the lines between voluntary data sharing and involuntary monitoring. When the tools of surveillance are also the tools of convenience and connection, the distinction becomes difficult to maintain, and the latter often overshadows the former in the public consciousness.

The sense of powerlessness also plays a critical role in fostering apathy. Faced with the immense capabilities of state and corporate surveillance apparatuses, many individuals may feel that their efforts to resist or even understand the scope of monitoring are futile. The sheer scale and technical sophistication of these systems can be intimidating, leading to a feeling that individual actions are insignificant in the face of such overwhelming power. This can manifest as a form of learned helplessness, where individuals conclude that resistance is pointless and that the best course of action is simply to adapt and comply. The complex legal frameworks and the often opaque nature of intelligence operations further exacerbate this feeling, creating a knowledge gap that makes meaningful public oversight incredibly challenging. Without clear understanding or a perceived ability to effect change, the natural inclination is to disengage.

The media's role in this normalization process is also multifaceted. While the initial revelations might be met with extensive coverage, the continuous, incremental expansion of surveillance capabilities often receives less sustained attention. The novelty wears off, and the public's attention span is often drawn to newer, more immediate concerns. Moreover, the framing of surveillance in the media can also contribute to its acceptance. If reporting consistently emphasizes the necessity of surveillance for national security, economic competitiveness, or public safety, without adequately exploring the counterarguments or the potential harms to civil liberties, it can reinforce the idea that such measures are simply the price of a secure and modern society. This can lead to a situation where the public is consistently exposed to the rationale for surveillance without a robust counter-narrative that emphasizes the value of privacy and the risks of unchecked monitoring.

The gradual desensitization that occurs through prolonged exposure is another significant factor. What once seemed alarming – the collection of metadata, the tracking of online activity, the use of facial recognition in public spaces – becomes part of the ambient background of digital existence. This process is akin to adapting to noise pollution; initially, it is disruptive, but over time, the brain learns to filter it out. For many, the constant, low-level awareness of being potentially monitored ceases to be a source of active concern and instead becomes a background hum, an accepted, if unacknowledged, condition of contemporary life. This desensitization can be particularly pronounced for younger generations who have grown up with pervasive digital tracking as a constant, rather than a newly introduced, element of their lives.

The fragmentation of public discourse also contributes to apathy. Instead of a unified public outcry against surveillance, opinions often become polarized along political or ideological lines. Those who are already skeptical of government power may remain vocal, while those who prioritize security or trust in authority may remain indifferent or even supportive of surveillance measures. This division prevents the formation of a broad, cohesive front demanding change. Without a unified public voice, the pressure on policymakers to enact significant reforms diminishes. The issue becomes another partisan talking point rather than a fundamental civil liberties concern that transcends political divides.

Furthermore, the commodification of data by private corporations, while distinct from government surveillance, often blurs the lines and contributes to the overall sense of being perpetually observed and cataloged. When personal data is routinely collected and monetized by companies for advertising, targeted marketing, and other commercial purposes, it normalizes the idea that personal information has become a public commodity. This widespread practice by the private sector can create a precedent and a psychological shift, making it easier for government surveillance programs to be perceived as just another iteration of this broader trend. If one's every click and purchase is already being tracked and analyzed for profit, the addition of government monitoring might seem like a less significant transgression, especially if it is couched in terms of public good. This blurring of public and private surveillance creates a complex environment where the boundaries of privacy are constantly shifting, and public vigilance can be easily diluted.

The legal and regulatory responses to surveillance also play a role in normalization. When legislative efforts to curb surveillance powers are perceived as weak, riddled with loopholes, or ultimately ineffective, it can breed cynicism and further entrench apathy. If citizens see that even well-intentioned reforms fail to produce meaningful change, they may conclude that the system is immutable and that their engagement will not alter the trajectory. This can lead to a feeling of disillusionment, where the energy required to remain informed and advocate for reform feels like a wasted effort. The cyclical nature of revelations, followed by limited reform, and then further expansion of surveillance capabilities, can create a sense of futility that erodes public engagement over time.

The economic pressures and the conveniences offered by surveillance-enabled technologies further entrench their acceptance. For instance, many essential services, from banking to healthcare to social interaction, are increasingly mediated by digital platforms that necessitate data sharing. Opting out of these systems entirely can impose significant social and economic costs, making it practically impossible for many individuals to fully escape the surveillance apparatus. This creates a situation where participation in modern society is, to a significant degree, conditional on accepting a certain level of monitoring. The implicit trade-off – security and convenience for privacy – becomes an increasingly difficult bargain to refuse when the alternatives are so unappealing.

Ultimately, the normalization of surveillance is not a sudden event but a gradual process of erosion. It is the quiet accumulation of small concessions, the slow acceptance of new norms, and the gradual dimming of public vigilance. While the initial shock of exposure might spark outrage and demand for change, the sustained effort required to maintain that vigilance against the backdrop of pervasive, often convenient, monitoring is a significant challenge. Without a constant stream of compelling revelations, without ongoing public education, and without robust, consistently enforced legal protections, the tide of apathy can easily reclaim the public consciousness, leaving surveillance as an accepted, though perhaps still disquieting, feature of our increasingly digital lives. This underscores the critical need for continuous dialogue, accessible information, and a steadfast commitment to defending privacy not just as an abstract ideal, but as a tangible necessity for a free and democratic society. The challenge for civil libertarians and advocates lies in combating this pervasive apathy, reigniting public concern, and ensuring that the normalization of surveillance does not become a permanent fixture of the 21st century.

9: RESISTANCE, ACTIVISM, AND ADVOCACY

The fight against pervasive surveillance and the erosion of civil liberties is not waged solely in the abstract realm of philosophical debate or through the isolated actions of whistleblowers. It is a sustained, multifaceted campaign orchestrated by dedicated civil liberties organizations. These groups serve as the vanguard, translating public concern into concrete action, challenging overreach in courtrooms, advocating for legislative reform, educating the populace, and even developing technological countermeasures. Their work is indispensable in a landscape where the tentacles of surveillance, both governmental and corporate, are constantly expanding.

Foremost among these defenders is the American Civil Liberties Union (ACLU). For over a century, the ACLU has stood as a bulwark for the Bill of Rights, and in the digital age, its focus on privacy and surveillance has become increasingly central. The organization employs a comprehensive strategy that encompasses litigation, lobbying, public awareness campaigns, and grassroots organizing. In the realm of surveillance, the ACLU has been at the forefront of challenging various government programs. Following the revelations of widespread electronic surveillance, the ACLU, often in conjunction with its state affiliates, has filed numerous lawsuits seeking to invalidate or at least curb the scope of these programs. These legal battles are not merely about privacy for privacy's sake; they are fundamentally about safeguarding the Fourth Amendment's protection against unreasonable searches and seizures, the First Amendment's guarantees of free speech and association, and the Fourteenth Amendment's promise of due process and equal protection.

One significant area of ACLU litigation has been against the National Security Agency's (NSA) bulk collection of phone metadata. These lawsuits have argued that such indiscriminate collection violates constitutional protections, as it allows the government to amass vast troves of information about the communications of millions of innocent Americans without any individualized suspicion of wrongdoing. While these cases have faced considerable legal hurdles, including government claims of national security secrecy and the classification of evidence, the ACLU's persistent efforts have kept these issues in the public eye and forced some degree of transparency and limited judicial review. The organization has also been instrumental in challenging the legal frameworks that underpin surveillance, such as arguing against the broad interpretations of statutes like the Foreign Intelligence Surveillance Act (FISA) that critics contend have been exploited to authorize expansive surveillance powers.

Beyond the courtroom, the ACLU actively engages in legislative advocacy. This involves meticulously researching surveillance technologies and government practices, drafting proposed legislation, and lobbying lawmakers at both federal and state levels. Their goal is to enact laws that strengthen privacy protections, increase transparency and accountability for surveillance programs, and rein in the executive branch's unilateral power in this domain. This legislative work often involves educating policymakers about the potential harms of unchecked surveillance, including its chilling effect on free speech, its disproportionate impact on marginalized communities, and its capacity to erode public trust. The ACLU's lobbying efforts are crucial in ensuring that the voices of privacy advocates are heard amidst the powerful narratives of national security that often dominate policy debates.

The Electronic Frontier Foundation (EFF) represents another critical pillar in the defense of civil liberties in the digital age. Founded in 1990, the EFF is a non-profit organization dedicated to defending and promoting civil liberties in connection with technology. Its approach is deeply rooted in technological understanding and a commitment to using legal and technological tools to protect user privacy, free expression, and innovation. The EFF's litigation strategy often focuses on novel legal theories and groundbreaking cases that establish important precedents for digital rights. They have been instrumental in cases challenging government access to digital communications, arguing for stronger protections for online speech, and advocating for transparency regarding government surveillance practices.

The EFF has been particularly active in pushing the boundaries of Fourth Amendment law in the context of modern technology. They have litigated cases involving the search of electronic devices at borders, the use of location data obtained from cell phone providers, and the government's ability to compel access to encrypted communications. A key aspect of the EFF's legal strategy is its emphasis on "privacy as a fundamental right," arguing that constitutional protections must evolve to keep pace with technological advancements. Their work has helped to clarify that digital data, like emails, text messages, and location records, is entitled to the same, if not greater, constitutional protection as physical possessions.

In addition to litigation, the EFF is a powerhouse of public education and digital advocacy. They produce extensive reports, white papers, and online resources that demystify complex surveillance technologies and legal issues for the general public and policymakers alike. The EFF's "Know Your Rights" guides, for example, provide practical advice to individuals about their privacy rights when interacting with law enforcement or government agencies. They also run high-profile public awareness campaigns, often leveraging social media and digital platforms, to galvanize public opinion and mobilize support for privacy-protective policies. This educational role is vital for countering the "nothing to hide" mentality and fostering a broader understanding of why privacy matters to everyone.

Furthermore, the EFF engages in crucial technological advocacy. This can involve developing tools and resources that help individuals protect their digital privacy, such as encryption guides or browser extensions that block trackers. They also play a significant role in commenting on proposed government regulations and industry standards, offering expert technical advice to ensure that new technologies and policies do not inadvertently undermine privacy rights. The EFF's commitment to technological solutions underscores the understanding that legal and policy advocacy must be complemented by practical, user-empowering tools.

The intersection of these two organizations, the ACLU and the EFF, along with numerous other smaller, specialized groups, forms a robust ecosystem of advocacy. They often collaborate on major legal challenges and legislative initiatives, pooling their resources and expertise. For instance, in landmark cases challenging government surveillance programs, it is common to see joint filings or amicus briefs submitted by multiple civil liberties organizations, amplifying their collective impact. This collaborative spirit is essential given the scale and complexity of the forces they are up against.

Other organizations also play vital roles. Groups like the Center for Democracy & Technology (CDT) focus on policy and advocacy, working to shape technology policy to promote democratic values and civil liberties. They engage extensively with policymakers, academics, and industry leaders to promote privacy-enhancing technologies and responsible data practices. The National Association of Criminal Defense Lawyers (NACDL) also champions Fourth Amendment rights, often highlighting how surveillance practices disproportionately affect those within the criminal justice system. In specific communities, local organizations often address the particular surveillance concerns that impact them most directly, such as racial profiling through facial recognition technology or invasive monitoring of public housing residents.

The strategies employed by these organizations are varied and adaptable. Litigation is a cornerstone, aiming to secure legal victories that can set binding precedents, compel changes in government behavior, and provide redress for those whose rights have been violated. These cases can span years, requiring immense dedication and resources, but their impact can be profound, shaping the legal landscape for decades to come.

Legislative advocacy is equally critical. It involves direct engagement with elected officials and government agencies to influence the creation and enforcement of laws and regulations. This can range from testifying at congressional hearings to drafting model legislation for states to adopt. The goal is to build a legal framework that respects individual privacy and limits government intrusion. This often involves a painstaking process of building consensus, educating diverse stakeholders, and countering well-funded opposition.

Public education campaigns are perhaps the most broadly

influential strategy. By informing the public about the realities of surveillance, the potential harms, and the importance of privacy, these organizations empower individuals to demand change. This includes creating accessible content, organizing public forums, and utilizing traditional and social media to reach a wide audience. The aim is to foster an informed citizenry that understands the stakes and is willing to engage in advocacy.

Technological solutions also represent a growing area of focus. Recognizing that legal and policy measures alone may not be sufficient, some organizations are involved in developing or promoting technologies that enhance privacy, such as encryption, anonymization tools, and secure communication platforms. This proactive approach seeks to empower individuals with the means to protect themselves in an increasingly monitored world.

The victories achieved by these organizations are significant, though often hard-won and subject to ongoing challenges. For instance, legal challenges have led to some degree of judicial scrutiny over certain NSA surveillance programs, forcing greater transparency and prompting legislative reforms, such as the USA FREEDOM Act of 2015, which aimed to end the NSA's bulk collection of U.S. telephone metadata. While the effectiveness and scope of such reforms are subjects of continuous debate, they represent tangible outcomes of sustained advocacy. Public awareness campaigns have undoubtedly raised the salience of privacy issues, making it more difficult for governments to expand surveillance without public scrutiny.

However, these organizations also face immense challenges. The sheer volume of data collected, the complexity of the technologies involved, and the secrecy that often shrouds national security operations create significant hurdles. Government claims of national security are frequently invoked to resist transparency and legal challenges, making it difficult to obtain evidence or to persuade courts to intervene. Furthermore, the economic incentives for data collection by private companies, while distinct from government surveillance, create a pervasive environment of monitoring that can desensitize the public and blur the lines of accountability.

The ongoing battles fought by civil liberties organizations are a testament to their unwavering commitment to fundamental freedoms. They are not merely reacting to revelations; they are proactively working to shape a future where technology serves humanity without compromising liberty. Their efforts highlight that the defense of privacy and civil liberties is not a passive endeavor but an active, ongoing struggle that requires constant vigilance, strategic engagement, and the collective will of concerned citizens and dedicated advocates. The sustained impact of these organizations in courtrooms, legislative chambers, and public consciousness underscores their indispensable role in preserving the core tenets of a free and democratic society in the face of evolving technological threats.

The fight for privacy and against overreaching surveillance has often found its most critical battleground in the courtroom. Civil liberties organizations, legal scholars, and affected individuals have consistently turned to the judiciary as a crucial avenue for challenging government practices that they deem unconstitutional or otherwise unlawful. These legal challenges are not merely abstract exercises in jurisprudence; they are direct confrontations with state power, aiming to secure tangible protections for fundamental rights in the face of evolving technologies and national security imperatives. The process is often arduous, marked by complex legal doctrines, significant resource disparities, and the inherent difficulty of litigating matters shrouded in secrecy.

At the heart of many of these legal battles lies the Fourth Amendment of the U.S. Constitution, which guards against unreasonable searches and seizures. As surveillance technologies have advanced, so too has the debate over what constitutes a "search" in the digital age and what level of suspicion is required to justify government intrusion. Litigants and their legal representatives have argued that the collection of vast amounts of digital data—whether it be phone metadata, internet browsing history, location information, or the contents of electronic devices—without a warrant based on probable cause, or at least without a clear legal framework, infringes upon constitutionally protected privacy interests. These arguments often center on the concept of reasonable expectation of privacy, a doctrine that has been continually re-examined and redefined in the context of new technologies. For instance, the Supreme Court's ruling in *Katz v. United States* (1967), which famously stated that "the Fourth Amendment protects people, not places," laid the groundwork for extending constitutional protections to intangible information, a principle that has been crucial in digital privacy litigation.

Beyond the Fourth Amendment, plaintiffs have also invoked other constitutional provisions. The First Amendment's guarantees of freedom of speech and association have been central to arguments that surveillance can have a chilling effect on political dissent and public discourse. When individuals believe their communications are being monitored, they may self-censor, refrain from participating in protests, or avoid associating with certain groups, thereby diminishing the robust exchange of ideas essential to a democratic society. The Fourteenth Amendment's Due Process Clause and Equal Protection Clause have also been invoked, particularly in cases where surveillance practices are alleged to disproportionately target certain communities, such as racial minorities or political activists, leading to discriminatory outcomes.

The types of lawsuits filed span a wide spectrum, reflecting the diverse ways in which surveillance can impact individuals and society. Some cases have directly challenged the legality of specific government surveillance programs, such as the bulk collection of telephone metadata by the National Security Agency (NSA), as revealed by former contractor Edward Snowden. Organizations like the ACLU and the EFF have been instrumental in filing and supporting such litigation, often seeking declaratory judgments that certain practices are unconstitutional or injunctions to halt them. Other cases have focused on more specific applications of surveillance, such as challenges to warrantless searches of electronic devices at the border, the use of cell-site location information to track individuals, or the government's ability to compel companies to decrypt user data.

A significant hurdle in many of these legal challenges is the classification of information and the invocation of national security privileges by the government. Defendants frequently argue that the details of surveillance programs are classified secrets, and that disclosing such information would jeopardize national security. This often leads to requests for dismissal of cases based on the state secrets privilege or other secrecy doctrines, making it exceedingly difficult for plaintiffs to gather evidence, understand the full scope of the government's actions, or even present their case effectively. Courts have had to grapple with balancing the need for transparency and accountability with the government's legitimate need to protect classified information. This tension has sometimes resulted in cases being dismissed on procedural grounds, even when there are significant concerns about the underlying surveillance practices.

The landmark case of *Klayman v. Obama* (later renamed *Clapper v. Amnesty International USA*), filed by Larry Klayman and later consolidated with other suits, sought to challenge the NSA's bulk metadata collection program. While the initial attempts to halt the program were met with significant legal obstacles, primarily concerning standing (the ability of plaintiffs to demonstrate they were directly harmed by the program), the legal arguments laid the groundwork for subsequent challenges. The Supreme Court's eventual ruling in *Clapper v. Amnesty International USA* (2013) denied standing to the plaintiffs, ruling that the alleged harm was too speculative. However, the case highlighted the profound difficulties individuals and organizations face in proving they have been directly and imminently harmed by secret surveillance programs, a barrier that has been a recurring theme in digital privacy litigation.

The subsequent revelations by Edward Snowden in 2013 provided concrete evidence of the NSA's extensive data collection activities, reinvigorating legal efforts. This led to new lawsuits, including *In re: National Security Agency Telecommunications Records Litigation*, where a federal judge in Maryland ruled in 2015 that the NSA's bulk collection of phone metadata was likely unconstitutional. Judge Richard J. Leon found that the program violated the Fourth Amendment's prohibition against unreasonable searches, stating, "This is not science fiction. This is the reality of the Fourth Amendment challenges of the 21st century." While this ruling was a significant victory for privacy advocates, it was later overturned on appeal by the U.S. Court of Appeals for the Fourth Circuit, which found that the plaintiffs lacked standing, a decision that was ultimately upheld by the Supreme Court, which declined to hear the case. This back-and-forth demonstrates the precarious nature of legal victories in this domain.

The passage of the USA FREEDOM Act in 2015, which aimed to reform the NSA's bulk metadata collection program by requiring the government to obtain specific court orders to access call detail records from telecommunications companies, was a direct consequence of these legal and public pressures. While proponents hailed it as a significant privacy enhancement, critics argued that it did not go far enough and that the government retained substantial surveillance powers. The ongoing interpretation and implementation of this legislation have continued to be a subject of legal scrutiny and advocacy.

Beyond bulk data collection, litigation has also focused on other aspects of surveillance. Cases challenging the government's ability to access location data from mobile phone providers, such as the Supreme Court's decision in *Riley v. California* (2014), which held that police generally need a warrant to search the digital contents of a cell phone seized from an arrested person, have been crucial in establishing digital privacy rights. In *Riley*, the Court recognized that cell phones contain vast amounts of personal information, far exceeding the scope of physical documents, and thus warrant strong Fourth Amendment protections. This ruling was a significant affirmation of privacy in the digital age and has had a broad impact on law enforcement practices.

Another area of intense legal contestation involves the government's access to data held by technology companies, often through requests under statutes like the Stored Communications Act (SCA) or through National Security Letters (NSLs). Organizations like the EFF have been active in challenging the broad interpretation of these authorities, arguing for greater transparency and stronger legal safeguards. Cases involving government demands for user data, including emails, cloud storage contents, and social media activity, have raised critical questions about privacy in the context of third-party data custodians and extraterritorial data access.

The judiciary's role in these legal battles is multifaceted. Courts are tasked with interpreting existing laws and constitutional principles in light of new technologies, balancing competing interests of national security and individual liberty, and providing a check on potential government overreach. However, the judiciary itself faces challenges. Judges must often make decisions based on incomplete information due to secrecy, navigate complex technical issues, and grapple with the political and societal implications of their rulings. The effectiveness of judicial review in safeguarding civil liberties against surveillance is therefore dependent on the judiciary's willingness and ability to rigorously scrutinize government claims and to uphold constitutional principles even when doing so may be unpopular or politically difficult.

The legal landscape is constantly evolving, with new technologies and new government surveillance methods emerging regularly. This necessitates a continuous cycle of legal challenges, advocacy, and judicial interpretation. While some legal battles may result in setbacks or be resolved through legislative compromises, the persistent engagement of civil liberties advocates and affected individuals in the courtroom remains a vital component of the broader resistance against pervasive surveillance. Each lawsuit, each argument, and each judicial decision contributes to the ongoing development of legal precedent and public understanding of the critical balance between security and freedom in the digital age. The courtroom, despite its inherent limitations and the formidable challenges it presents, remains an indispensable arena for defending the privacy rights of citizens in an increasingly monitored world.

The struggle against pervasive government surveillance has not been confined solely to the hallowed halls of the judiciary. While legal challenges represent a critical front, the arena of legislative reform has also served as a vital, albeit often more complex and politically fraught, battleground. Recognizing that laws, not just court interpretations, define the boundaries of government power, a persistent wave of activism has targeted the very statutes that enable and regulate surveillance activities. These efforts aim to not only dismantle problematic existing frameworks but also to proactively establish new legal protections for privacy and civil liberties in an era of rapidly advancing technology. The legislative path, however, is a winding one, often characterized by intense political wrangling, the influence of powerful special interests, and the inherent difficulty of achieving consensus on matters that touch upon national security, technological capabilities, and fundamental rights.

One of the most significant legislative targets has been the Foreign Intelligence Surveillance Act (FISA) of 1978. Enacted in the wake of revelations about abuses of power by intelligence agencies during the Vietnam War and Watergate eras, FISA sought to provide a legal framework and judicial oversight for certain types of electronic surveillance for foreign intelligence purposes. However, as surveillance technologies evolved and the nature of national security threats shifted, key provisions of FISA, particularly Section 702 concerning the collection of foreign intelligence information, became subjects of intense debate and scrutiny. Section 702 allows the U.S. government to compel U.S. telecommunications companies to assist in the acquisition of foreign intelligence information concerning non-U.S. persons located outside the United States. While intended for foreign targets, the program's broad collection methods have raised concerns about the incidental collection of communications data belonging to U.S. citizens. This incidental collection, often referred to as "backdoor searches," occurs when queries are made of the data held by the government using identifiers that may belong to Americans, without a warrant based on probable cause.

The debate over Section 702 has spurred numerous legislative reform efforts. For years, civil liberties organizations, digital rights advocates, and a bipartisan coalition of lawmakers have called for significant reforms. These proposed changes have often focused on several key areas: requiring warrants based on probable cause for any search of Section 702 data that targets U.S. persons; enhancing transparency regarding the scope and application of the program; and strengthening oversight by both the Foreign Intelligence Surveillance Court (FISC) and Congress. For instance, the USA FREEDOM Act of 2015, while primarily addressing the NSA's bulk collection of telephone metadata under Section 215 of the PATRIOT Act, also included some reforms aimed at the FISA process, albeit not directly targeting Section 702's controversial querying practices. The USA FREEDOM Act's passage marked a significant legislative effort to reform intelligence gathering in the wake of the Snowden revelations, shifting from bulk collection to a more targeted approach for certain types of data.

More recently, as Section 702 has come up for reauthorization, more aggressive legislative proposals have emerged. These have included amendments that would explicitly prohibit the querying of Section 702 data for information on Americans without a warrant. Advocates for these reforms argue that the current practice fundamentally undermines the Fourth Amendment's protections against unreasonable searches. They contend that allowing the government to search vast databases of intercepted communications for information about its own citizens, even if incidentally collected, without individualized suspicion and a warrant, represents a profound erosion of privacy. The political battles surrounding these reauthorization efforts have been fierce. Proponents of reform have marshaled public opinion, presented expert testimony, and engaged in extensive lobbying to persuade lawmakers. They have pointed to instances where Section 702 data has been queried for information on Americans, sometimes for purposes unrelated to national security, as evidence of the program's potential for abuse.

However, these reform efforts have consistently faced formidable opposition. Intelligence agencies and their allies in Congress argue that Section 702 is an indispensable tool for identifying and disrupting foreign terrorist plots, cyberattacks, and other national security threats. They maintain that the ability to query the data without a warrant for U.S. persons is crucial for national security because an American citizen might be involved in a plot or communication with foreign adversaries, and delays caused by obtaining a warrant could have dire consequences. Critics of the proposed warrant requirement, meanwhile, argue that it would create an untenable burden and significantly hamper the effectiveness of the program, potentially allowing threats to materialize. They also point to existing oversight mechanisms, such as the FISC's review and the FBI's internal compliance protocols, as sufficient safeguards. The tension between these competing arguments has made achieving consensus on substantial legislative reform incredibly challenging.

Beyond FISA, other legislative avenues have been explored to curb government surveillance powers. Following the revelations of widespread NSA data collection programs, there was a significant push for amendments to the PATRIOT Act and its successor legislation, such as the USA PATRIOT

Improvement and Reauthorization Act. These efforts aimed to narrow the scope of Section 215, which allowed the government to compel the production of "any tangible things" relevant to an investigation, and to increase transparency and accountability for the use of National Security Letters (NSLs) and other national security authorities. For instance, the PATRIOT SUNRISE Act and the USA RIGHTS Act were examples of legislative proposals introduced in Congress that sought to roll back some of the expansive surveillance authorities granted in the post-9/11 era and to reassert congressional oversight. These bills often proposed to reinstate higher standards for government access to data, enhance judicial review, and provide greater protections for civil liberties.

The role of advocacy groups and civil liberties organizations has been paramount in driving these legislative reform efforts. Groups like the American Civil Liberties Union (ACLU), the Electronic Frontier Foundation (EFF), the Center for Democracy & Technology (CDT), and many others have been instrumental in educating the public, testifying before congressional committees, drafting legislative proposals, and lobbying lawmakers. They have meticulously analyzed government surveillance programs, highlighted potential abuses, and advocated for specific statutory changes. Their work has often involved building broad coalitions, bringing together diverse stakeholders—including privacy advocates, technologists, academics, and even some business groups concerned about data security—to present a united front. These organizations frequently engage in public awareness campaigns, utilizing media outreach, social media, and grassroots organizing to generate public pressure on elected officials.

The political challenges inherent in legislative reform are substantial. The national security apparatus, including the intelligence community and the Department of Defense, often presents a unified and powerful voice in Washington, advocating for the maintenance or expansion of surveillance capabilities. This is often amplified by bipartisan consensus on the necessity of robust national security tools in the face of perceived threats. Lawmakers who champion surveillance reform often find themselves navigating a political landscape where expressing concern about government power can be framed as being soft on terrorism or national security. Furthermore, the complexity of surveillance technologies and legal frameworks can make it difficult for many legislators to fully grasp the implications of the laws they are debating, making them more reliant on the assessments provided by executive branch agencies.

The process of legislative change is also slow and incremental. Bills can languish in committee for years, and even when they gain traction, they are often subject to intense negotiation and compromise, which can dilute their intended impact. The reauthorization cycles of key surveillance statutes, such as FISA, have become regular flashpoints for legislative debate and reform attempts. For example, the periodic reauthorization of Section 702 of FISA has consistently provided an opportunity for lawmakers to attach amendments aimed at curbing perceived abuses or enhancing privacy protections. These moments are critical, as failure to reauthorize the law can have significant implications for national security operations, creating a powerful incentive for compromise, though not always in favor of stronger privacy protections.

Successes in legislative reform have been hard-won and often partial. The USA FREEDOM Act of 2015 is often cited as a significant legislative achievement. It responded directly to public outcry and legal challenges concerning the NSA's bulk telephone metadata collection program. The Act ended the government's bulk collection of these records under Section

215, replacing it with a system where telecommunications companies retain the data and the government must obtain specific court orders from the FISC to access it, based on specific selectors. This was a direct legislative intervention that fundamentally altered a major surveillance practice. Similarly, reforms aimed at National Security Letters (NSLs) have been enacted, increasing some transparency and requiring the Attorney General to review their use.

However, these successes are often juxtaposed with significant failures or incomplete reforms. Many proposed amendments to FISA, particularly those seeking to impose a warrant requirement for querying Section 702 data concerning Americans, have failed to pass Congress. Similarly, efforts to repeal or significantly curtail other national security surveillance authorities have generally stalled. The ongoing debate around the PATRIOT Act and its successors demonstrates a continuous tension between those who advocate for stronger privacy protections and those who prioritize enhanced national security capabilities. The balance of power in these legislative debates often leans towards the latter, making comprehensive reform a persistent challenge.

Moreover, the legislative landscape is dynamic. As new technologies emerge and new national security threats are identified, new surveillance tools and authorities are developed, creating a constant need for legislative adaptation and oversight. This cyclical process means that the fight for privacy through legislation is never truly over. It requires sustained engagement, vigilance, and a willingness to adapt strategies as the technological and political environments shift. The impact of lobbying by various interest groups— both those advocating for greater surveillance powers and those championing civil liberties—plays a crucial role in shaping legislative outcomes. The resources and influence of the intelligence community and its contractors, for instance, are significant, often countered by the grassroots activism and advocacy of civil liberties organizations.

The effectiveness of legislative reform is also contingent on robust oversight mechanisms. Even when laws are enacted, their implementation and adherence require constant monitoring. Congressional committees responsible for oversight, such as the House and Senate Intelligence Committees, play a vital role, but their effectiveness can be hampered by partisan politics, a lack of access to complete information, or a perceived conflict of interest in simultaneously overseeing and supporting national security agencies. Therefore, efforts to strengthen congressional oversight itself, through increased transparency, better-resourced staff, and a greater willingness to challenge executive branch assertions, are also a critical component of legislative reform.

In essence, legislative reform efforts represent a crucial, ongoing battle to rebalance the scales between government surveillance powers and individual privacy rights. While court decisions can strike down specific practices, legislation can establish broad new frameworks, define clear rules, and ensure ongoing accountability. The path is fraught with political obstacles, the influence of powerful interests, and the inherent difficulty of legislating in areas as complex and sensitive as national security and technology. Yet, the persistent advocacy, the tireless work of civil liberties groups, and the engagement of concerned citizens demonstrate a profound commitment to ensuring that technological advancements do not outpace fundamental protections for privacy and freedom. Each legislative proposal, each debate, and each enacted or rejected amendment contributes to the ongoing evolution of surveillance law and its impact on the democratic society.

The technological landscape, often perceived as a double-edged sword, undeniably facilitates the very surveillance it is now increasingly being harnessed to counter. While governments and corporations have leveraged advancements in data processing, artificial intelligence, and network infrastructure to expand their monitoring capabilities, a parallel and equally dynamic movement has emerged, dedicated to developing and deploying technological solutions that empower individuals to reclaim their digital privacy. This is not merely a reactive response, but an active, ongoing innovation cycle, a digital arms race where privacy-enhancing technologies (PETs) are constantly being refined and adopted to stay ahead of evolving surveillance methods. At the heart of this technological counter-offensive lies the principle of making surveillance more difficult, costly, and ultimately, less effective, by providing individuals with the tools to secure their data and communications.

One of the most fundamental and impactful technological solutions for privacy protection is **encryption**. Encryption, in essence, is the process of encoding information in such a way that only authorized parties can access it. When applied to digital communications and data storage, it renders information unintelligible to anyone who intercepts it without the correct decryption key. This is particularly crucial in the context of preventing mass surveillance. End-to-end encryption (E2EE), for instance, is a game-changer. In an E2EE system, data is encrypted on the sender's device and can only be decrypted by the intended recipient's device. This means that even the service provider facilitating the communication, or any intermediary network operator, cannot read the content of the messages. This principle is the bedrock of many secure messaging applications that have gained immense popularity among privacy-conscious users. Applications like Signal, WhatsApp (with its E2EE implementation), and Telegram (with its optional "secret chats") have become vital conduits for private communication, ensuring that conversations remain confidential from prying eyes, including those of governments or malicious actors. The strength of these encryption protocols, such as the Signal Protocol used by Signal and WhatsApp, is a subject of rigorous cryptographic research, aiming to provide a robust defense against even sophisticated decryption attempts.

Beyond messaging, encryption plays a critical role in protecting data at rest. Full-disk encryption, for example, encrypts the entire contents of a hard drive or solid-state drive. This means that if a device is lost, stolen, or seized by authorities, the data stored on it remains inaccessible without the user's password or encryption key. Operating systems like macOS and Windows offer built-in full-disk encryption (FileVault and BitLocker, respectively), making it a readily available feature for users to enable. Furthermore, secure cloud storage services are increasingly offering client-side encryption, where files are encrypted on the user's device *before* being uploaded to the cloud, ensuring that the cloud provider itself cannot access the plaintext data. Services like Sync.com and Proton Drive are pioneers in this area, offering a higher degree of assurance against unauthorized access to personal files stored remotely.

Another significant technological avenue for privacy protection lies in the development of **secure communication platforms and software**. This encompasses a range of tools designed to facilitate private interactions and shield users from being tracked or identified. **Virtual Private Networks (VPNs)** have become a cornerstone of digital privacy for many. A VPN creates an encrypted tunnel between a user's device and a VPN server, routing all internet traffic through that server. This masks the user's real IP address, making it appear as though they are browsing from the server's location. Furthermore, it encrypts the data traffic between the user and the VPN server, protecting it from interception by Internet Service Providers (ISPs), network administrators, or anyone monitoring the local network. The effectiveness of a VPN hinges on the trustworthiness of the VPN provider, their logging policies (ideally, no-logs policies), and the strength of their encryption protocols. Choosing a reputable VPN provider is crucial, as a compromised or untrustworthy VPN can be worse than no VPN at all.

The evolution of web browsers has also seen significant strides in privacy features. **Privacy-focused browsers** like Brave, Firefox (with enhanced tracking protection), and Tor Browser go beyond basic browsing security. Brave, for instance, includes built-in ad and tracker blockers, preventing websites from collecting user data through cookies and other tracking mechanisms. It also offers HTTPS Everywhere by default, ensuring that connections to websites are encrypted whenever possible. The Tor Browser, designed for maximum anonymity, routes internet traffic through a volunteer overlay network consisting of thousands of relays. Each relay decrypts only the layer of encryption needed to identify the next hop, effectively obscuring the user's true location and IP address from any single point in the network. While Tor offers a high degree of anonymity, it can come at the cost of slower browsing speeds, making it a tool best suited for situations where anonymity is paramount.

Beyond these established tools, the realm of **privacy-enhancing software** is constantly expanding. **Password managers** are essential for digital security, allowing users to create and store strong, unique passwords for each online account. By generating complex passwords that are difficult to guess or crack, and by securely storing them in an encrypted vault, password managers significantly reduce the risk of account compromise through weak or reused credentials. Tools like Bitwarden, LastPass, and 1Password are widely used for this purpose.

Furthermore, the concept of **decentralization** is emerging as a powerful technological countermeasure to centralized surveillance. In a decentralized system, data and control are distributed across a network of users, rather than being concentrated in a single entity. This makes it much harder for any single party, including governments or large corporations, to gain access to or control over the entire network or the data within it. **Decentralized communication platforms**, built on peer-to-peer (P2P) architectures, are gaining traction. These platforms often bypass central servers altogether, connecting users directly, which can enhance privacy and security. Projects exploring decentralized social media and file storage aim to create environments where user data is not a commodity to be harvested by a central authority.

Anonymization tools represent another critical layer of digital defense. While VPNs mask IP addresses, tools like the **Tor network** offer a more profound level of anonymity by obscuring the user's identity and location across multiple hops. However, anonymity is not a binary state and can be difficult to achieve perfectly. Emerging technologies are exploring more sophisticated anonymization techniques, such as **zero-knowledge proofs**, which allow one party to prove the truth of a statement to another party without revealing any information beyond the bare fact of its truth. While still largely in the research and development phase for widespread consumer use, zero-knowledge proofs hold immense potential for enabling private transactions and verifications without exposing sensitive personal data.

The ongoing "cat and mouse" game between surveillance and privacy is perhaps most evident in the evolution of **metadata collection and protection**. While encryption can protect the content of communications, metadata – the information about who communicated with whom, when, and for how long – has often been a prime target for mass surveillance. However, technologies are emerging that can help obscure or minimize the collection of such metadata. **Encrypted DNS (DoH/DoT)**, for example, encrypts DNS queries, preventing ISPs from seeing which websites users are visiting based on their DNS lookups. This adds a layer of privacy to the initial stages of internet browsing. Similarly, advancements in secure multi-party computation (SMPC) and homomorphic encryption are being explored to enable data analysis and aggregation without revealing the underlying raw data or individual identities, offering a path towards data utility while preserving privacy.

It is crucial to acknowledge that no single technology offers absolute privacy. The effectiveness of these tools often depends on their proper implementation, user vigilance, and the ongoing evolution of the technologies themselves. For instance, **secure operating systems**, like Tails or Qubes OS, are designed from the ground up with privacy and security in mind, offering features that isolate different activities and make it more difficult for malware or surveillance tools to gain a foothold. These systems often incorporate Tor or VPN integration by default and are designed to leave minimal traces on the host machine.

Moreover, the very act of adopting and promoting these technologies represents a form of **digital activism**. By choosing to use privacy-respecting software and platforms, individuals demonstrate demand for these products, encouraging further innovation and adoption. Open-source software, in particular, plays a vital role. The transparent nature of open-source code allows for independent security audits, building trust and enabling the community to identify and fix vulnerabilities. Projects like Signal, VeraCrypt (for encrypted storage), and LibreWolf (a privacy-focused fork of Firefox) are testaments to the power of collaborative development in creating robust privacy tools.

The challenge, however, is not solely about technological

capability but also about **accessibility and usability**. For privacy technologies to be truly effective in empowering citizens, they must be accessible to the average user, not just the tech-savvy elite. Developers are continuously working to improve user interfaces and simplify complex processes, making tools like VPNs, encrypted messaging, and secure browsers easier to install and use. The widespread adoption of two-factor authentication (2FA) is another example of a security measure that has become more accessible, significantly enhancing account security against unauthorized access.

The ongoing tension between surveillance and privacy is a dynamic one. As governments and other entities develop new methods for data collection and analysis, privacy advocates and technologists respond with innovative solutions. This includes not only encrypting communications but also developing techniques for **de-identifying data, anonymizing browsing habits**, and **securing personal information against breaches**. The proliferation of smart devices and the Internet of Things (IoT) presents new frontiers for surveillance, necessitating the development of IoT-specific privacy solutions, such as secure device management, data minimization protocols, and user control over data sharing.

Ultimately, the technological solutions for privacy protection are not just tools; they are enablers of autonomy and resistance in the digital age. They empower individuals to make informed choices about their data and communications, to resist unwarranted monitoring, and to participate in the digital sphere without constant fear of surveillance. The continuous innovation in encryption, secure platforms, anonymization techniques, and decentralized systems represents a vital and evolving front in the broader struggle for privacy and civil liberties. This technological counter-movement is a testament to human ingenuity and the persistent desire to maintain control over one's personal information in an increasingly data-driven world. The ongoing development and adoption of these digital defenses are crucial for fostering a future where privacy is not a luxury, but a fundamental right, secured by robust and accessible technology.

The effectiveness of technological safeguards and legal frameworks designed to protect privacy is intrinsically linked to the public's understanding of the issues at play. Without a well-informed populace, even the most robust privacy-enhancing technologies can remain underutilized, and legislative efforts can falter due to a lack of public mandate. This is where the crucial role of public education and awareness campaigns comes into sharp focus, acting as the bedrock upon which any meaningful resistance, activism, and advocacy for digital civil liberties must be built.

Advocacy groups, investigative journalists, academic researchers, and grassroots movements have collectively undertaken the monumental task of illuminating the often-opaque world of government and corporate surveillance for the average citizen. Their efforts aim not merely to inform, but to empower, fostering a critical understanding of how data is collected, analyzed, and utilized, and the profound implications this has for individual freedoms and democratic societies. This process begins with demystifying complex technological concepts and legal frameworks. For instance, explaining the nuances of data retention policies, the capabilities of facial recognition technology, or the implications of Section 702 of the Foreign Intelligence Surveillance Act (FISA) requires translating technical jargon and legalistic language into accessible terms. Organizations like the Electronic Frontier Foundation (EFF), the American Civil Liberties Union (ACLU), and Privacy International have been at the forefront of this educational charge, producing a wealth of accessible materials, from in-depth reports and white papers to simple infographics and explainer videos. These resources break down intricate issues like the Fourth Amendment's application in the digital age, the privacy implications of smart home devices, or the potential for algorithmic bias in law enforcement tools.

Journalists play an indispensable role in this ecosystem of awareness. Through meticulous investigative reporting, they uncover instances of overreach, expose the mechanisms of surveillance, and bring to light the human stories behind the data. Whistleblowers, often risking significant personal consequences, have been pivotal in initiating public discourse. The revelations from Edward Snowden, for example, about the global scale of intelligence gathering by agencies like the NSA, fundamentally shifted the public conversation around surveillance. His disclosures, amplified by media outlets worldwide, spurred widespread debate about the balance between national security and individual privacy, leading to legislative reviews and increased scrutiny of surveillance programs. This kind of journalism is not merely about reporting facts; it is about contextualizing those facts, connecting them to fundamental rights, and illustrating the potential for chilling effects on free speech and association. The ability of journalists to provide this critical context, often under significant pressure, is vital for building an informed public opinion.

Academics and researchers contribute by providing rigorous, evidence-based analysis. They conduct studies on the psychological impact of constant surveillance, the economic incentives driving data collection, and the societal consequences of unchecked surveillance capitalism. Their work often forms the intellectual backbone of advocacy efforts, providing the data and theoretical frameworks necessary to argue for policy changes. University-led research centers, think tanks, and independent academic publications serve as crucial spaces for developing and disseminating this knowledge. For example, research into the accuracy and biases of predictive policing algorithms can directly inform policy debates and public understanding of the risks associated with relying too heavily on such technologies. Similarly, studies on the effectiveness and proportionality of mass surveillance programs provide essential data for evaluating their necessity and impact.

Grassroots movements and community organizers often excel at reaching specific demographics and mobilizing local action. They translate abstract concerns about privacy into tangible impacts on people's daily lives. Campaigns focused on issues like local police use of surveillance technologies, such as Stingrays or body cameras without adequate privacy protections, can galvanize community opposition and demand greater transparency and accountability. These movements often leverage social media and community forums to share information, organize protests, and lobby local officials. Their strength lies in their ability to connect with people on a personal level, illustrating how surveillance policies can affect school districts, public transportation, or neighborhood safety initiatives. By fostering local conversations and organizing collective action, they build a groundswell of support for broader reforms.

The impact of these collective efforts is manifold. Firstly, they significantly raise public consciousness. Before concerted awareness campaigns, discussions about digital privacy were often confined to niche tech communities or academic circles. Now, thanks to the persistent efforts of these groups, issues like data breaches, online tracking, and government surveillance are topics of mainstream public concern. This heightened awareness is the first crucial step in demanding change. A public that is unaware of the extent of surveillance or the implications of data collection is unlikely to advocate for stronger protections.

Secondly, these initiatives promote critical thinking about technology and its societal implications. It's no longer enough to simply accept new technologies at face value. The educational efforts encourage citizens to ask questions: Who is collecting this data? Why are they collecting it? How is it being used? What are the risks? This critical engagement is essential for moving beyond passive consumption of technology to active, informed participation in shaping its development and deployment. For instance, understanding how cookies track online behavior or how targeted advertising works empowers individuals to make more informed choices about the websites they visit and the services they use. Similarly, understanding the concept of metadata and why it is valuable to intelligence agencies helps people appreciate the privacy implications of their phone calls and internet activity, even if the content is encrypted.

Thirdly, and perhaps most importantly, these awareness campaigns mobilize support for policies that protect civil liberties in the digital age. An informed citizenry is more likely to support legislative reforms, demand greater transparency from government agencies and corporations, and hold elected officials accountable for privacy-related issues. This can manifest in various ways, such as increased public pressure for encryption mandates, stronger data protection laws like GDPR or CCPA, or calls for independent oversight of surveillance programs. The shift in public opinion following the Snowden revelations, for example, contributed to legislative efforts in the United States to reform surveillance practices, such as the USA FREEDOM Act, which placed some limits on the government's ability to collect bulk metadata.

The empowerment of the citizen is the ultimate goal of public education and awareness efforts. By providing individuals with the knowledge and tools to understand their digital rights, they are better equipped to protect themselves and to advocate for collective change. This includes understanding how to use privacy-enhancing technologies discussed previously, such as VPNs, encrypted messaging, and secure browsers. But it also extends to understanding their rights in interactions with law enforcement, the implications of terms of service agreements, and how to petition their representatives for change.

This educational endeavor is an ongoing process. The technological landscape is constantly evolving, with new surveillance capabilities emerging and existing ones becoming more sophisticated. Therefore, continuous public education is necessary to keep pace with these changes. Advocacy groups must remain agile, adapting their messaging and outreach strategies to address new threats and opportunities. This includes not only focusing on government surveillance but also on the increasing power of large technology corporations and their role in data collection and utilization, a phenomenon often referred to as "surveillance capitalism."

Furthermore, the effectiveness of these campaigns relies on reaching diverse segments of the population, including those who may not typically engage with civil liberties issues. This requires tailoring messages to different cultural contexts, age groups, and socioeconomic backgrounds. For example, digital literacy programs in underserved communities can be integrated with privacy education, ensuring that the benefits of digital citizenship are accessible to all. Similarly, engaging with younger generations through platforms they actively use, such as TikTok or Instagram, is crucial for instilling a lifelong commitment to privacy and digital rights.

The challenge is also one of overcoming apathy and the perception that surveillance is an inevitable or acceptable cost of modern life. Countering the narrative that "if you have nothing to hide, you have nothing to fear" is a persistent hurdle. Educational efforts must highlight that privacy is not about hiding wrongdoing, but about fundamental autonomy, the freedom to explore ideas, to form associations, and to express oneself without the chilling effect of constant monitoring. It is about the right to control one's own narrative and to define the boundaries of personal information.

In conclusion, public education and awareness campaigns are indispensable components of any successful strategy for resistance, activism, and advocacy in the digital age. They illuminate the hidden mechanisms of surveillance, foster critical engagement with technology, and build the public will necessary to enact meaningful reforms. By empowering citizens with knowledge, these efforts lay the groundwork for a society where digital privacy is not an abstract concept, but a tangible reality, protected by both robust technological solutions and an informed, engaged populace. The ongoing dialogue, the meticulous reporting, the rigorous research, and the persistent grassroots organizing all contribute to a vital, collective effort to ensure that technological progress serves, rather than undermines, fundamental human rights and freedoms.

10: THE GLOBAL DIMENSION OF SURVEILLANCE

The digital age has irrevocably blurred the lines of national sovereignty, transforming borders into porous conduits for information. The vast quantities of data generated by individuals and entities daily do not remain confined within the geographical confines of their origin. Instead, they traverse the globe with astonishing speed, facilitated by the interconnected nature of the internet and the globalized operations of technology companies. This incessant, dynamic flow of data across international borders forms the bedrock of modern digital economies, fuels global communication, and, crucially for our discussion, underpins many of the expansive surveillance capabilities wielded by states. Understanding this phenomenon is paramount to grasping the global dimension of surveillance, as it reveals how intelligence agencies and law enforcement bodies can reach beyond their territorial jurisdictions to access information, and conversely, how foreign entities might gather intelligence on a nation's own citizens.

At the heart of this intricate web lies the concept of data localization versus data free flow. While some nations advocate for stringent data localization laws, requiring data pertaining to their citizens or economic activities to be stored and processed exclusively within their borders, others champion the free flow of data, essential for international commerce, cloud computing, and the seamless operation of global digital services. This tension creates a complex legal and political landscape, shaping how data is accessed, protected, and governed internationally. For instance, a social media user in Europe might have their data stored on servers located in the United States, processed by a company headquartered in Ireland, and accessed by an intelligence agency in a third country under specific legal provisions. This multi-jurisdictional reality means that the privacy protections afforded to that user are not uniform and depend heavily on the laws of each nation involved in the data's journey.

The United States, with its significant presence in the global technology sector and its expansive national security apparatus, plays a pivotal role in this global data exchange. Many of the world's largest technology companies, which host and process vast amounts of personal data, are U.S.-based. This provides U.S. authorities with a degree of access to data concerning individuals worldwide, even when those individuals reside outside the United States. This access is often predicated on legal frameworks that, while intended for domestic application, have extraterritorial reach. The mechanisms employed can range from direct requests to U.S.-based companies for data held by them, to more coercive measures authorized under national security statutes. The CLOUD Act (Clarifying Lawful Overseas Use of Data Act), for instance, allows U.S. law enforcement to compel U.S. technology companies to provide requested data stored on servers regardless of whether the data is stored in the U.S. or abroad. While ostensibly aimed at facilitating cross-border investigations for U.S. law enforcement, it also signifies the U.S. government's ability to reach into the data of non-U.S. persons located outside the U.S., raising significant concerns among privacy advocates and foreign governments about sovereignty and due process.

Conversely, foreign governments and intelligence agencies are equally interested in data concerning U.S. citizens. The digital trail left by Americans as they interact with global online services can be a rich source of intelligence for other nations. In an era where information is a critical component of statecraft and economic competition, foreign intelligence services may seek to gather data on U.S. individuals, businesses, or government activities. This can occur through various means, including exploiting vulnerabilities in networks, compelling foreign-based service providers to disclose data, or through international agreements that facilitate information sharing. The challenge for the U.S. lies in protecting its citizens' data from foreign surveillance while simultaneously engaging in necessary international cooperation for national security and law enforcement.

The legal frameworks governing these cross-border data flows are a complex patchwork of bilateral and multilateral agreements, domestic laws with extraterritorial implications, and international norms that are still evolving. Treaties such as Mutual Legal Assistance Treaties (MLATs) provide a formal mechanism for countries to request and receive assistance in gathering and transferring information for criminal investigations. However, the effectiveness and speed of MLATs have often been criticized as too slow and cumbersome for the pace of digital investigations. This has led to a proliferation of more direct, often less transparent, methods of data acquisition.

The post-9/11 era saw a significant increase in international cooperation on intelligence sharing, driven by shared security concerns. Agreements between countries, often referred to as "intelligence sharing agreements" or "five eyes" (or more expanded variations), facilitate the exchange of signals intelligence, human intelligence, and other forms of information. While these agreements are crucial for combating terrorism and transnational crime, they can also create pathways for governments to circumvent domestic privacy protections by obtaining information through a partner nation that might be more difficult to obtain directly. For example, if Country A has stricter privacy laws that limit its ability to surveil its own citizens, it might request Country B to conduct surveillance and share the resulting data. If Country B then shares that data with Country A, the original privacy protections are effectively bypassed.

The legal basis for U.S. agencies accessing data stored abroad or data of non-U.S. persons abroad is often rooted in statutes like the Foreign Intelligence Surveillance Act (FISA), particularly Section 702, which allows for the targeting of non-U.S. persons reasonably believed to be located outside the United States for foreign intelligence purposes. While this section explicitly targets non-U.S. persons, the data collected can inevitably include communications of U.S. citizens who are communicating with those foreign targets. The "incidental collection" of U.S. person data under Section 702 has been a significant point of contention, leading to debates about the scope of U.S. surveillance authority and the adequacy of protections for both domestic and foreign individuals.

The concept of "digital sovereignty" has gained traction as a response to these challenges. Nations are increasingly asserting their right to control data generated within their borders and to protect their citizens' data from unauthorized access by foreign entities. This can manifest in regulations requiring data localization, imposing strict conditions on cross-border data transfers, or enacting robust data protection laws that mirror or exceed international standards. The European Union's General Data Protection Regulation (GDPR) is a prime example, establishing stringent rules for the processing of personal data and imposing significant restrictions on transfers of data outside the EU unless adequate levels of protection are ensured. This has created a complex compliance landscape for global technology companies and has also led to legal challenges, such as the invalidation of the EU-U.S. Privacy Shield framework by the Court of Justice of the European Union in the Schrems II decision. The court found that U.S. surveillance laws did not provide adequate protection for EU citizens' data transferred to the U.S., highlighting the fundamental differences in legal approaches to privacy and government access to data between the EU and the U.S.

These legal battles underscore the profound disagreements on how to balance national security imperatives with fundamental privacy rights in the digital age. The U.S. often frames its surveillance powers within the context of national security and the necessity of combating global threats, while many other nations, particularly in Europe, prioritize individual privacy rights and due process as foundational democratic principles. The extraterritorial reach of surveillance laws, therefore, becomes a point of friction, as a nation's exercise of its sovereign right to gather intelligence can be perceived by another nation as an infringement on its own sovereignty and the rights of its citizens.

The implications for international privacy standards are far-reaching. As data flows globally, the lowest common denominator of protection can inadvertently become the de facto standard, eroding privacy protections worldwide. Conversely, strong domestic privacy laws, like the GDPR, can create upward pressure, forcing international companies and governments to adapt their practices to meet higher standards. This creates a dynamic where legal and regulatory frameworks are constantly tested and challenged by the realities of global data flows and the evolving capabilities of surveillance technologies.

Furthermore, the lack of a universally agreed-upon

framework for international data access and surveillance creates significant legal uncertainty. Companies operating globally must navigate a complex and often contradictory set of laws, risking penalties or legal action in multiple jurisdictions. Governments struggle to balance the need for international cooperation in law enforcement and intelligence gathering with their obligations to protect citizens' privacy and uphold national sovereignty. Investigative journalists and civil liberties advocates face the daunting task of tracking and exposing surveillance activities that transcend national boundaries, often relying on fragmented information and challenging legal interpretations.

The global dimension of surveillance is not merely a technical or legal issue; it is deeply political. Treaties and agreements are negotiated against a backdrop of international relations, geopolitical alliances, and differing philosophical approaches to the role of the state versus the individual. The desire of some nations to expand their surveillance reach, often justified by security concerns, clashes with the desire of others to shield their citizens from such intrusion. This ongoing tension shapes the evolution of international data governance and the future of privacy in an increasingly interconnected world. As technology continues to advance, and as the value and volume of data grow, the challenges of managing cross-border data flows and ensuring adequate privacy protections will only become more complex, demanding continuous international dialogue, robust legal reform, and a vigilant public awareness of how data moves across our interconnected planet and who might be watching. The effectiveness of any national privacy law is inherently limited by the data practices and legal frameworks of other nations, making international cooperation and the establishment of clear, rights-respecting global norms not just desirable, but essential for safeguarding digital freedoms worldwide.

The intricate tapestry of global surveillance is woven with threads of international cooperation, a complex interplay of alliances and agreements that allow intelligence agencies to extend their reach far beyond their national borders. While the previous context highlighted the inherent data flows across jurisdictions, this section delves into the formalized mechanisms by which nations actively collaborate in intelligence gathering, particularly focusing on the extensive partnerships involving the United States and its allies. At the forefront of these collaborations is the "Five Eyes" alliance, a post-World War II intelligence-sharing arrangement initially between the United States, the United Kingdom, Canada, Australia, and New Zealand. This pact, born out of shared security concerns and a desire for reciprocal intelligence advantages, has evolved into a sophisticated network for collecting, analyzing, and disseminating vast amounts of signals intelligence (SIGINT). The operationalization of this alliance means that data collected by one member nation, often targeting individuals or communications originating in another member's territory or concerning citizens of a third country, can be readily shared.

The sheer scale of this cooperation is staggering. Through programs and agreements, often operating under different monikers and with varying degrees of public visibility, these nations engage in a constant exchange of intelligence. This can encompass everything from raw data streams harvested from telecommunications networks and internet traffic to highly analyzed reports on specific threats or individuals. The rationale, consistently presented, is the paramount need to combat transnational threats such as terrorism, cybercrime, and the proliferation of weapons of mass destruction. However, the practical effect is that citizens of allied nations, including those within the United States and its partner countries, can find their communications and digital activities subject to surveillance by foreign intelligence agencies, with the collected information then being fed back into the collaborative intelligence pool.

This reciprocal nature of intelligence sharing raises profound questions about privacy and oversight. When U.S. intelligence agencies collaborate with their counterparts, they are not merely sharing information that originates within their own borders or targets individuals they are legally permitted to surveil domestically. Instead, they are often beneficiaries of intelligence collected by other states under different legal frameworks, which may have varying standards of privacy protection. For instance, a communication from a U.S. citizen to a national in a country with less stringent privacy laws might be intercepted by that country's intelligence service. Under the umbrella of intelligence-sharing agreements, this data can then be provided to U.S. agencies, effectively bypassing the domestic legal safeguards that would typically govern the surveillance of U.S. persons. This dynamic has been characterized by critics as a form of "outsourcing" surveillance, where one nation might undertake surveillance activities that its partner nation is legally or politically constrained from performing on its own citizens.

The legal justifications for this cross-border data acquisition are often couched in terms of national security and the necessity of international cooperation. Statutes like Section 702 of the Foreign Intelligence Surveillance Act (FISA) in the United States play a crucial role in enabling the collection of foreign intelligence information. While Section 702 is designed to target non-U.S. persons reasonably believed to be located outside the United States, the communications of U.S. citizens with these foreign targets can be "incidentally" collected. When this incidentally collected U.S. person data is then shared with foreign intelligence partners, or when U.S. agencies receive data collected by foreign partners that may contain information about U.S. citizens, the complexities multiply. The legal protections for U.S. persons in such scenarios are often less robust than those that would apply to direct domestic surveillance.

Beyond the Five Eyes, the United States maintains a broad network of intelligence cooperation with numerous other countries. These relationships, often formalized through Memoranda of Understanding or less formal agreements, facilitate the exchange of intelligence derived from various sources, including signals intelligence, human intelligence, and open-source intelligence. The scope and legal underpinnings of these partnerships vary significantly. Some are deeply integrated, mirroring the extensive data sharing seen within the Five Eyes, while others may be more narrowly focused on specific threats or operational areas. Regardless of their depth, these collaborations contribute to a global intelligence apparatus where data flows across borders, and the surveillance of individuals, regardless of their nationality or location, becomes a shared endeavor.

The ethical and legal quandaries presented by this level of international cooperation are substantial. One primary concern is the potential for circumventing domestic legal protections. If a nation's laws prohibit certain types of surveillance or data collection on its own citizens, or even on foreign nationals within its borders, the ability to obtain similar intelligence from a foreign partner can serve as a backdoor. This raises questions about accountability and the rule of law. When intelligence is collected under foreign legal frameworks, which may offer fewer protections or different due process guarantees, and then shared with domestic agencies, the oversight mechanisms designed to protect civil liberties within the originating nation may be rendered ineffective.

Furthermore, the notion of "sovereignty" becomes a complex factor in these collaborations. While nations engage in intelligence sharing to protect their own interests, the act of collecting intelligence on the citizens of allied or neutral nations can be viewed as an intrusion into their sovereignty. Conversely, allowing foreign intelligence agencies unfettered access to data pertaining to one's own citizens, even under the guise of mutual security, can be seen as a dereliction of a government's primary duty to protect its populace. The balancing act between national security imperatives and the protection of individual privacy rights, particularly in an international context, is a perpetual challenge.

The legal basis for accessing and sharing data across borders is a constantly evolving landscape. Mutual Legal Assistance Treaties (MLATs) represent a formal, albeit often slow, mechanism for judicial cooperation in criminal matters, allowing for the exchange of evidence and other forms of assistance. However, the speed and volume of digital information often outstrip the capabilities of traditional MLAT processes. This has led to the development of more direct and often less transparent channels for intelligence sharing, particularly for national security purposes, which may not be subject to the same level of judicial oversight.

The debate over the legality and ethics of these international surveillance collaborations is often framed by differing legal traditions and cultural values regarding privacy. In many European countries, for example, privacy is considered a fundamental human right, with strong legal protections enshrined in national constitutions and supranational frameworks like the GDPR. In contrast, while the U.S. has privacy laws, the emphasis on national security and the expansive interpretation of executive powers, particularly in the post-9/11 era, have sometimes led to a different approach to balancing security and privacy. This divergence in legal philosophies can create friction when intelligence is shared between countries with such different approaches, leading to concerns that data collected under one set of rules might be used in ways that would be impermissible under another.

The implications of these extensive cooperation agreements extend to the global digital economy. Technology companies, operating across multiple jurisdictions, find themselves caught in the middle of these complex legal and intelligence-sharing frameworks. They are often compelled by law in one country to provide data that may concern citizens of another country, while simultaneously being subject to privacy regulations in the latter that restrict such disclosures. This creates significant compliance challenges and legal uncertainty for businesses, and it underscores the need for clearer international norms and agreements that respect both national security needs and fundamental privacy rights.

For investigative journalists and civil liberties advocates, unravelling the specifics of these international intelligence-sharing arrangements is a formidable task. Information is often classified, and the legal frameworks are intricate and opaque. However, evidence has emerged over the years, through leaks, declassified documents, and legal challenges, that sheds light on the depth and breadth of this cooperation. These revelations often point to a system where data is gathered broadly and shared widely among allied nations, with the ultimate goal of enhancing collective security capabilities.

The continued expansion of digital technologies and the increasing reliance on globalized communication networks mean that the scope and impact of international intelligence cooperation are likely to grow. As states continue to grapple with evolving threats and the challenges of operating in a borderless digital environment, the mechanisms for sharing intelligence will undoubtedly adapt and expand. This necessitates a sustained public discourse and rigorous oversight to ensure that such collaborations, while serving legitimate security purposes, do not erode the fundamental rights and freedoms of individuals caught within the vast web of global surveillance. The critical question remains: how can nations cooperate to enhance security without compromising the privacy and civil liberties that underpin democratic societies? This ongoing tension shapes the future of data governance and the very nature of privacy in an interconnected world, demanding constant vigilance and a commitment to upholding democratic values even in the face of perceived threats. The interconnectedness that facilitates global trade and communication also creates unprecedented opportunities for intelligence agencies to collaborate, collect, and analyze information on an immense scale, underscoring the profound impact of these alliances on individual privacy worldwide.

The United States, a global leader in technological innovation, has also become a significant exporter of surveillance technologies. This export market, often driven by national security interests and lucrative commercial opportunities, plays a crucial role in the global proliferation of tools that can be used for mass monitoring and the suppression of dissent. U.S. companies, fueled by government policies and often operating within a framework that prioritizes technological advancement and market share, develop and market sophisticated hardware and software capable of intercepting communications, tracking individuals, and analyzing vast datasets. These technologies, ranging from sophisticated spyware and facial recognition systems to advanced network surveillance platforms, find their way into the hands of governments worldwide, including those with deeply concerning human rights records.

The apparatus for this export is multifaceted. It involves direct sales by American corporations, partnerships with foreign entities, and, in some cases, indirect facilitation through government-approved channels or by turning a blind eye to the end-use of certain technologies. The U.S. government, while occasionally expressing concerns about human rights abuses, has historically maintained policies that encourage the export of dual-use technologies, including those with surveillance capabilities. This can be attributed to a complex web of geopolitical considerations, economic incentives, and a perceived need to equip allied nations with the tools to combat terrorism and other transnational threats. However, the reality on the ground is that these same tools are frequently repurposed for domestic repression, creating a moral and ethical quandary that deeply implicates the U.S. in the human rights violations committed by recipient regimes.

Consider the case of advanced mobile phone interception technology. Companies, often based in the U.S. or with significant U.S. investment and R&D, develop sophisticated systems that can intercept calls, SMS messages, and even access data on smartphones remotely. These systems can be deployed by law enforcement and intelligence agencies. When exported to countries where due process is weak, and the rule of law is selectively applied, such technologies become potent instruments for targeting political opponents, journalists, activists, and minority groups. Reports have consistently emerged detailing the use of such exported U.S. technologies by authoritarian governments to monitor dissidents, gather intelligence on opposition movements, and silence critical voices. The sale of these tools, even when ostensibly for "legitimate" law enforcement purposes, overlooks the corrupting influence of unchecked power and the propensity for these systems to be abused.

Furthermore, the export of facial recognition technology,

an area where U.S. companies have been at the vanguard, presents another alarming dimension. These systems, capable of identifying individuals in crowds or from existing databases with increasing accuracy, are being adopted by governments globally. In countries with a history of human rights abuses, facial recognition deployed in public spaces can create an environment of constant surveillance, chilling legitimate public assembly and free expression. The U.S. government's role in this export ecosystem, by not imposing stringent controls on the sale of such technologies to regimes known for human rights abuses, effectively sanctions their use in ways that are antithetical to democratic values. The argument that these are purely commercial transactions, free from government complicity, is disingenuous when the government's regulatory framework allows for, and at times implicitly encourages, such exports.

The business model for these surveillance technology exports is often built on a foundation of cutting-edge research and development, much of which benefits from the very ecosystem of innovation fostered within the United States. U.S. universities, research institutions, and venture capital firms play a role in nurturing the companies that eventually produce these technologies. While this innovation is celebrated domestically, its international application raises serious ethical questions. The data privacy and civil liberties concerns that are increasingly debated within the United States are often disregarded when these same technologies are sold to foreign governments with fewer safeguards and a different set of priorities. This creates a perverse incentive structure where the very advancements intended to enhance security or efficiency are weaponized against fundamental human rights in other parts of the world.

The regulatory landscape governing the export of surveillance technology from the United States is notoriously complex and, critics argue, deeply inadequate. While certain export controls exist, they often struggle to keep pace with the rapid evolution of technology. Furthermore, the classification of these technologies as "dual-use" products—meaning they have both civilian and military or intelligence applications —allows them to navigate a less stringent export control regime. The Commerce Department, through its Bureau of Industry and Security (BIS), manages many of these controls, but the focus has often been on preventing the proliferation of weapons technology rather than the broader impact of surveillance tools on human rights. This oversight gap allows a significant volume of sophisticated surveillance equipment to flow to countries that may use it to facilitate widespread human rights abuses.

The mechanisms of export are varied. Direct sales by U.S. companies to foreign governments or their security agencies are common. These transactions can be lucrative, and companies are eager to tap into international markets. However, the ultimate destination and use of the technology are not always rigorously monitored. Even when sales are subject to review, the sheer volume of requests and the classification of information often mean that potential human rights implications are not fully assessed. Moreover, U.S. companies may establish subsidiaries or partnerships in other countries, allowing them to circumvent certain direct export regulations, or they may sell components and source code to foreign integrators who then assemble and market the final surveillance systems. This diffusion of responsibility makes accountability incredibly difficult to establish.

The human rights implications are profound and well-documented. In numerous instances, surveillance technologies exported from the U.S. have been implicated in the persecution of journalists, human rights defenders, and political activists. For example, reports have detailed how governments have used sophisticated spyware to infiltrate the devices of opposition leaders, extract sensitive information, and track their movements. This digital intrusion not only compromises the safety and security of individuals but also creates a chilling effect on free speech and association, as people become aware that their communications and activities are being constantly monitored. The U.S. government's role in enabling this is not always direct, but its failure to implement robust export controls and its promotion of a global surveillance infrastructure undeniably contribute to these abuses.

The concept of "safe harbor" for technology exports is often invoked by companies and governments, arguing that the responsibility for misuse lies with the recipient nation, not the exporter. However, this argument fails to acknowledge the foreseeable consequences of selling powerful surveillance tools to regimes known for their authoritarian tendencies. When a company or a government knowingly facilitates the transfer of technology that is almost certain to be used for repression, they share a degree of culpability. The U.S. government's own policies and export licensing decisions directly influence which technologies are available to which countries, and therefore, it bears a significant responsibility for the human rights outcomes that result from these transfers.

The financial aspect of the surveillance export economy is also substantial. Many of these technologies are highly specialized and command premium prices. This economic incentive can sometimes override ethical considerations, as companies prioritize profitability over the potential for misuse. The U.S. government, while advocating for democratic values abroad, also benefits indirectly through the growth and success of its technology sector. This creates a conflict of interest, where the promotion of American innovation and economic competitiveness can come at the expense of human rights in other nations. The lack of transparency surrounding many of these export licenses and sales makes it difficult for the public to fully understand the scale of this trade and its implications.

Moreover, the proliferation of these technologies contributes to a global "race to the bottom" in terms of privacy standards. As more countries acquire and deploy advanced surveillance capabilities, the pressure increases on others to do the same, often leading to the normalization of intrusive monitoring practices. The U.S. itself, while often seen as a proponent of openness, has significantly expanded its own surveillance capabilities in recent decades. This domestic expansion can create a tacit endorsement of such capabilities, making it harder to argue against their export. The global reach of American technology companies means that policies and practices developed or tolerated within the U.S. can have a direct and profound impact on citizens in countries with far weaker democratic institutions and legal protections.

The specific types of technologies exported are diverse. They include:

Spyware and Malware: Tools that can be covertly installed on devices to extract data, monitor communications, and track user activity. These often exploit software vulnerabilities.

Network Interception Systems: Devices and software capable of tapping into telecommunications networks, internet traffic, and cellular communications, allowing for mass collection of data.

Facial Recognition and Biometric Surveillance: Systems that use advanced algorithms to identify and track individuals based on their physical characteristics, often deployed in public spaces or integrated with surveillance camera networks.

Data Analytics and Social Media Monitoring Tools: Software that can sift through vast amounts of data, including social media posts, to identify patterns, individuals of interest, and potential threats.

Location Tracking and Geolocation Services: Technologies that enable the tracking of individuals' movements through their mobile devices or other networked electronics.

The U.S. government's export licensing process, particularly for items controlled under the International Traffic in Arms Regulations (ITAR) and the Export Administration Regulations (EAR), is meant to prevent the proliferation of technologies that could undermine U.S. national security or foreign policy interests. However, the interpretation and application of these regulations to advanced surveillance technologies have been a subject of considerable debate and criticism. Advocates for stronger controls argue that the existing framework is insufficient to address the human rights risks associated with the export of such powerful tools. They call for greater transparency in the licensing process, more rigorous end-use monitoring, and a stronger emphasis on human rights considerations in export decisions.

The impact on democratic movements and civil society in countries that import these technologies is often devastating. When governments can effectively monitor and suppress dissent, the space for political opposition, independent journalism, and advocacy shrinks dramatically. Activists may be arrested, intimidated, or forced into exile based on information gathered through these surveillance systems. The psychological effect of constant monitoring can also be profound, leading to self-censorship and a general atmosphere of fear. This makes the role of U.S. companies and government policies in facilitating this trade a critical issue in the global struggle for human rights and democratic freedoms. The very technologies that are a hallmark of American innovation are, in many instances, being turned into instruments of oppression abroad, a direct consequence of export policies that have not adequately prioritized human rights. The ongoing development of more sophisticated and harder-to-detect surveillance tools only exacerbates these concerns, creating an escalating challenge for those seeking to protect privacy and civil liberties in an increasingly monitored world. The export economy of surveillance is not merely a commercial transaction; it is a significant factor in the global balance of power between states and their citizens, and the U.S. plays a pivotal, often enabling, role.

The United States, often perceived as a pioneer in technological advancement, finds its domestic approach to surveillance and privacy increasingly contrasted with the legal frameworks adopted by other developed nations. While the U.S. grapples with balancing national security imperatives with civil liberties, many countries, particularly in Europe and Canada, have enacted more stringent and comprehensive data protection laws that significantly alter the landscape of digital surveillance. This divergence in legal philosophy creates a complex web of challenges for global technology companies, necessitates international cooperation, and raises critical questions about whether American privacy norms are indeed lagging behind evolving global benchmarks.

The European Union's General Data Protection Regulation (GDPR), which came into effect in May 2018, stands as a landmark piece of legislation in this regard. Unlike the sector-specific, and often fragmented, approach to privacy in the United States, the GDPR offers a unified, rights-based framework for the processing of personal data of all EU citizens. It establishes a broad definition of personal data, encompasses a wide array of data processing activities, and grants individuals significant rights, including the right to access, rectification, erasure, and data portability. For businesses operating globally, compliance with the GDPR means implementing robust data protection measures, obtaining explicit consent for data processing, and adhering to strict rules regarding data breaches and cross-border data transfers. This regulation has had a profound ripple effect, compelling companies worldwide to re-evaluate their data handling practices, even those outside the EU, if they process the data of EU residents. The extraterritorial reach of the GDPR means that a company in the United States, processing the personal data of a French citizen, for example, is subject to its provisions.

This regulatory environment in Europe directly impacts surveillance practices. The GDPR places significant restrictions on the collection and processing of personal data by both public authorities and private entities. While it allows for data processing for legitimate purposes, such as national security, it mandates that such processing must be necessary, proportionate, and subject to safeguards. This contrasts sharply with the broader powers often afforded to intelligence agencies in the United States under legislation like the Foreign Intelligence Surveillance Act (FISA), particularly Section 702, which has been used to collect communications data of non-U.S. persons overseas on a massive scale, often without specific warrants targeting individuals. European courts and data protection authorities have been more assertive in scrutinizing government surveillance programs, often deeming broad, indiscriminate data collection to be incompatible with fundamental rights. The Schrems II ruling by the Court of Justice of the European Union, for instance, invalidated the EU-U.S. Privacy Shield framework, highlighting concerns about U.S. surveillance laws and their compatibility with EU data protection standards. This decision underscored that U.S. intelligence agencies' access to data transferred from the EU was not subject to adequate protections, effectively creating a significant hurdle for transatlantic data flows.

Canada, while not possessing a single overarching data protection law like the GDPR, has a robust federal and provincial legal framework for privacy. The Personal Information Protection and Electronic Documents Act (PIPEDA) at the federal level, and similar provincial legislation in Alberta, British Columbia, and Quebec, govern the collection, use, and disclosure of personal information by commercial organizations. These laws are generally based on principles of consent, accountability, and limited collection. Furthermore, Canada has specific legislation governing government surveillance, such as the **Criminal Code** and the **Privacy Act**, which regulate lawful access to data by law enforcement and intelligence agencies. While these laws do grant powers for surveillance in the context of criminal investigations, they also impose requirements for judicial authorization and specify limitations on the types of data that can be accessed and the purposes for which it can be used. The Canadian approach tends to emphasize transparency and accountability, requiring government agencies to report on their surveillance activities.

Compared to the United States, where a significant portion of government surveillance powers, particularly concerning foreign intelligence, is exercised through secret court orders and executive branch interpretations of statutes, Canada and many European nations have placed a greater emphasis on judicial oversight and public accountability for such activities. This difference is particularly evident in the realm of bulk data collection. While the U.S. has engaged in extensive programs collecting metadata on a vast scale, often justified by national security, many European jurisdictions have viewed such practices with considerable skepticism, often requiring demonstrable suspicion or specific warrants for the collection of personal communications data. The legal justifications for surveillance in the U.S., such as those derived from the interpretation of the Fourth Amendment in the digital age, have often been viewed by international observers as providing less robust protection for individual privacy compared to European constitutional rights and specific data protection statutes.

The complexities arising from these differing legal standards have a tangible impact on global technology companies. These companies, operating across multiple jurisdictions, must navigate a patchwork of regulations, each with its own requirements for data collection, storage, processing, and transfer. For a company like Google, Meta, or Apple, this means adhering to GDPR in Europe, PIPEDA in Canada, and various state-level privacy laws in the U.S., alongside federal laws like the Electronic Communications Privacy Act (ECPA). The cost and effort involved in ensuring compliance across such diverse legal landscapes are substantial. Furthermore, these differences create an uneven playing field, potentially disadvantaging companies that are more diligent in their privacy practices if competitors operate under less stringent regulations.

The issue of international data transfers is particularly contentious. The GDPR, for instance, restricts the transfer of personal data outside the EU unless the receiving country ensures an adequate level of data protection. This has led to ongoing legal battles and the need for complex data transfer mechanisms, such as Standard Contractual Clauses (SCCs) and Binding Corporate Rules (BCRs), which require rigorous due diligence and ongoing assessments to ensure compliance. The invalidation of the EU-U.S. Privacy Shield and the continued scrutiny of SCCs highlight the deep-seated concerns in Europe regarding the U.S. government's access to personal data of its citizens, often through mechanisms like FISA Section 702 and Executive Order 12333, which are perceived as lacking adequate safeguards and judicial oversight from a European perspective.

This divergence also raises questions about the adequacy of U.S. privacy norms. While the U.S. has strong privacy protections in specific sectors, such as healthcare (HIPAA) and finance (GLBA), it lacks a comprehensive federal privacy law akin to the GDPR or Canada's PIPEDA that would provide a baseline of protection across all sectors. The U.S. approach has often been criticized as being too industry-friendly and reactive, relying on sector-specific legislation and self-regulation rather than a proactive, rights-based approach. This has led to a situation where, on the global stage, U.S. privacy standards are often seen as weaker, particularly concerning government access to data and the breadth of permissible commercial data collection and use.

The implications of this legal disparity extend to the very nature of surveillance. In nations with strong data protection laws, there is a greater emphasis on proportionality and necessity in surveillance measures. This means that government surveillance must be targeted, justified, and limited in scope. In contrast, the legal framework in the United States has, at times, allowed for more expansive surveillance operations, particularly in the context of national security and counter-terrorism, often based on broad interpretations of existing statutes and executive authority. The debate over the U.S. government's ability to compel tech companies to provide access to encrypted data or user information, even when such data is stored abroad, further exemplifies these differing legal philosophies.

Moreover, the global expansion of American technology companies means that U.S. legal norms and practices regarding surveillance can inadvertently influence or become the de facto standard in other countries, even those with more protective laws. When U.S. companies design their services and data handling policies, they often do so with U.S. legal requirements in mind. This can lead to a "California effect" or "U.S. effect" where U.S. privacy standards, which are arguably less protective than those in Europe, become embedded in global digital infrastructure. This is a concern for privacy advocates and regulators in other nations who fear that U.S. legal practices might erode hard-won privacy protections elsewhere.

The differing legal approaches also manifest in the mechanisms of oversight and enforcement. In the EU, data protection authorities have significant powers to investigate breaches, impose substantial fines, and halt data processing activities. In Canada, privacy commissioners have similar investigative and enforcement powers. While the U.S. has agencies like the Federal Trade Commission (FTC) and state Attorneys General that enforce privacy laws, their powers and scope can be more limited, and the penalties may not always act as a sufficient deterrent against widespread data misuse or intrusive surveillance. The U.S. also has a complex system of congressional oversight, inspector generals, and internal agency reviews for government surveillance, but the effectiveness and transparency of these mechanisms are often debated.

The global nature of the internet and digital services means that these legal divergences are not merely academic exercises; they have real-world consequences for individuals and businesses. Citizens in countries with stronger privacy laws may find their data is less protected when it is processed by U.S. companies, especially if that data is transferred to or accessed by U.S. authorities. Businesses, particularly small and medium-sized enterprises, may struggle to understand and comply with the myriad of international privacy regulations, potentially hindering their ability to compete globally.

The ongoing evolution of technology, including advancements in artificial intelligence, facial recognition, and the Internet of Things (IoT), further complicates this regulatory landscape. As new forms of data collection and analysis emerge, existing legal frameworks are constantly tested. Nations are grappling with how to regulate the use of AI in surveillance, the collection of biometric data, and the potential for mass surveillance through interconnected devices. The differing approaches to these emerging technologies reflect the underlying legal philosophies regarding privacy and state power. Some jurisdictions are opting for a precautionary approach, seeking to establish strong safeguards before widespread deployment, while others are adopting a more permissive stance, prioritizing innovation and economic development.

In conclusion, the comparison of privacy laws and surveillance practices between the United States and other developed nations, particularly in Europe and Canada, reveals a significant divergence in approach. While the U.S. has a complex and evolving legal framework, many other nations have adopted more comprehensive, rights-based data protection laws that impose stricter limits on government and corporate surveillance. These differing legal standards create substantial complexities for global technology companies, impact international data flows, and raise pertinent questions about whether U.S. privacy norms are keeping pace with global expectations. The challenges of harmonizing these disparate legal requirements are immense, but they are essential for fostering trust in the digital economy, protecting fundamental human rights, and ensuring accountability in an increasingly monitored world. The legal frameworks in place—or not in place—in the United States have a direct bearing on the privacy rights of citizens globally, especially as American technology companies shape the digital infrastructure used by billions. The global dimension of surveillance is therefore not just a matter of how foreign governments conduct their operations, but also how the United States' own legal traditions and commercial practices influence the global privacy landscape, often setting a lower bar than many international counterparts aspire to reach.

The extensive reach of surveillance capabilities, particularly those revealed through disclosures like the Snowden revelations, has fundamentally altered the landscape of international relations, casting a long shadow over diplomatic trust and fostering a climate of suspicion between nations, even those historically considered close allies. What was once understood as a necessary, albeit sometimes covert, function of national security has, in many instances, been perceived by other sovereign states as an overreach, an infringement on their own sovereignty, and a profound betrayal of shared intelligence understandings. This perception has not remained confined to diplomatic communiqués; it has translated into tangible shifts in how countries interact, cooperate, and even negotiate the very terms of their global engagement.

The immediate aftermath of the revelations saw a palpable chill descend upon relationships between the United States and its traditional partners. Countries like Germany, France, and Brazil expressed profound indignation, feeling blindsided by the extent to which their own leaders and citizens were allegedly being monitored. Chancellor Angela Merkel of Germany, a staunch ally, found herself on the receiving end of alleged U.S. surveillance, including the monitoring of her personal mobile phone. This personal violation, coupled with the systemic revelation of widespread data collection on German citizens, led to public outcry and a significant strain on bilateral relations. Germany, along with other European nations, began to re-evaluate the nature of intelligence sharing and cooperation with the U.S., questioning the reciprocity and mutual respect that underpin such partnerships. The very foundations of trust, upon which alliances are built and shared security interests are pursued, were shaken.

This erosion of trust had direct implications for the intricate web of intelligence sharing that is crucial for combating transnational threats like terrorism, organized crime, and cyber warfare. When allies begin to question the motives and methods of their partners, the seamless flow of critical intelligence can falter. Countries became more hesitant to share sensitive information, fearing that it might be indiscriminately collected, stored, or even exploited for purposes beyond the agreed-upon cooperative efforts. This hesitancy created significant operational challenges, potentially leaving all parties more vulnerable to the very threats they sought to counter. The paradox was stark: the very tools of surveillance, ostensibly deployed for security, were inadvertently undermining the collaborative security efforts that are vital in a globalized world.

The issue of data protection and privacy, once primarily a domestic concern or a matter of bilateral negotiation, rapidly escalated to become a significant point of international contention. The differing legal and cultural attitudes towards privacy, as discussed previously, became a focal point of diplomatic friction. European nations, with their robust data protection laws and a historical sensitivity to state surveillance, viewed the U.S. approach as alarmingly intrusive and lacking in adequate safeguards. The decisions by the Court of Justice of the European Union (CJEU), such as the Schrems II ruling which invalidated the EU-US Privacy Shield, were not merely legal pronouncements; they were potent signals of disapproval from a major global bloc regarding the adequacy of U.S. privacy protections. These rulings effectively hampered transatlantic data flows, impacting countless businesses and raising questions about the future of digital trade.

Global commerce and digital trade, inherently reliant on the free and secure flow of data across borders, were significantly affected. Technology companies, the backbone of the modern global economy, found themselves caught in the crossfire. As U.S. surveillance practices came under scrutiny, the perceived security of data handled by American tech giants became a concern for international users and governments alike. Companies were forced to navigate a complex and often contradictory regulatory landscape, attempting to comply with differing data protection standards in various jurisdictions while also facing potential demands from their own government for data access. This created uncertainty, increased compliance costs, and, in some cases, led to a competitive disadvantage for U.S. firms compared to their counterparts in regions with less stringent government data access demands.

The debate over data localization—the requirement for data to be stored within the borders of the country where it is collected—gained traction in many nations. This was partly a reaction to concerns about foreign government access to data, driven by revelations of extensive surveillance. While data localization can offer some measure of control, it also presents significant challenges for the global digital economy, potentially fragmenting the internet and hindering innovation. It also raises questions about its effectiveness, as data can often be accessed or routed through other jurisdictions regardless of its primary storage location. Nevertheless, the push for data localization underscored the deep-seated anxieties about data sovereignty and the perceived erosion of national control in the digital age.

Beyond the realm of intelligence and commerce, surveillance practices also influenced international cooperation on a broader spectrum of global challenges. Cybersecurity initiatives, for instance, require a high degree of trust and information sharing between nations to effectively combat state-sponsored attacks and criminal hacking operations. When allies suspect each other of engaging in covert surveillance, the willingness to collaborate on sensitive cybersecurity matters can diminish. Similarly, efforts to counter illicit financial flows, combat drug trafficking, or enforce international sanctions often rely on the ability of countries to share financial and communication data. A breakdown in trust, stemming from surveillance revelations, can impede these vital international efforts, making the world a less secure place for everyone.

The geopolitical ramifications extended to the formation of new alliances and the reshaping of existing power dynamics. Countries that felt particularly aggrieved by U.S. surveillance practices began to look towards alternative partnerships and frameworks for digital governance. Some nations sought to establish their own independent data infrastructure and surveillance capabilities, further fragmenting the global digital landscape. Others advocated for stronger international norms and treaties governing surveillance and data protection, aiming to create a more level playing field and hold all nations accountable to common standards. The discussion around multilateral agreements on cyber norms and data governance, which had previously been niche, gained significant urgency and visibility on the international stage.

The concept of "reciprocity" became a recurring theme in diplomatic discussions. Allies who felt they were being monitored without their knowledge or consent began to demand that the U.S. adhere to the same standards of transparency and mutual respect that they themselves applied in their intelligence relationships. This often translated into calls for reciprocal data access limitations, demands for greater transparency in intelligence operations conducted on their soil, and a push for more formal agreements that would clearly delineate the boundaries of intelligence cooperation and data sharing. The U.S. government faced pressure to acknowledge these concerns and to demonstrate a tangible commitment to rebuilding trust.

The influence of U.S. surveillance policy also extended to the development of international standards for emerging technologies. As countries grappled with regulating artificial intelligence, facial recognition, and the Internet of Things, the U.S. approach to data collection and privacy often served as a point of reference, either as a model to emulate or, more frequently, as an example of what to avoid. Nations seeking to protect their citizens' privacy often looked to more stringent regulatory frameworks, like those in Europe, as benchmarks for their own domestic policies. This created a dynamic where U.S. practices could inadvertently set a lower global standard if not carefully managed, potentially leading to a "race to the bottom" in terms of privacy protections. Conversely, the U.S. also faced international pressure to adopt more robust privacy norms to facilitate global digital commerce and maintain its leadership role in technological innovation.

The revelations also highlighted the significant power imbalance inherent in global surveillance capabilities. Nations with advanced technological infrastructure and extensive data collection programs, such as the United States, possessed a distinct advantage. This advantage, however, was not always perceived as legitimate or equitable by other nations, particularly when it appeared to be used for economic or political gain rather than purely for mutual security. The perception that surveillance was being used to gain an unfair advantage fueled resentment and led to calls for greater international oversight and accountability mechanisms that would apply universally, regardless of a nation's technological prowess.

Furthermore, the legal justifications used by the U.S. for its surveillance programs, often rooted in national security imperatives and broad interpretations of existing statutes, were frequently viewed with skepticism by international legal scholars and human rights advocates. The emphasis on proportionality and necessity, which is a cornerstone of data protection in many other jurisdictions, seemed to be applied more loosely in some U.S. programs. This divergence in legal philosophy not only created friction between governments but also influenced the broader international discourse on human rights, privacy, and the legitimate scope of state power in the digital age.

The long-term implications for global governance are profound. The ability of nations to trust each other, to cooperate effectively, and to establish common rules for the digital world are all contingent on a shared understanding of acceptable practices. When surveillance practices create deep divisions and erode trust, the very fabric of international cooperation is weakened. This can hinder progress on a wide range of global issues, from climate change to public health, all of which increasingly rely on international data sharing and coordinated action. The challenge for the United States, and indeed for all nations, is to find a way to balance legitimate national security interests with the need to maintain trust, uphold fundamental rights, and foster a global digital environment that is both secure and respects the privacy of individuals and the sovereignty of nations. The ongoing dialogue and negotiation over these issues will continue to shape the geopolitical landscape for years to come, underscoring that surveillance policy is no longer merely a technical or domestic issue, but a fundamental component of international diplomacy and global power dynamics. The path forward requires a renewed commitment to transparency, accountability, and a genuine effort to understand and respect the differing legal and cultural norms that govern privacy and surveillance across the globe, ensuring that the pursuit of security does not inadvertently dismantle the architecture of international trust.

11: THE FUTURE OF SURVEILLANCE: EMERGING TRENDS

T he integration of Artificial Intelligence (AI) and Machine Learning (ML) into surveillance systems represents a profound evolution, fundamentally augmenting the capabilities and reach of monitoring technologies. These advanced computational tools are not merely incremental improvements; they are transformative forces, enabling surveillance to move beyond simple data collection and storage towards sophisticated analysis, prediction, and even autonomous decision-making. The efficiency gains are undeniable, allowing for the processing of vast datasets at speeds and scales previously unimaginable, thereby accelerating the very nature of how individuals and their activities can be tracked and understood by state and corporate entities. This shift portends an era of unprecedented surveillance intensity, where patterns are identified, deviations are flagged, and potential future actions are anticipated with increasing accuracy, all powered by algorithms learning from colossal volumes of information.

At the forefront of this technological shift is the widespread application of AI in facial recognition systems. Gone are the days when manual comparison of images was the norm. Modern AI-powered facial recognition can scan through millions of images and videos, identifying individuals with remarkable speed and often with high degrees of accuracy. This technology is deployed in diverse environments, from public spaces equipped with CCTV networks to digital platforms where user photos are uploaded. The ability to instantaneously match a face to a database of known individuals – whether they are suspected criminals, persons of interest, or even simply registered users – transforms passive video feeds into active identification tools. This pervasive identification capability is a cornerstone of many modern surveillance architectures, allowing for the rapid tracking of movements and associations in both physical and digital realms. The underlying ML models are trained on massive datasets of facial images, learning to recognize subtle variations in features, lighting conditions, and angles, continuously refining their accuracy through exposure to more data.

Beyond mere identification, AI is also instrumental in behavioral analysis. Machine learning algorithms can be trained to recognize patterns of behavior that are deemed anomalous or suspicious. This can range from tracking unusual movement patterns in public spaces – such as loitering in specific areas for extended periods, or frequent and unpredictable changes in direction – to analyzing communication patterns, such as the frequency and timing of calls or messages between individuals. By sifting through vast quantities of data, AI can identify subtle deviations from established norms, flagging individuals or groups for further scrutiny. This predictive element is particularly concerning, as it shifts the focus of surveillance from observing past actions to anticipating future ones. The algorithms learn what constitutes "normal" behavior for a given context and then flag any departure from that norm. This can be applied to a wide array of scenarios, from monitoring public transport hubs for potential security threats to analyzing online activity for signs of radicalization.

The concept of predictive policing, heavily reliant on AI and ML, exemplifies this forward-looking surveillance. By analyzing historical crime data, including location, time, and offender characteristics, ML models can forecast where and when crimes are most likely to occur. This allows law enforcement agencies to allocate resources more strategically, deploying officers to high-risk areas proactively. While proponents argue this approach can reduce crime rates by deterring potential offenders and enabling quicker response times, critics raise significant concerns about fairness and bias. The algorithms are trained on past data, which may reflect existing societal biases and discriminatory policing practices. If certain neighborhoods or demographic groups have historically been over-policed, the AI will likely identify these areas and groups as high-risk, perpetuating a cycle of surveillance and enforcement that disproportionately affects marginalized communities. This creates a self-fulfilling prophecy, where increased police presence in these areas leads to more arrests, which in turn further reinforces the AI's prediction, regardless of the actual underlying crime rate.

Furthermore, the automation of data processing is a significant outcome of AI integration. Traditional surveillance involved human analysts manually reviewing footage, analyzing reports, and cross-referencing information. AI, however, can automate many of these labor-intensive tasks. Algorithms can process and categorize vast amounts of unstructured data – such as text from social media, audio from intercepted communications, or video footage from ubiquitous cameras – identifying keywords, sentiment, and key entities. This allows for the rapid identification of relevant information within immense datasets, a task that would be practically impossible for human analysts alone. This automation significantly increases the sheer volume of data that can be processed and the speed at which insights can be extracted, thereby expanding the effective scope of surveillance operations. It means that more data can be continuously collected, analyzed, and acted upon, creating a feedback loop that intensifies the surveillance apparatus.

The potential for AI-driven surveillance to operate with minimal human intervention is a particularly salient ethical consideration. As algorithms become more sophisticated, they are increasingly capable of making decisions and taking actions without direct human oversight. This can manifest in various ways, from automatically flagging individuals based on behavioral analysis to even, in extreme hypothetical scenarios, triggering automated responses. This escalating autonomy raises profound questions about accountability and responsibility. If an AI system incorrectly identifies an individual as a threat or makes a flawed prediction that leads to an adverse outcome, who is responsible? The programmer? The deploying agency? The AI itself? The lack of clear lines of accountability can create a significant governance gap, where powerful surveillance tools operate with reduced human judgment and oversight, increasing the risk of errors and abuses.

Algorithmic bias is another critical concern that permeates AI-driven surveillance. ML models learn from the data they are trained on, and if this data reflects existing societal biases, the algorithms will invariably reproduce and often amplify these biases. For instance, facial recognition systems have historically shown higher error rates when identifying individuals from minority ethnic groups or women, due to underrepresentation or biased representation in training datasets. Similarly, predictive policing algorithms can be biased against certain socioeconomic groups if the historical crime data used for training reflects discriminatory enforcement patterns. This means that AI-powered surveillance systems are not neutral observers; they are inherently shaped by the data they consume, potentially leading to unfair targeting, wrongful accusations, and the exacerbation of existing inequalities. The concept of "black box" AI, where the internal workings of complex algorithms are opaque and difficult to interpret, further complicates the identification and mitigation of such biases. Understanding *why* an AI made a particular decision is crucial for ensuring fairness and accountability, but this is often challenging with advanced ML models.

The capacity of AI to analyze sentiment and infer intent from online communications or public expressions is another area where surveillance capabilities are significantly enhanced. Natural Language Processing (NLP) techniques, a subfield of AI, allow systems to analyze text and speech for emotional tone, opinion, and even potential threats. This can be applied to social media monitoring, the analysis of public forums, or even the processing of intercepted communications to gauge public mood, identify dissent, or detect individuals expressing extremist views. While such analysis can be a legitimate tool for intelligence gathering, it also raises concerns about freedom of speech and the potential for misinterpretation or overreach. Analyzing sentiment can be subjective, and algorithms may struggle to understand context, sarcasm, or cultural nuances, leading to the mischaracterization of harmless expressions as malicious.

The sheer scalability of AI-powered surveillance means that it can be deployed on a global scale, creating interconnected networks of monitoring that span vast geographical areas. This is particularly evident in the context of smart cities, where AI is being integrated into urban infrastructure to manage traffic, optimize energy consumption, and enhance public safety. While these applications offer potential benefits, they also create unprecedented opportunities for pervasive surveillance. Every connected device, every camera, every sensor can become a node in a vast data collection network, feeding information into AI systems that build comprehensive profiles of citizens' lives. This creates a chilling effect on individual freedoms, as people may self-censor their behavior or expressions knowing that they are constantly being monitored and analyzed. The ubiquity of such systems raises fundamental questions about privacy as a right and the potential for a society where every action, every interaction, is subject to algorithmic scrutiny.

Moreover, the data required to train and operate these sophisticated AI surveillance systems is immense, leading to an ever-increasing appetite for data collection. This can incentivize entities to collect more data than is strictly necessary, and to find new ways to acquire it, often through increasingly intrusive means. The aggregation of data from disparate sources – social media, financial transactions, location data from mobile phones, public CCTV, smart home devices, and more – creates incredibly detailed and intimate portraits of individuals. AI algorithms excel at finding connections and patterns within this fused data that would be invisible to human analysts, revealing insights into relationships, habits, and even vulnerabilities. This granular understanding of individuals can be used for a variety of purposes, ranging from targeted advertising to more concerning applications like social scoring or political profiling.

The ethical dilemmas surrounding AI-driven surveillance are multifaceted and complex, touching upon issues of privacy, fairness, accountability, and autonomy. The ability of AI to analyze and predict behavior raises questions about pre-crime interventions and the potential for individuals to be penalized or restricted based on what an algorithm *predicts* they might do, rather than what they have actually done. This shifts the paradigm of justice and accountability, potentially infringing upon fundamental rights to due process and presumption of innocence. The lack of transparency in many AI systems, coupled with the potential for embedded biases, means that individuals may be subjected to surveillance and its consequences without understanding why or having any meaningful recourse to challenge the decisions made by the algorithms.

The development of AI in surveillance is an ongoing and rapidly evolving field. As AI capabilities continue to advance, so too will the sophistication and pervasiveness of surveillance systems. This necessitates a continuous societal dialogue about the ethical boundaries of these technologies, the appropriate legal frameworks for their deployment, and the mechanisms for ensuring accountability and mitigating harm. The efficiency gains offered by AI must be carefully weighed against the potential erosion of civil liberties and the risk of creating a society under constant, automated scrutiny. The challenge lies in harnessing the power of AI for legitimate security purposes without sacrificing the fundamental rights and freedoms that define a democratic society. As AI-powered surveillance becomes more deeply embedded in our lives, understanding its mechanics, its limitations, and its ethical implications is no longer an academic exercise but a critical imperative for safeguarding individual privacy and societal values. The future of surveillance is inextricably linked to the advancement of AI, presenting both unprecedented opportunities for efficiency and deeply concerning challenges for human rights and autonomy, demanding careful consideration and robust oversight.

The pervasive integration of the Internet of Things (IoT) into the fabric of our daily lives marks a new frontier in ubiquitous data collection, extending the reach of surveillance far beyond traditional cameras and digital communications. As 'smart' devices proliferate, transforming our homes, our commutes, and even our bodies into interconnected data points, the volume and intimacy of the information being gathered have escalated exponentially. These devices, often designed for convenience and efficiency, inadvertently become sophisticated sensors, constantly streaming an array of personal data to manufacturers, service providers, and potentially, to wider surveillance networks.

Consider the modern smart home. Devices like connected thermostats learn our daily routines, adjusting temperatures based on our presence and preferences, thereby mapping our comings and goings with remarkable precision. Smart speakers, always listening for wake words, can inadvertently record and transmit conversations, offering a window into private domestic life. Connected refrigerators track our grocery habits, while smart televisions log our viewing preferences, creating detailed profiles of our consumption patterns and leisure activities. Even seemingly innocuous devices, such as smart light bulbs that can be controlled remotely, collect data on when lights are on or off, contributing to a granular understanding of occupancy and activity within a household. This constant, often unarticulated, exchange of data transforms the home, traditionally a sanctuary of privacy, into a densely monitored environment.

Wearable technology, from fitness trackers to smartwatches, represents an even more direct and intimate form of data collection. These devices monitor heart rate, sleep patterns, activity levels, and even blood oxygen saturation. They collect biometric data that is inherently personal and can reveal a great deal about an individual's health, lifestyle, and physical condition. For instance, a fitness tracker might log when a person wakes up, how vigorously they exercise, and where they go for their runs. A smartwatch could record pauses in activity, or the duration of a particular social interaction based on elevated heart rate and proximity to other devices. This data, often shared with third-party apps and cloud services for analysis and personalized feedback, can paint an incredibly detailed picture of an individual's physical and even emotional state.

The connected car further extends this data-gathering ecosystem into our public and private transit. Modern vehicles are equipped with an array of sensors that track driving behavior, speed, braking patterns, routes taken, and even in-car conversations or music preferences. Telematics systems, often installed for insurance purposes or vehicle diagnostics, can provide a constant stream of location data and operational information. Furthermore, the infotainment systems within these cars can collect data on destinations, frequently visited locations, and even the user's interactions with connected services, effectively creating a mobile surveillance unit that follows our every journey. The data captured here is not just about the mechanics of driving; it's about where we go, who we're with (through paired phone contacts or in-car communication), and what we do while on the move.

The sheer volume and variety of data generated by these interconnected devices create a rich tapestry of personal information. Unlike traditional surveillance that might focus on specific activities or locations, IoT data provides a continuous, multi-dimensional view of an individual's life. It's not just about what you do, but *when* you do it, *how* you do it, *where* you do it, and even *how you feel* while doing it. This granular level of detail allows for sophisticated profiling, enabling entities to infer not just habits but also intentions, relationships, and vulnerabilities. For commercial entities, this data is gold, used to refine targeted advertising, develop new products, and personalize user experiences. For government surveillance programs, however, it presents an unprecedented opportunity to monitor populations with unparalleled depth and breadth.

A significant concern arising from the proliferation of IoT devices is their inherent security vulnerability. Many of these devices are designed with a focus on functionality and affordability, often at the expense of robust security measures. Default passwords that are rarely changed, unencrypted data transmission, and infrequent or non-existent software updates leave many IoT devices susceptible to hacking. This creates a wide-open door for malicious actors to gain unauthorized access to sensitive personal data. Exploits can range from hijacking smart cameras to gain visual access to homes, to intercepting data streams from wearables, or even taking control of connected vehicles.

These security lapses are not merely theoretical. Numerous reports and cybersecurity analyses have highlighted the ease with which many IoT devices can be compromised. Botnets, like the infamous Mirai botnet, have leveraged weakly secured IoT devices – particularly those with default credentials – to launch massive Distributed Denial of Service (DDoS) attacks. However, the implications extend beyond service disruption. Compromised smart home devices can allow attackers to spy on occupants, while hacked medical IoT devices could potentially expose sensitive health records or even be manipulated to cause harm. The interconnected nature of IoT means that a vulnerability in one device could potentially serve as an entry point into an entire network of devices, be it in a home, a vehicle, or a larger infrastructure.

The data collected by IoT devices, even when not directly accessed through hacking, is often shared with third parties, including manufacturers, app developers, and data brokers, sometimes with very limited user consent or transparency. Privacy policies for IoT devices can be lengthy, complex, and vague, making it difficult for consumers to understand precisely what data is being collected, how it is being used, and with whom it is being shared. This opaque data ecosystem means that information gathered in the privacy of one's home can end up in the hands of entities with commercial interests, or potentially be requested or compelled by government agencies through legal means.

From a surveillance perspective, this aggregated data from IoT devices offers a powerful tool for building comprehensive profiles of individuals and populations. Imagine a scenario where location data from a connected car, combined with activity logs from a fitness tracker, and smart home usage patterns, is fed into an AI analysis system. Such a system could infer not only where and when a person is active, but also their general health status, their social habits, their daily routines, and even their likely presence at home or away. This level of detailed, continuous monitoring makes it possible to track individuals' movements, activities, and associations with unprecedented accuracy, blurring the lines between public and private life.

The aggregation of data across multiple IoT devices and other digital footprints creates a mosaic of personal information that can be remarkably revealing. A connected thermostat can indicate when you are likely at home, a smart wearable can suggest your activity levels and even stress indicators, and your connected car's GPS history can map your daily commute and weekend excursions. When combined with social media activity, online browsing history, and financial transactions, this IoT data contributes to an extraordinarily detailed digital dossier for each individual. This dossier can be used for highly personalized advertising, but it also provides fertile ground for more intrusive forms of surveillance, enabling authorities or other entities to reconstruct an individual's life with remarkable clarity.

The potential for government agencies to access this vast troves of IoT data raises significant civil liberties concerns. While warrants and legal processes are generally required for accessing data held by telecommunications companies or internet service providers, the landscape for IoT data can be more ambiguous. Data stored by third-party manufacturers or service providers might be subject to different legal frameworks, and the sheer volume and constant generation of IoT data can create opportunities for mass data collection and analysis without specific individual suspicion. Furthermore, the reliance on data brokers who aggregate and sell personal information further complicates the privacy landscape, as this data can be acquired by various entities, potentially including government agencies operating under different legal authorities.

The concept of "dataveillance," or surveillance through the systematic collection and analysis of data, is amplified manifold by the IoT. Every connected device becomes a potential sensor, a data point contributing to a sprawling surveillance infrastructure. Unlike traditional surveillance methods that are often reactive or targeted, IoT-enabled surveillance can be pervasive, continuous, and proactive. It allows for the identification of patterns and deviations from those patterns across entire populations, enabling a form of social sorting and control that is both automated and deeply integrated into everyday objects.

Consider the implications for public spaces. Smart city initiatives, while promising efficiency and improved services, also deploy networks of interconnected sensors. Traffic sensors, environmental monitors, smart streetlights with integrated cameras, and public Wi-Fi hotspots all contribute to a city-wide data collection grid. When combined with the data streams from personal IoT devices carried by citizens, the potential for comprehensive tracking of urban populations becomes immense. An individual's movement through a city can be mapped not only by public CCTV but also by their connected car, their smartphone, their wearable, and even smart transit cards or payment systems.

The economic incentives driving the proliferation of IoT also play a role in this data collection landscape. Many IoT devices are offered at low cost, with the underlying business model relying on the ongoing collection and monetization of user data. This creates a system where users are, in essence, trading their personal data for the convenience or functionality of the device. This data is then used to build detailed user profiles, which are highly valuable for targeted marketing and other commercial applications. However, this data is also a valuable asset for surveillance, as it provides a rich and continuous stream of information about individuals' lives, habits, and preferences.

The security challenges are compounded by the 'lifecycle' of IoT devices. Many are deployed with the expectation of long-term use, but without adequate provisions for ongoing security updates and maintenance. This means that devices that are perfectly secure at the time of purchase can become increasingly vulnerable over time as new exploits are discovered and as the manufacturers cease to provide security patches. This creates a persistent risk of data compromise and potential surveillance over the lifespan of the device, which can often span many years.

Furthermore, the interoperability and standardization challenges within the IoT ecosystem can also contribute to security weaknesses. The lack of universal security standards means that devices from different manufacturers may not be able to communicate securely, or that insecure devices can be integrated into otherwise more secure networks. This can create attack vectors that are difficult to anticipate and mitigate, as the security of the entire system can be compromised by the weakest link in the chain.

The ethical implications of this ubiquitous data collection

are profound. It raises questions about consent, particularly when data collection is implicit or bundled within complex terms of service. It challenges traditional notions of privacy, as more aspects of our lives become digitized and accessible. It also raises concerns about autonomy, as the insights derived from this data could be used to influence behavior, manipulate choices, or even restrict opportunities based on predicted actions or inferred characteristics.

The aggregation of IoT data into large, centralized databases, often managed by a few dominant tech companies or cloud service providers, creates single points of failure and attractive targets for sophisticated attacks. A breach of such a database could expose the personal information of millions of individuals, including intimate details about their health, habits, and daily lives. This concentration of data also provides governments with a more efficient target for surveillance requests, as a single legal order could potentially yield vast amounts of information about a large number of people.

The challenge, therefore, lies in balancing the undeniable benefits of convenience and innovation offered by IoT devices with the fundamental right to privacy and security. As more devices become connected, and as the data they collect becomes more intimate and comprehensive, the imperative to develop robust security measures, transparent data handling practices, and clear legal frameworks governing data access becomes increasingly urgent. Without such safeguards, the Internet of Things risks evolving into an unprecedented surveillance apparatus, subtly and pervasively monitoring every aspect of our lives, transforming our homes, our bodies, and our movements into data points within a vast, interconnected web of observation. The future of privacy hinges on our ability to understand and effectively govern the data streams generated by this ever-expanding ecosystem of connected devices, ensuring that technological advancement does not come at the cost of fundamental freedoms.

The evolution of surveillance is not solely confined to the digital realm of interconnected devices and data streams; it is increasingly rooted in the biological distinctiveness of human beings themselves. Biometric data, encompassing the unique physical and behavioral characteristics that define an individual, has emerged as a potent and increasingly ubiquitous tool for identification and tracking. This encompasses a spectrum of identifiers, from the instantly recognizable fingerprints and DNA, to the more subtle yet equally distinctive patterns found in iris scans, facial geometry, and even the unique rhythm of a person's gait. The ability to capture, store, and analyze these inherent traits offers a level of persistent identification that transcends the limitations of traditional identifiers like passwords, ID cards, or even digital footprints, which can be lost, stolen, or obfuscated.

The collection of biometric data is no longer the exclusive domain of high-security facilities or forensic investigations. It has permeated everyday life through a variety of channels. Smartphones, now equipped with fingerprint sensors and facial recognition technology, routinely capture and store these unique biological markers for user authentication. Retailers are exploring the use of facial recognition in stores to identify known shoplifters or to personalize customer experiences. Airports and border crossings have long utilized fingerprint and iris scans for passenger processing, but these systems are becoming more sophisticated and widespread, aiming for faster and more seamless identification. Even public health initiatives, such as contact tracing during a pandemic, have raised the specter of DNA or other biological data being linked to individuals' movements and interactions.

At the heart of this trend lies the creation and maintenance of vast biometric databases. These digital repositories consolidate fingerprints, DNA profiles, facial templates, and other identifying biological information. For law enforcement and national security agencies, these databases have become invaluable for identifying suspects, solving crimes, and monitoring individuals deemed to be a risk. However, the expansion of these databases beyond traditional criminal justice contexts to include information on the general populace presents significant privacy challenges. The linking of biometric data to other forms of surveillance, such as CCTV footage, location data from mobile devices, or online activity, creates a powerful and potentially inescapable system of identification and tracking.

The implications for personal privacy and anonymity are profound. Unlike a forgotten password or a lost ID card, one's biological identity is inextricably linked to their existence. Biometric tracking offers a highly reliable method for identifying and monitoring individuals, potentially enabling persistent tracking across different contexts and locations without relying on traditional identifiers. This means that an individual's movements, associations, and activities can be pieced together with a degree of certainty that was previously unimaginable. If a person's gait is recognized by sensors in one location, and their fingerprint is scanned at another, and their face is captured by a camera linked to a facial recognition system in a third, a coherent and continuous record of their presence can be constructed. This capability moves beyond simple identification to a form of persistent, passive surveillance that can follow individuals wherever they go, irrespective of whether they have consciously disclosed their identity.

Consider the growing sophistication of gait analysis. This emerging field focuses on identifying individuals based on their unique walking patterns – the rhythm, stride length, arm swing, and even the subtle way they land their feet.

As wearable technology becomes more advanced and sensors become more integrated into our environments (smart floors, street furniture, public transport), the ability to passively identify individuals by their gait becomes increasingly feasible. This form of biometric tracking operates in the background, requiring no active participation or consent from the individual, and can track someone's movement through public spaces with remarkable accuracy, even if their face is obscured or their digital devices are off.

The collection and storage of DNA, once primarily confined to forensic science and medical diagnostics, is also entering the realm of broader surveillance. Direct-to-consumer genetic testing services have put DNA analysis within reach of millions, creating massive, privately held databases of genetic information. While often voluntarily submitted for personal ancestry or health insights, this data holds the potential for re-identification and tracking. Regulatory frameworks around the ownership, sharing, and security of this genetic data are still nascent, leaving open the possibility of it being accessed by law enforcement or other entities, either through legal means, data breaches, or by purchase from data brokers. The inherent immutability of DNA means that once collected and linked to an individual, it provides a permanent identifier that can be used for tracking and monitoring throughout their life.

Facial recognition technology has seen exponential growth in its deployment, moving from niche applications to widespread use in public spaces, law enforcement, and even consumer electronics. High-resolution cameras, coupled with advanced algorithms, can identify individuals in crowds with increasing accuracy, even under challenging lighting conditions or from oblique angles. This technology allows for the real-time tracking of individuals as they move through monitored areas, linking their physical presence to their identity. When combined with extensive databases of known individuals (such as driver's license photos or mugshots), facial recognition effectively transforms public spaces into a network of identifying sensors, capable of flagging and following specific persons of interest. The ability to link facial recognition data with other surveillance streams, such as license plate readers or location data from cell towers, creates a comprehensive picture of an individual's movements and activities.

The implications of this pervasive biometric tracking extend to the erosion of anonymity in public life. In societies where biometric identification is routinely employed, the ability to move through public spaces without being identified and potentially cataloged diminishes significantly. This can have a chilling effect on freedom of assembly and expression, as individuals may be hesitant to participate in protests or associate with certain groups if they know their presence is being meticulously recorded and attributed to their identity. The fear of being permanently linked to a particular location or event, even if their actions were lawful, can stifle dissent and encourage self-censorship.

Furthermore, the potential for mission creep is a significant concern. Systems initially designed for specific security purposes, such as identifying known terrorists at airports, can be gradually expanded to encompass broader categories of individuals or to monitor everyday activities. What begins as a tool for identifying threats can evolve into a system for tracking citizens, enforcing minor infractions, or profiling behavior based on perceived risk factors. The sheer accuracy and persistence of biometric identification make it an attractive tool for any entity seeking to maintain order or gather intelligence, potentially leading to a society where every movement is observed and every identity is logged.

The ethical considerations surrounding biometric data

collection are also complex. Issues of consent are paramount. While individuals may consent to fingerprint scanning for a job or facial recognition for unlocking their phone, the collection of their biometric data in public spaces is often done without their explicit knowledge or consent. The opacity of many surveillance systems means that individuals may not even be aware that their unique biological markers are being captured and analyzed. This lack of transparency undermines the fundamental principle of informed consent in data collection.

Moreover, the potential for bias in biometric systems, particularly facial recognition, is a well-documented problem. Studies have shown that these algorithms can be less accurate for women and people of color, leading to higher rates of misidentification. This can have serious consequences, particularly in law enforcement contexts, where a false match could lead to unwarranted suspicion, detention, or even wrongful arrest. The reliance on flawed technology to identify and track individuals raises serious questions about fairness and equity in the application of surveillance measures.

The concentration of such sensitive personal data in centralized databases also presents significant security risks. Biometric databases are highly attractive targets for hackers, as a successful breach could expose the immutable identities of millions of individuals. Such a breach could have catastrophic consequences, leading to identity theft, impersonation, and other forms of malicious activity that are difficult to rectify. The long-term security and integrity of these databases are therefore of paramount importance, yet the history of data breaches suggests that no system is entirely impervious to attack.

The interoperability of different biometric systems and databases also raises concerns. As various agencies and private entities deploy their own biometric identification tools, there is a growing trend towards data sharing and integration. This interconnectedness, while potentially offering efficiencies, also creates a more complex and vulnerable surveillance ecosystem. A compromise in one system could potentially cascade to others, creating widespread risks. The lack of standardized protocols for data sharing and security across these diverse systems exacerbates these vulnerabilities.

The notion of a "digital identity" is increasingly being conflated with one's biological identity through the application of biometric tracking. In a future where our faces, fingerprints, or gaits can be used to authenticate us across a myriad of services and locations, the very concept of being anonymous or unidentifiable in public may cease to exist. This shift has far-reaching implications for individual freedom and the nature of public space. It is a subtle but profound transformation, one that necessitates a thorough understanding of the technologies involved and a robust public discourse on the ethical and societal trade-offs being made. The ability to be identified by one's very being, rather than by a token or a digital credential, represents a new and powerful frontier in surveillance, one that demands careful scrutiny and robust safeguards to protect fundamental liberties. The ease with which these identifiers can be captured and the permanence of the data collected necessitate a proactive approach to regulation and oversight, ensuring that these powerful tools are used responsibly and ethically, without eroding the essential freedoms that underpin a democratic society. The future of personal privacy hinges on our collective ability to comprehend and control how our most fundamental, immutable characteristics are harnessed in the ever-expanding landscape of surveillance.

The relentless march of technological innovation continues to redefine the boundaries of surveillance, moving beyond the passive observation of physical spaces or the analysis of digital footprints. A nascent yet profoundly transformative frontier is emerging with the integration of augmented reality (AR) into our everyday lives, and with it, the potential for entirely new dimensions of monitoring and data collection. As AR devices, from smart glasses and contact lenses to enhanced smartphone interfaces, become more sophisticated, they are poised to seamlessly blend digital information with our perception of the physical world. This overlay, while offering immense potential for utility and enriched experiences, simultaneously creates a fertile ground for unprecedented surveillance capabilities. The very fabric of reality, as we perceive it, can become a dynamic canvas for tracking, identification, and analysis, fundamentally altering the nature of privacy and public interaction.

Imagine a scenario where individuals walking down a street are instantly recognized and identified through their AR interfaces, with real-time data about them – their social media profiles, known affiliations, purchase history, or even their current mood indicators derived from subtle physiological cues captured by sensors – being superimposed onto their visual field. This isn't the realm of science fiction; it's a plausible extrapolation of current technological trajectories. Governments and law enforcement agencies could equip officers with AR devices that, when focused on a person, instantly access and display relevant information from existing databases, facilitating rapid identification and risk assessment. Similarly, private entities, such as retailers or security firms, could leverage AR to monitor customer behavior, identify potential threats, or even personalize interactions based on real-time data feeds about individuals.

The implications of such pervasive AR-driven surveillance are multifaceted and far-reaching. Public spaces, traditionally zones where a degree of anonymity was assumed, could transform into intensely monitored environments. Every individual could become a data point, their presence, movements, and interactions continuously logged and analyzed. This persistent layer of digital observation, woven directly into our visual experience, raises significant concerns about the erosion of privacy and the chilling effect it might have on behavior. If individuals know that their every public appearance can be instantly linked to their digital identity and potentially scrutinized, it could lead to a society characterized by self-censorship and a reluctance to engage in activities that might draw attention, even if those activities are perfectly legal and benign.

Furthermore, the development of AR surveillance layers moves beyond simply identifying individuals. It can extend to the tracking of interactions and associations. AR systems, equipped with sophisticated object and people recognition capabilities, could potentially log who an individual speaks to, where they go, and for how long. This granular level of data collection, when aggregated, could build an intricate and highly personal map of an individual's life – their social networks, their routines, their preferred haunts, and their daily activities. For authoritarian regimes, this presents an unparalleled tool for social control, enabling the identification and suppression of dissent or any deviation from prescribed norms. Even in democratic societies, the potential for misuse, data breaches, or unauthorized access to such intimate information is a grave concern, threatening to undermine personal autonomy and freedom of association.

The concept of "contextual surveillance" becomes acutely relevant here. AR devices, constantly processing visual and auditory information from their surroundings, can interpret and act upon the context of a situation. A person attending a political rally, for instance, might be flagged by an AR system that cross-references their face with watchlists or known associations. Their presence at such an event, rendered visible and datafied through AR, could then trigger further scrutiny or documentation, regardless of their individual actions. This ability to imbue physical presence with contextual data allows for a far more sophisticated form of monitoring than traditional CCTV, which typically relies on human operators to interpret feeds or on more rudimentary motion detection. AR systems, by contrast, can perform complex analysis in real-time, making connections and drawing inferences that were previously impossible.

Consider the privacy implications for everyday interactions. If AR glasses become commonplace, allowing users to see enhanced information about the world around them, what prevents the development of AR applications that provide real-time "social scores" or "risk assessments" of individuals encountered? Such systems could draw upon vast datasets, including online activity, public records, and even data gleaned from previous AR interactions, to offer a curated, and potentially biased, impression of any given person. This could fundamentally alter social dynamics, leading to a society where trust is replaced by algorithmic judgment and where individuals are prejudged based on data that may be incomplete, inaccurate, or discriminatory. The ability to meet someone without an immediate, algorithmically generated profile being available could become a relic of the past.

The creation of these "surveillance layers" is not confined to governmental or security applications. The commercial sector is equally poised to harness AR for monitoring and data exploitation. Retailers might deploy AR systems that track customer movement within stores, analyze dwell times in front of specific products, and even infer purchasing intent. This data could be used for highly targeted advertising delivered directly through AR interfaces, creating an immersive and potentially coercive shopping experience. Imagine walking through a mall and having virtual advertisements or special offers pop up, precisely tailored to your inferred interests, based on your AR device's constant awareness of your location, your gaze, and your past behavior. While seemingly convenient, this represents a powerful form of persuasive technology that blurs the lines between organic discovery and algorithmic manipulation.

Moreover, the data generated by AR surveillance layers could be aggregated and cross-referenced with other data sources, creating incredibly detailed profiles of individuals. A person's gait, as captured by AR sensors, could be linked to their facial recognition data from a public camera, which in turn could be connected to their online browsing history or their smartphone's location data. This triangulation of information, facilitated by the ubiquitous nature of AR, creates a comprehensive digital twin of an individual, vulnerable to analysis and exploitation by any entity that possesses the means to access and process it. The inherent difficulty in opting out of such systems, especially when they become deeply embedded in social and professional life, makes the protection of privacy a monumental challenge.

The development of AR as a surveillance tool also raises questions about the ownership and control of the data it generates. Who owns the data captured by AR devices as users navigate public spaces? Is it the user, the AR platform provider, the government, or the entity whose premises the user is visiting? The current legal and ethical frameworks surrounding data ownership are still grappling with the implications of cloud computing and the Internet of Things; they are wholly unprepared for the deluge of hyper-contextual data that AR promises to unleash. The potential for data brokers to acquire, analyze, and sell this rich, real-time information about individuals' lives presents a new and alarming market for personal data.

The very concept of public space is at stake. Traditionally, public spaces were characterized by a degree of shared experience and, for the most part, an expectation of being unobserved in one's ordinary behavior. AR surveillance layers threaten to transform these spaces into highly curated and policed environments, where every interaction and movement is potentially recorded and assessed. This could lead to a gradual erosion of spontaneity and freedom in public life. The ability to simply "be" in public, without the constant awareness of being monitored or cataloged, might become a luxury, accessible only to those who can afford to disconnect or who live in areas not yet saturated with AR surveillance.

Furthermore, the algorithms driving AR surveillance systems are not neutral. They are designed by humans, trained on data, and therefore susceptible to biases. If AR systems are trained on datasets that reflect existing societal prejudices, they are likely to perpetuate and amplify those biases in their surveillance activities. This could lead to disproportionate scrutiny of certain demographic groups, reinforcing cycles of discrimination. For example, an AR system designed to identify "suspicious behavior" might be more prone to flagging individuals from minority communities if its training data overrepresents such groups in contexts of law enforcement intervention. The opacity of many machine learning algorithms makes it difficult to identify and rectify these biases, creating a "black box" effect where unfair surveillance practices can persist unseen.

The potential for "surveillance creep" is also a significant concern. Systems initially deployed for narrowly defined security purposes could gradually expand their scope. For instance, AR devices used by transit authorities to monitor passenger flow might evolve to track individuals' movements across different modes of transportation, building a comprehensive travel itinerary. Similarly, AR systems for retail analytics could be repurposed for law enforcement investigations, with data being shared between private companies and public agencies without explicit consent. This gradual, almost imperceptible expansion of surveillance capabilities, often justified by evolving security threats or the promise of greater efficiency, is a hallmark of how invasive technologies become normalized.

The development of interoperable AR surveillance systems is another critical consideration. As various entities deploy their own AR monitoring solutions, the push towards interoperability and data sharing will inevitably grow. This interconnectedness, while offering potential benefits for coordinated responses, also creates a more complex and vulnerable surveillance architecture. A vulnerability in one AR system could potentially be exploited to gain access to data from numerous other systems, creating cascading security risks. The lack of standardized security protocols and data governance frameworks across the AR ecosystem exacerbates these vulnerabilities.

The very nature of human interaction could be fundamentally altered. If AR overlays provide constant social profiling, it could lead to a world where authentic, unmediated encounters are replaced by interactions filtered through algorithmic judgment. The ability to connect with others based on genuine human empathy and understanding could be overshadowed by data-driven assessments of their perceived value or threat level. This potential for AR to create social stratification and to pre-emptively categorize individuals based on data risks fostering a more divided and less compassionate society. The spontaneous act of striking up a conversation with a stranger, for example, could become fraught with the anxiety of what one's AR profile might reveal, or what the stranger's AR profile might imply about them.

Moreover, the increasing sophistication of AR in mimicking real-world environments raises the possibility of AR being used to create "digital decoys" or to manipulate perceived reality for surveillance purposes. Imagine AR systems that can digitally alter the appearance of individuals in real-time, creating false identities or disguising recognized individuals for tracking purposes. This could be used to mislead surveillance systems, but also by surveillance systems themselves to create false trails or to obscure the identities of their own operatives. The lines between the real and the augmented, the observed and the observer, become increasingly blurred in such scenarios.

The proactive design of AR systems with privacy and security at their core, rather than as an afterthought, is therefore paramount. This necessitates robust regulatory frameworks that govern the collection, use, and retention of data generated by AR devices. It requires transparent communication with the public about how AR surveillance technologies are being deployed and what data is being collected. Furthermore, it demands the development of strong encryption methods and secure data storage practices to protect sensitive AR-generated information from unauthorized access and misuse. Without such safeguards, the widespread adoption of AR could inadvertently usher in an era of unprecedented and inescapable surveillance, fundamentally reshaping our understanding of freedom, privacy, and the very nature of reality itself. The integration of AR into our perception is not merely an enhancement of our senses; it is a potential gateway for a new, all-encompassing layer of observation that requires careful consideration and deliberate action to ensure it serves humanity rather than enslaves it to constant monitoring.

The very essence of digital anonymity, once a nascent concept nurtured in the early days of the internet, now finds itself under siege. As the capabilities of surveillance expand, encompassing not just the digital realm but also the augmented layers of our perceived reality, the struggle to remain unseen, uncataloged, and unidentifiable has intensified. This is not merely an abstract debate for technologists or a niche concern for privacy zealots; it is a fundamental battle for the preservation of individual liberty, free expression, and the very autonomy of thought and action in an increasingly interconnected world. The sophisticated, pervasive surveillance mechanisms that have been outlined demand a robust and multifaceted response, one that seeks to reclaim and protect the right to digital anonymity.

At the forefront of this defense are dedicated privacy advocates, forward-thinking technologists, and courageous activists who understand the profound implications of pervasive digital observation. Their work is a constant, often uphill, battle against well-resourced state and corporate actors who possess the infrastructure and the motivation to erode anonymity. These groups are not simply reacting to threats; they are proactively developing, refining, and disseminating the tools and methodologies that can empower individuals to resist invasive monitoring. This involves a deep dive into the technical underpinnings of digital communication, network infrastructure, and data collection, seeking out vulnerabilities and designing solutions that can shield users from prying eyes.

One of the foundational pillars of this fight is the promotion and development of strong encryption. End-to-end encryption, for instance, ensures that only the sender and intended recipient can read messages, rendering them unintelligible to intermediaries, including internet service providers, governments, or the platforms themselves. While widely adopted for messaging applications, the ongoing debate surrounding "backdoors" – cryptographic keys or vulnerabilities that would allow law enforcement or intelligence agencies access – represents a significant legal and technical battleground. Advocates argue vehemently that any weakening of encryption, even for ostensibly legitimate purposes, creates systemic vulnerabilities that can be exploited by malicious actors, thereby undermining the security and privacy of all users. The push for zero-knowledge proofs and other advanced cryptographic techniques further aims to enable verification and transaction without revealing underlying data, offering a more sophisticated form of privacy preservation.

Beyond encryption, the development and promotion of anonymizing networks form another critical component. Technologies like Tor (The Onion Router) have been instrumental in this regard. Tor works by routing internet traffic through a worldwide network of volunteer relays, encrypting the data at each step and stripping off successive layers of encryption as it passes from node to node. This process makes it extremely difficult to trace the origin of the traffic, effectively anonymizing the user's IP address and their online activities. However, Tor is not without its challenges. Its reliance on volunteer relays can lead to slower speeds, and the exit nodes, where traffic leaves the Tor network and enters the public internet, can potentially be monitored. Consequently, there is ongoing research into enhancing the robustness and usability of such networks, exploring decentralized alternatives and more resilient routing mechanisms.

The concept of digital self-sovereignty is also gaining traction within these advocacy circles. This involves empowering individuals with greater control over their digital identities and the data they generate. Decentralized identity solutions, built on blockchain technology or similar distributed ledger systems, aim to shift control away from central authorities and towards individuals. Instead of relying on platforms to manage identities, users can store and manage their own verified credentials, choosing precisely what information to share and with whom. This not only enhances privacy but also reduces the risk of mass data breaches and identity theft that plague centralized systems. The challenge lies in making these complex technologies accessible and user-friendly for the general public, ensuring that privacy-enhancing tools do not become the exclusive domain of the technically proficient.

The legal and legislative arenas are equally crucial battlegrounds. Privacy advocates tirelessly work to influence policy, lobbying for stronger data protection laws, challenging surveillance programs through legal means, and raising public awareness about privacy rights. This includes fighting against legislation that expands government surveillance powers, arguing for the necessity of warrants based on probable cause for accessing digital communications, and demanding transparency regarding data collection practices. The legal framework surrounding digital privacy is constantly evolving, often lagging behind technological advancements. This necessitates continuous engagement to ensure that legal protections keep pace with the increasing sophistication of surveillance. Landmark cases, such as those challenging mass metadata collection or advocating for stronger protections against digital searches, are vital in shaping the future of digital anonymity.

The role of whistleblowers and investigative journalists is inextricably linked to the fight for anonymity. Individuals who expose wrongdoing, whether by governments or corporations, often rely on anonymized communication channels and secure data handling to protect themselves from retribution. Similarly, journalists require anonymity to protect their sources, enabling them to report on sensitive issues without jeopardizing the safety of those who provide vital information. The very act of journalism and dissent is often predicated on the existence of protected private spaces where information can be shared and investigated without fear of immediate identification and reprisal. Efforts to weaken anonymity directly threaten the ability of these crucial societal actors to perform their functions effectively.

However, the path to digital anonymity is fraught with significant hurdles. The sheer ubiquity of data collection, often enabled by the very devices we use for convenience and connectivity, creates a pervasive environment where anonymity is difficult to achieve. Smart devices, social media platforms, and even the underlying internet infrastructure are designed to collect and process vast amounts of personal data. This data is then often aggregated, analyzed, and used for various purposes, including targeted advertising, behavioral profiling, and, increasingly, surveillance. The economic incentives for data collection are immense, creating a powerful counterforce to privacy advocacy. Companies benefit from understanding consumer behavior, and governments often cite national security or public safety as justifications for broad data access.

The concept of "friction" is also a major obstacle. Privacy-enhancing tools often introduce a degree of friction into online interactions, whether it's slower browsing speeds, the need to manage encryption keys, or the effort required to understand and configure privacy settings. For many users, the perceived benefits of anonymity may not outweigh the inconvenience. This highlights the need for privacy technologies to be not only effective but also seamless and intuitive, integrating into daily life without creating undue burdens. The challenge is to make privacy the default, rather than an opt-in feature that requires technical expertise and conscious effort.

Furthermore, the evolving landscape of augmented reality presents new frontiers for surveillance that directly challenge anonymity. As discussed previously, AR systems can overlay digital information onto our perception of the physical world, potentially identifying individuals in real-time and linking them to vast databases. This creates a scenario where even physical presence can be instantly associated with a digital identity, eroding the traditional anonymity afforded by public spaces. The fight for digital anonymity must therefore expand to encompass these emerging technologies, demanding privacy-by-design principles in AR development and advocating for regulations that prevent the creation of pervasive AR-based surveillance layers that strip away any semblance of unobserved existence.

The battle for anonymity also extends to legal and jurisdictional complexities. The global nature of the internet means that data can traverse multiple borders, making it difficult to apply consistent privacy protections. Different countries have varying laws regarding data retention, surveillance, and privacy rights, creating a patchwork of regulations that can be exploited by those seeking to circumvent privacy measures. International cooperation and the establishment of harmonized privacy standards are therefore essential, though achieving consensus among nations with diverse legal traditions and geopolitical interests is a formidable task.

The technical sophistication of surveillance also requires continuous innovation in anonymization techniques. As governments and corporations develop more advanced methods for data analysis, tracking, and identification, privacy advocates and technologists must constantly adapt. This includes researching and implementing new forms of encryption, exploring decentralized communication protocols, and developing methods to obfuscate digital footprints. The cat-and-mouse game between surveillance and privacy is an ongoing one, demanding persistent vigilance and a commitment to staying ahead of emerging threats.

The social and cultural dimensions of anonymity are also critical. There is a growing awareness of the importance of privacy, but this is often counterbalanced by a societal embrace of transparency and sharing, particularly through social media. The normalization of constant digital presence and the public sharing of personal information can, paradoxically, create an environment where genuine anonymity is viewed with suspicion. Overcoming this cultural shift requires education and a clear articulation of why anonymity is not inherently about hiding wrongdoing, but about safeguarding fundamental rights, enabling free thought, and protecting individuals from undue influence and control. Anonymity is crucial for protecting vulnerable populations, including victims of abuse, political dissidents, and individuals seeking to explore sensitive aspects of their identity without public scrutiny.

The economic model of the internet, largely driven by advertising and data monetization, presents a fundamental conflict with the principles of anonymity. As long as companies profit from collecting and analyzing personal data, there will be a constant incentive to erode privacy. This necessitates exploring alternative economic models for online services, such as subscription-based platforms or publicly funded digital infrastructure, that do not rely on invasive data practices. The transition to such models is challenging, requiring significant societal and political will.

The ongoing development of artificial intelligence further complicates the fight for anonymity. AI algorithms can process and analyze data at unprecedented scales, identifying patterns and making inferences that were previously impossible. This includes advanced facial recognition, behavioral analysis, and even the potential to infer sensitive personal characteristics from seemingly innocuous data. The challenge for anonymity advocates is to ensure that AI is developed and deployed ethically, with robust safeguards against its misuse for mass surveillance and the erosion of privacy. This may involve advocating for regulatory frameworks that govern the use of AI in surveillance contexts, promoting transparency in AI algorithms, and developing AI-powered tools that can protect user privacy.

Ultimately, the fight for digital anonymity is a continuous struggle to maintain space for individual freedom in an increasingly monitored world. It requires a confluence of technical innovation, legal advocacy, public education, and a fundamental commitment to the principles of privacy and autonomy. The tools and techniques developed by privacy advocates, from strong encryption and anonymizing networks to decentralized identity solutions, are vital components in this effort. However, their effectiveness depends on their widespread adoption and the creation of a legal and cultural environment that values and protects digital anonymity. As surveillance technologies continue to evolve, so too must the strategies and the resolve of those who believe that the right to be unobserved, to express oneself freely, and to exist in the digital realm without constant scrutiny is a cornerstone of a free and democratic society. The maintenance of private digital spaces is not a luxury, but a necessity for a healthy public sphere and for the protection of fundamental human rights in the digital age.

12: REFORMING THE SURVEILLANCE STATE

The escalating sophistication and pervasiveness of government surveillance necessitate a robust re-evaluation and strengthening of our legal frameworks. While technological advancements have outpaced legislative action for years, a concerted effort to update and reform existing laws is paramount to safeguarding individual privacy and civil liberties in the digital age. This involves not only scrutinizing and amending statutes that grant broad surveillance powers but also introducing new legislation that anticipates future technological developments and enshrines fundamental privacy rights into law.

A critical area for reform lies in the Foreign Intelligence Surveillance Act (FISA) and its associated amendments. Originally designed to govern electronic surveillance for foreign intelligence purposes, FISA has, over time, been expanded and interpreted in ways that arguably extend its reach into the communications of American citizens, often under the guise of national security. The Patriot Act, enacted in the wake of the September 11th attacks, significantly broadened the government's surveillance capabilities, including Section 215, which allowed the government to compel the production of "any tangible things" relevant to foreign intelligence or terrorism investigations. This provision, famously used to collect the metadata of millions of phone calls, demonstrated a significant gap in privacy protections, as it permitted the collection of vast amounts of data without the individualized suspicion typically required by the Fourth Amendment.

Strengthening legal safeguards requires a multi-pronged approach, starting with the reining in of what is often termed "bulk data collection." The practice of acquiring and retaining massive datasets of communications metadata—such as who called whom, when, and for how long—without a warrant predicated on probable cause, represents a significant intrusion into the private lives of ordinary citizens. Such metadata, while not containing the content of conversations, can reveal intimate details about an individual's relationships, habits, and associations. Proposals to amend FISA and other relevant statutes must explicitly require a warrant issued by an independent court, based on probable cause, for the acquisition of such metadata, even when collected in bulk. This would align the collection of digital metadata with the constitutional standard for physical searches, acknowledging that in the digital realm, metadata itself can be highly revealing.

Furthermore, the scope of Section 702 of FISA, which allows for the warrantless surveillance of non-U.S. persons located outside the United States, has become a point of contention. While ostensibly targeting foreign nationals, the "incidental collection" of communications data belonging to American citizens who communicate with these foreign targets—often referred to as "backdoor searches"—raises serious privacy concerns. Reforms should mandate that any such incidentally collected data pertaining to U.S. persons can only be accessed by the FBI through a Section 702-specific query process that requires a warrant or court order, thereby preventing warrantless backdoor searches. The ability of the government to conduct these types of searches without a warrant, even if the initial target is foreign, effectively bypasses the Fourth

Amendment protections that would normally apply to U.S. citizens.

Transparency in surveillance operations is another crucial pillar for strengthening legal safeguards. The secrecy that often surrounds intelligence gathering activities, while sometimes necessary for operational security, can also serve to shield overreach and abuse from public scrutiny. Legislative reforms should mandate greater transparency regarding the scope and nature of surveillance programs. This could include requiring the government to publicly disclose aggregate data on the number of warrants sought and granted under various surveillance statutes, the types of data collected, and the general categories of individuals or groups targeted. While specific operational details must remain classified, a higher level of aggregate transparency would allow for more informed public debate and legislative oversight.

The Patriot Act's authority for the FBI to issue National Security Letters (NSLs) also warrants closer examination and reform. NSLs are administrative subpoenas that do not require prior judicial approval and can compel the production of certain types of sensitive information, such as subscriber information and toll billing records. While they have legitimate uses, the broad application and the gag orders often attached to them have raised concerns about due process and First Amendment rights. Legislators should consider implementing stricter controls on the issuance of NSLs, requiring judicial review in more circumstances, limiting the scope of information that can be obtained, and providing clearer avenues for challenging gag orders.

Beyond amending existing legislation, the creation of new legal mechanisms tailored to emerging surveillance technologies is essential. As mentioned previously, augmented reality (AR) systems and the increasing integration of AI into surveillance capabilities present new challenges. Laws need to be developed to address the collection and use of data generated by AR devices, including biometric data and location information, ensuring that such collection is subject to stringent legal standards and that individuals have control over their digital presence in augmented environments. Similarly, regulations governing the use of AI in surveillance must be established, focusing on preventing discriminatory practices, ensuring algorithmic transparency, and safeguarding against the creation of ubiquitous, automated surveillance systems.

The concept of "purpose limitation" should be more rigorously applied to government data collection. This principle dictates that data collected for a specific, authorized purpose should not be repurposed for other, unrelated investigations without further legal authorization. Current legal frameworks sometimes permit the retention and analysis of data collected for foreign intelligence purposes to be used in domestic law enforcement investigations, often through mechanisms like the "}+1 rule" (Section 105(d)(1) of FISA) which allows for the dissemination of incidentally collected information to law enforcement if it suggests a crime may have been committed. Strengthening legal safeguards would involve clearly defining and enforcing purpose limitations, ensuring that data collected under one legal authority is not implicitly re-authorized for use under another without meeting the latter's stricter standards.

Furthermore, the judicial oversight of surveillance must be strengthened. The FISA Court, while intended to be an independent arbiter, operates largely in secret and without adversarial representation. While maintaining some level of secrecy is unavoidable in national security matters, reforms could explore mechanisms to introduce more robust adversarial perspectives into its proceedings, perhaps through the appointment of amici curiae (friends of the court) who are tasked with representing the interests of civil liberties. Increased judicial training on the nuances of digital privacy and the capabilities of modern surveillance technologies would also be beneficial.

The ability for individuals to seek redress for unlawful surveillance must also be enhanced. Currently, proving that one has been subjected to unlawful surveillance can be exceedingly difficult, especially given the classified nature of many intelligence programs. Legislators should consider creating clearer legal pathways for individuals to challenge surveillance practices and seek damages when their rights have been violated. This might involve establishing specialized courts or tribunals to handle such cases or creating mechanisms for supervised declassification of evidence in specific instances to allow for fair adjudication.

The debate surrounding encryption and government access is another area where legal safeguards need clarification. While many privacy advocates argue against any form of "backdoor" access to encrypted communications, citing the potential for abuse and the weakening of security for all, governments often contend that such access is necessary for investigating serious crimes. Legislation could explore a middle ground, perhaps by strengthening legal requirements for data preservation orders that require companies to maintain unencrypted data for a limited period before it is encrypted, or by facilitating secure, lawful access to data with appropriate judicial authorization, while simultaneously enacting strong prohibitions against mandated vulnerabilities in encryption systems. This is a complex area, but a complete lack of legislative direction leaves a critical gap.

The accountability of intelligence agencies and law enforcement for surveillance practices is also a key component of strengthening legal safeguards. Existing oversight mechanisms, such as those provided by Inspectors

General and Congressional committees, are important but can sometimes be limited by the information they receive and the political pressures they face. Reforms could include granting these oversight bodies greater independent authority, ensuring broader access to information, and establishing clearer lines of accountability for any misuse of surveillance powers. Whistleblower protections are also vital; ensuring that individuals within these agencies can report suspected abuses without fear of retaliation is critical for maintaining integrity.

Legislative action must also keep pace with the rapid evolution of technology. This necessitates a more agile and forward-looking approach to lawmaking, rather than constantly playing catch-up. Establishing independent bodies composed of legal experts, technologists, ethicists, and civil liberties advocates that can provide ongoing recommendations to lawmakers on how to adapt surveillance laws to new technologies would be a proactive step. Such bodies could help identify potential privacy risks before they become widespread problems and offer informed policy suggestions.

Ultimately, the strengthening of legal safeguards is not merely about restricting government power; it is about striking a balance that allows for legitimate security needs to be met while rigorously protecting the fundamental right to privacy. The current legal landscape, shaped by legislation often enacted in response to past threats, is increasingly inadequate for the challenges of digital-age surveillance. A comprehensive legislative agenda that prioritizes warrants for data collection, limits bulk data acquisition, enhances transparency, strengthens judicial oversight, and adapts to new technologies is essential for preserving civil liberties and maintaining public trust in the institutions that are tasked with protecting national security. This is an ongoing process, requiring continuous vigilance and a commitment to adapting our laws to the realities of the 21st century. The goal is to ensure that the tools used to protect society do not inadvertently erode the very freedoms that society is meant to uphold. This requires a proactive, rather than reactive, approach to legal reform.

The opaque nature of modern surveillance operations presents a significant impediment to democratic oversight and public trust. To counter this, a fundamental shift towards enhanced transparency and accountability is not merely desirable but essential. This involves establishing clear, actionable mechanisms that illuminate the scope and application of government surveillance, thereby fostering informed public discourse and holding those responsible to account. A cornerstone of this reform would be the implementation of mandatory public reporting requirements for all surveillance activities. These reports, generated on a regular basis, should provide granular yet anonymized statistical data concerning the types of surveillance employed, the number of authorizations sought and granted, the categories of individuals or entities targeted, and the general nature of the investigations for which surveillance was deemed necessary. Such comprehensive reporting would move beyond broad generalizations and offer concrete insights into the practical application of surveillance powers, allowing citizens and their representatives to assess whether these powers are being exercised judiciously and in accordance with legal and ethical boundaries.

Crucially, these public reports must be accessible and understandable to the general populace. The technical jargon and legal complexities often associated with surveillance legislation can render even well-intentioned transparency efforts obscure. Therefore, reports should be presented in a clear, concise manner, accompanied by explanatory glossaries and summaries that demystify the data for a lay audience. Furthermore, the information should be published in readily searchable and downloadable formats, enabling researchers, journalists, and civil society organizations to conduct independent analyses and identify potential trends or anomalies. This level of public visibility is vital for cultivating a shared understanding of the surveillance state's footprint and for empowering citizens to engage meaningfully in the ongoing debate about its appropriate role and limits. Without such transparency, the risk of overreach and the erosion of civil liberties under the veil of national security remains unacceptably high.

Beyond statistical reporting, the establishment of robust, independent oversight bodies is paramount. These bodies must be vested with genuine investigative powers, unfettered by political influence or the very agencies they are tasked with scrutinizing. Current oversight mechanisms, often housed within executive branches or relying on self-reporting from intelligence agencies, frequently suffer from inherent conflicts of interest and a lack of true autonomy. True independence would necessitate that these oversight entities are composed of individuals with diverse expertise, including legal scholars, civil liberties advocates, technologists, and former judges, selected through a transparent and merit-based process that insulates them from partisan pressures. Their mandates should include the authority to conduct unannounced audits of surveillance programs, access all relevant documentation, interview personnel at all levels, and, critically, to compel the production of information.

The investigative powers of these independent oversight bodies must extend to the examination of specific surveillance operations when warranted, whether triggered by credible allegations of misconduct, statistical anomalies, or significant shifts in technological application. Furthermore, these bodies should be empowered to recommend systemic reforms, policy changes, and disciplinary actions when abuses or inefficiencies are identified. Their findings and recommendations, while acknowledging the need for appropriate redactions to protect legitimate national security secrets, should be made public, ensuring that the insights gained from their investigations contribute to broader public understanding and accountability. The effectiveness of such oversight hinges on its perceived and actual independence and its ability to act as a credible check on executive power.

A critical component of this oversight framework involves empowering individuals with mechanisms to access and correct personal data held by surveillance agencies. In an era where vast amounts of information are collected and analyzed, individuals must have a meaningful right to know what data exists about them, how it is being used, and the ability to challenge its accuracy or legality. This data access and correction right, often referred to as a "right to be forgotten" or a right to access one's digital footprint, would necessitate the creation of clear procedures for individuals to submit requests for information and to have those requests addressed in a timely and transparent manner. Agencies would be obligated to respond to these requests, providing access to the collected data unless doing so would demonstrably jeopardize an ongoing investigation or compromise a clearly defined national security interest.

When inaccuracies or illegalities are identified in the data, individuals should have a clear pathway to request corrections or deletions. The process for challenging data must be fair and accessible, potentially involving an independent review by the oversight bodies mentioned previously, particularly in cases where agency responses are unsatisfactory or deemed inadequate. This right of access and correction serves a dual purpose: it empowers individuals to maintain control over their personal information and acts as a powerful deterrent against the careless or malicious collection and retention of data. It acknowledges that in a surveillance-driven society, personal data is not merely an administrative record but a fundamental extension of an individual's identity and privacy.

Moreover, the concept of "data minimization" should be rigorously enforced, aligning with the principles of privacy by design. Surveillance agencies should be legally compelled to collect only the data that is strictly necessary for the stated purpose of their investigations, and to establish clear policies for the timely and secure deletion of data that is no longer required. This proactive approach to limiting data accumulation reduces the potential for misuse, breaches, and the chilling effect that the awareness of pervasive data collection can have on freedom of expression and association. Public reporting should also include statistics on data retention periods and the volume of data routinely purged, demonstrating a commitment to responsible data stewardship.

The judiciary also plays a crucial role in enhancing accountability. While the Foreign Intelligence Surveillance Court (FISC) operates in secrecy, reforms can be implemented to increase transparency and adversarial input within its proceedings. The appointment of independent special advocates or amici curiae (friends of the court) to represent the interests of civil liberties during ex parte warrant applications could provide a vital counterpoint to government arguments, ensuring that privacy concerns are rigorously debated. Furthermore, while operational details must remain classified, the FISC could be mandated to release declassified, redacted versions of its significant rulings and legal interpretations, providing greater insight into the evolving legal landscape of surveillance and fostering a more informed public debate about judicial oversight.

Whistleblower protections are another indispensable element in the architecture of accountability. Individuals within intelligence agencies or law enforcement who come forward with credible information about illegal surveillance activities or systemic abuses often face severe retaliation, including job loss, prosecution, and reputational damage. Strengthening whistleblower protections would involve establishing secure, confidential channels for reporting wrongdoing, granting enhanced legal immunities for those who report in good faith, and ensuring robust investigation and redress for any retaliatory actions. By fostering an environment where individuals feel safe to speak out against potential abuses, these protections serve as an essential early warning system and a critical mechanism for uncovering and rectifying misconduct that might otherwise remain hidden.

The effectiveness of accountability mechanisms is also directly tied to the public's awareness and engagement. Educational initiatives aimed at informing the public about their privacy rights, the nature of surveillance technologies, and the existing legal frameworks are crucial. When citizens are informed, they are better equipped to advocate for their rights, demand accountability from their elected officials, and participate meaningfully in shaping surveillance policy. This could involve public awareness campaigns, accessible online resources, and civic education programs that demystify the complexities of the surveillance state.

Ultimately, the success of any reforms aimed at enhancing transparency and accountability hinges on a fundamental shift in governmental culture – a move from secrecy as the default to openness as a guiding principle, balanced appropriately with legitimate national security needs. It requires a recognition that in a democracy, the power to surveil must be accompanied by a profound responsibility to be transparent and accountable to the people. The establishment of independent oversight, mandatory public reporting, rights of data access and correction, and robust whistleblower protections are not merely bureaucratic adjustments; they are essential pillars for safeguarding democratic values and ensuring that the tools of national security do not become instruments of oppression. This commitment to transparency and accountability is the bedrock upon which public trust in governmental institutions can be rebuilt and maintained in the face of increasingly pervasive technological capabilities. Without these measures, the surveillance state risks operating beyond the reach of democratic control, eroding the very freedoms it purports to protect. The ongoing evolution of technology necessitates a dynamic and adaptive approach to these reforms, ensuring that transparency and accountability remain at the forefront of policy discussions and legislative action. This continuous process of scrutiny and adaptation is vital to prevent the normalization of unchecked surveillance and to uphold the principles of a free and open society.

The existing architecture of oversight for surveillance activities, while formally established, often falters in practice due to structural limitations, resource constraints, and a persistent imbalance of information in favor of the executive branch. Congressional oversight, primarily vested in intelligence committees within both the House and Senate, operates under the premise that elected representatives are best positioned to scrutinize the actions of the national security apparatus. However, these committees are frequently hamstrung by their own memberships, which often include individuals with deeply ingrained pro-intelligence or national security leanings, potentially compromising their ability to provide genuinely critical appraisal. Moreover, the sheer volume and complexity of classified information presented to these committees can be overwhelming, demanding significant resources and specialized expertise that are not always adequately allocated. The information asymmetry is particularly acute; agencies can readily provide curated briefings and documentation, while committees struggle to independently verify or deeply probe the operational realities. To rectify this, a multi-pronged approach to strengthening congressional oversight is essential. This includes ensuring that members of intelligence committees possess a foundational understanding of technological capabilities and legal frameworks governing surveillance, potentially through mandatory specialized training. Furthermore, the professional staffs supporting these committees must be significantly enhanced, both in number and in their technical and legal acumen, to allow for more granular analysis of agency proposals and activities. These staffs should have direct, unfettered access to declassified information, enabling them to perform independent research and verification without relying solely on agency-provided materials. The practice of assigning committee members based on seniority alone should be re-evaluated, with a greater emphasis placed on demonstrated interest and expertise in intelligence oversight. Furthermore, the frequency and depth of committee briefings need to be increased, moving beyond perfunctory updates to substantive engagements that allow for genuine interrogation of intelligence gathering and analytic processes. The committees' power to conduct oversight is also directly linked to their subpoena power and their ability to compel testimony; these powers must be unequivocally asserted and utilized to ensure full accountability, even when faced with executive branch resistance.

The judicial branch, particularly through the Foreign Intelligence Surveillance Court (FISC), plays a critical role in authorizing surveillance operations. However, the FISC's historically ex parte and largely secret nature raises profound questions about the robustness of this oversight. While the court's mandate is to ensure compliance with legal standards, the absence of adversarial representation during warrant applications means that only the government's perspective and legal arguments are typically presented. This inherent imbalance can lead to a situation where the government's interpretation of legal boundaries and the necessity of surveillance is accepted with insufficient challenge. To bolster the effectiveness of judicial oversight, reforms should focus on introducing a more robust adversarial process. This could involve the mandatory appointment of independent special advocates, who are not affiliated with any intelligence agency and possess a strong background in civil liberties and privacy law, to represent the interests of the public and constitutional rights during FISA warrant applications. These special advocates would be empowered to review the government's submissions, conduct their own independent assessments of necessity and proportionality, and present counterarguments to the court. Their role would be to ensure that all potential legal and constitutional implications are thoroughly examined, thereby providing the court with a more balanced perspective. Additionally, the FISC should be empowered to proactively seek out and appoint amici curiae (friends of the court) in cases involving novel legal interpretations or significant technological shifts in surveillance capabilities. These amici could bring specialized knowledge in areas such as cryptography, data analytics, or constitutional law to bear, enriching the court's understanding and decision-making process. Furthermore, while the operational specifics of surveillance must remain classified, the FISC should be mandated to publish declassified and redacted versions of its significant legal interpretations and rulings. This transparency, even in a limited form, would allow for a greater public understanding of how surveillance laws are being applied and interpreted, fostering more informed debate and enabling legal scholars and civil liberties advocates to identify trends or potential overreach. The selection process for FISC judges should also be reviewed to ensure that appointees possess a demonstrable commitment to civil liberties and a nuanced understanding of the Fourth Amendment in the digital age, perhaps through a nomination process that explicitly considers these attributes alongside judicial experience.

Beyond the established congressional and judicial channels, a critical need exists to fortify the independent oversight functions of Inspectors General (IGs) and government auditors, and to explore innovative approaches to surveillance oversight. Inspectors General within intelligence agencies and related departments are tasked with conducting audits and investigations into waste, fraud, and abuse. However, their independence and effectiveness can be compromised if they are too closely aligned with the agencies they oversee or if their mandates are not sufficiently broad to encompass the ethical and constitutional implications of surveillance. To enhance their capabilities, IGs should be granted greater autonomy in selecting their staff and in initiating audits and investigations without requiring prior approval from agency leadership. Their reports and findings, particularly those pertaining to systemic issues or potential abuses in surveillance practices, should be made publicly accessible in a declassified form, perhaps with the same level of redaction as FISC rulings to protect sensitive operational details. This public reporting would serve as a crucial mechanism for accountability, allowing Congress and the public to assess the efficacy of internal controls and identify areas requiring external intervention. Similarly, government auditors, such as those at the Government Accountability Office (GAO), play a vital role in assessing the efficiency and effectiveness of government programs. The GAO should be routinely tasked with conducting comprehensive reviews of surveillance technologies, programs, and their associated costs, examining whether these investments are yielding intended results and whether they are being implemented in a manner that is proportionate to the threats they are designed to address. These GAO reports should explicitly evaluate compliance with legal requirements, privacy policies, and ethical considerations, providing an independent, data-driven assessment of the surveillance state's performance.

Innovative oversight models that incorporate elements of public input or external expert consultation could also prove invaluable in adapting oversight to the rapidly evolving landscape of surveillance technology and practice. One such model could involve the establishment of an independent, non-governmental advisory council composed of leading experts in technology, law, ethics, and civil liberties. This council would be tasked with providing regular, independent assessments of emerging surveillance technologies and their potential implications for privacy and civil rights. The council's reports and recommendations, while not binding, would offer crucial insights to policymakers and oversight bodies, helping to anticipate challenges and proactively shape policy and legislation. Such a council could also play a role in advising on best practices for data security, retention, and minimization, ensuring that technological advancements are accompanied by robust ethical frameworks. Another avenue for enhanced oversight could involve the creation of specialized, technology-focused review boards within congressional intelligence committees, staffed by individuals with deep technical expertise in areas such as artificial intelligence, encryption, and network surveillance. These boards would work in tandem with the broader committee structure, providing the technical depth necessary to critically evaluate the government's use of sophisticated surveillance tools. Furthermore, the concept of "sunset clauses" for broad surveillance authorities could be more rigorously applied. These clauses, which require legislative reauthorization of surveillance powers after a set period, provide recurring opportunities for Congress to review and debate the necessity and scope of these powers, ensuring that they do not become permanently entrenched without ongoing justification. The debate surrounding these reauthorizations must be informed by robust public discourse and accessible data, moving beyond the often-closed-door deliberations that have characterized past renewals. The very process of oversight needs to be imbued with a greater sense of urgency and adaptability, recognizing that the state's surveillance capabilities are in a constant state of flux, driven by technological innovation. Without commensurate adaptation in our oversight mechanisms, the balance between security and liberty will inevitably tilt further towards the erosion of the latter. The current system, while containing the building blocks of oversight, often lacks the teeth, the independence, and the transparency required to effectively govern the modern surveillance state. Strengthening these disparate elements and fostering a more integrated and dynamic approach to oversight is not merely a matter of bureaucratic refinement; it is a fundamental necessity for preserving democratic principles and public trust in an age of pervasive digital monitoring. The ability of oversight bodies to access, analyze, and disseminate information, to compel testimony, and to make independent judgments is paramount. This requires a conscious and sustained effort to overcome ingrained secrecy, to resource oversight adequately, and to empower those tasked with ensuring that surveillance powers are exercised judiciously and in strict accordance with the law and the Constitution. The ultimate goal is to ensure that the tools designed to protect society do not inadvertently undermine the freedoms and values that define it.

The architecture of our digital lives, from the applications we use daily to the very hardware that connects us, is not a neutral landscape. It is sculpted by design choices, often made with efficiency, functionality, and profitability at the forefront, but with a regrettable disregard for the fundamental right to privacy. The current paradigm often treats privacy as an afterthought, a feature to be bolted on if consumer demand or regulatory pressure dictates, rather than an intrinsic value to be protected from the genesis of a technology. This approach has inadvertently facilitated the pervasive surveillance capabilities that define the modern state, embedding intrusive data collection and retention into the very fabric of the systems we rely upon. To truly reform the surveillance state, a radical shift in how technology is conceived, developed, and deployed is not just desirable; it is imperative. This shift must be anchored in the principle of "privacy by design," a proactive and foundational approach that integrates privacy considerations into every stage of the technological lifecycle.

Privacy by design is not merely a set of guidelines; it is a philosophy that compels engineers, product managers, and corporate strategists to prioritize user privacy as a core design objective. It mandates that privacy be embedded into the system's architecture and operational practices from the earliest conceptualization, rather than being addressed as a mere compliance issue or an add-on feature. This philosophy, first articulated by Dr. Ann Cavoukian, former Information and Privacy Commissioner of Ontario, Canada, emphasizes seven foundational principles: proactive not reactive; privacy as the default setting; privacy embedded into design; full functionality—the "win-win" scenario; end-to-end security— life cycle protection; visibility and transparency; and respect for user privacy—keeping it user-centric. Adopting these principles means moving away from a model of data maximization, where services are built around the premise of collecting as much user data as possible, towards a model that prioritizes data minimization and user control.

For software engineers and hardware manufacturers, this

translates into a fundamental rethinking of their development processes. It means scrutinizing every data point collected, asking not only "can we collect this data?" but more critically, "should we collect this data?" and "what is the absolute minimum data required to achieve the intended functionality?" For instance, a social media platform could be designed to require minimal personal information for account creation, opting for pseudonymous or anonymous access where feasible. Instead of collecting precise location data by default for every interaction, the application could prompt users to grant location access only when specific location-based features are actively used, and even then, allow for generalized or temporary location sharing. Similarly, the data retention policies must be re-evaluated. Technologies should be built with automatic data deletion or anonymization mechanisms, ensuring that personal information is not stored indefinitely without a clear, ongoing justification. This proactive approach to data minimization reduces the potential for misuse and significantly limits the scope of data that can be accessed by state actors, whether through lawful requests or unauthorized breaches.

The implementation of privacy-enhancing technologies (PETs) is a cornerstone of privacy by design. These technologies are specifically engineered to protect personal information and can be integrated into the core functionality of digital products and services. Examples include end-to-end encryption, which ensures that only the sender and intended recipient can read messages, rendering them unreadable to intermediaries, including the service provider and potentially, state surveillance apparatuses. Differential privacy, a more advanced technique, allows for the analysis of large datasets to reveal patterns and trends without exposing the identity or specific data of any individual within the dataset. This is crucial for statistical analysis and machine learning models that might otherwise require the aggregation of sensitive personal information. Homomorphic encryption, a nascent but promising technology, enables computations to be performed on encrypted data without decrypting it first, offering a powerful way to process sensitive information while maintaining its confidentiality. These are not mere theoretical concepts; they are increasingly practical tools that can be woven into the design of everything from messaging apps and cloud storage services to online advertising platforms and government databases.

Hardware manufacturers also have a critical role to play. The physical devices we use are the gateways to our digital lives, and their design can either facilitate or hinder privacy. For example, smartphones could be designed with hardware switches that physically disable cameras and microphones, providing users with undeniable assurance that they are not being recorded. Secure enclaves within processors can be utilized to isolate sensitive data and cryptographic keys, making them more resilient to software-based attacks. The manufacturing process itself can be scrutinized for vulnerabilities that might be exploited for surveillance purposes, such as the introduction of backdoors or embedded surveillance hardware. Furthermore, the lifecycle management of devices needs to consider privacy. This includes secure wiping of data when devices are repurposed or disposed of, and providing clear, accessible information to users about how their data is handled throughout the device's lifespan. The increasing sophistication of IoT (Internet of Things) devices presents a particularly challenging frontier. Many of these devices, from smart speakers and thermostats to connected appliances and wearables, are designed with minimal security and privacy considerations, often collecting vast amounts of intimate personal data with little transparency or user control. Adopting privacy by design principles in the development of IoT devices is paramount to prevent these ubiquitous technologies from becoming pervasive surveillance instruments.

Service providers, encompassing everything from internet service providers (ISPs) to cloud hosting companies and software-as-a-service (SaaS) providers, are custodians of user data and play a pivotal role in its protection. Their business models often rely on the aggregation and monetization of user data. A paradigm shift towards privacy by design would necessitate a fundamental restructuring of these models, moving away from a reliance on extensive data harvesting towards service offerings that genuinely prioritize user privacy. This could involve offering tiered service models, where a premium price is paid for enhanced privacy protections, or developing data-neutral infrastructure that does not actively monitor or log user activities beyond what is strictly necessary for service delivery and security. For instance, ISPs could offer "privacy-focused" browsing plans that do not collect or sell user browsing history. Cloud providers could offer end-to-end encrypted storage by default, with robust key management systems that place control firmly in the hands of the user. Transparency in data handling practices is also crucial. Users should be provided with clear, concise, and easily accessible privacy policies that detail what data is collected, why it is collected, how it is used, and with whom it is shared. Moreover, service providers should implement robust data breach notification protocols and provide mechanisms for users to access, correct, and delete their personal data, aligning with principles like those found in the GDPR.

The influence of the technology industry's design choices extends beyond individual products and services to shape the very infrastructure of mass surveillance. When companies prioritize data collection for targeted advertising, machine learning training, or predictive analytics, they are inadvertently creating vast repositories of information that are attractive targets for government data requests and surveillance programs. The "move fast and break things" ethos, while driving innovation, has often resulted in the creation of systems with inherent privacy vulnerabilities. The challenge is to instill a culture within the tech industry where privacy is not a liability to be managed, but a core competency and a competitive advantage. This requires educating engineers and designers about privacy principles and legal requirements, fostering internal champions for privacy, and establishing strong ethical review processes for new technologies and features. It also means resisting the pressure to incorporate intrusive functionalities simply because they are technically feasible or could potentially generate revenue, and instead, actively seeking out privacy-preserving alternatives.

Consider the evolution of digital assistants and smart home devices. Initially marketed for convenience, their underlying design often involves continuous audio recording and sophisticated data analysis to understand user behavior, preferences, and even emotional states. While some of this data is used to improve service, a significant portion is likely retained and analyzed for other purposes, creating a detailed surveillance footprint of individuals' private lives. A privacy-by-design approach would dictate that these devices should only record and process audio when explicitly activated by a wake word, and that all collected data should be anonymized or deleted locally on the device whenever possible, with explicit user consent required for any off-device transmission or retention. Furthermore, the algorithms used to process this data should be designed to be as transparent as possible, allowing users to understand how their data is being interpreted and to challenge any inaccuracies or inferences.

The development of artificial intelligence (AI) and machine learning (ML) presents both significant opportunities and profound risks for privacy. These technologies are data-hungry, often requiring massive datasets for training. Without careful design, the collection and processing of this data can lead to significant privacy intrusions. However, AI and ML can also be leveraged to enhance privacy. Techniques such as federated learning allow models to be trained on decentralized data residing on user devices, without the data ever leaving the device itself. This dramatically reduces the privacy risks associated with centralized data aggregation. Generative AI models, while capable of producing sophisticated outputs, also raise concerns about the potential for bias embedded in training data and the generation of misinformation or deepfakes that can be used for malicious purposes. Designing these models with privacy and ethical considerations at their core means carefully curating training data to avoid bias, implementing safeguards against malicious use, and providing users with transparency about the generative process.

The legal and regulatory frameworks surrounding technology development often lag behind the pace of innovation. While laws like the GDPR in Europe have made strides in emphasizing privacy, the fragmented nature of global regulations and the emphasis on a consent-based model, which can be easily circumvented by complex, opaque terms of service, demonstrate the limitations of current approaches. A more effective strategy is to mandate privacy by design and by default. This means that privacy considerations should not be optional or subject to user opt-in for core functionalities. Instead, privacy-preserving settings should be the default, and any deviation from these defaults should require an explicit, informed, and affirmative choice by the user. This shifts the burden from the user having to navigate complex privacy settings to the technology provider having to justify any less-private design choices.

The economic incentives within the technology sector

also need to be realigned. The dominant business model of "surveillance capitalism," as described by Shoshana Zuboff, thrives on the commodification of personal data. To foster a technology ecosystem that respects privacy, alternative business models that do not rely on mass data collection must be encouraged and supported. This could involve public funding for privacy-preserving technologies, tax incentives for companies that adopt strong privacy practices, and regulatory measures that penalize data-hoarding and privacy violations more effectively. Consumer demand can also be a powerful driver. As users become more aware of the privacy implications of the technologies they use, they can exert pressure on companies to adopt more responsible design practices. Supporting and promoting technologies and companies that prioritize user privacy can help to create a virtuous cycle, where privacy becomes a key differentiator and a driver of market success.

Ultimately, the transition to a technology landscape where privacy is intrinsically valued requires a concerted effort from multiple stakeholders: engineers and designers must be trained and incentivized to prioritize privacy; hardware manufacturers must build privacy into the very foundation of their devices; service providers must rethink their business models to de-emphasize data exploitation; policymakers must create robust legal frameworks that mandate privacy by design and by default; and consumers must demand greater respect for their personal information. This is not merely a technical challenge; it is a societal one that calls for a re-evaluation of our relationship with technology and a commitment to building a digital future that is not only innovative and convenient but also secure, ethical, and respectful of fundamental human rights. Without embedding privacy into the DNA of our technologies, the surveillance state will continue to evolve, its reach expanding with every new data point collected and every new algorithm deployed, eroding the very freedoms it purports to protect. The path forward requires a conscious, deliberate, and sustained effort to design technology not for surveillance, but for liberation and autonomy.

The digital realm, by its very nature, transcends national borders. Data flows ceaselessly across continents, carried by invisible currents of information. This inherent borderlessness of the internet and the globalized nature of the technology industry present a formidable challenge to any nation seeking to unilaterally protect its citizens' privacy. What one country enacts in terms of data protection can be undermined by the laxer standards or different priorities of another. Consequently, addressing the pervasive issue of state and corporate surveillance in the digital age necessitates a robust framework of international cooperation. Without it, efforts to reform national surveillance states risk being piecemeal, easily circumvented, and ultimately insufficient.

The necessity for global collaboration on privacy standards stems from several interconnected factors. Firstly, cross-border data flows are fundamental to modern commerce, communication, and governance. Personal data is regularly transferred between individuals, businesses, and governments located in different jurisdictions. This creates a complex web where data processed in one country might be stored in another, analyzed by entities in a third, and potentially accessed by governments in multiple locations. If each nation operates under a vastly different set of privacy rules, it becomes incredibly difficult to ensure consistent protection for individuals whose data is in transit or stored abroad. For instance, a European citizen whose data is processed by a US-based company might find their privacy rights significantly diminished if US surveillance laws permit broader access to that data than the General Data Protection Regulation (GDPR) allows. This jurisdictional arbitrage effectively creates privacy loopholes that can be exploited by both states and private entities.

Secondly, the architecture of surveillance itself is increasingly globalized. Intelligence-sharing agreements between nations, such as those that have come under scrutiny following revelations about programs like PRISM, demonstrate a coordinated effort by states to collect and analyze data on a global scale. These arrangements often operate in a legal and technical grey area, with differing national laws on data access and retention creating complexities. Harmonizing legal frameworks for surveillance, therefore, becomes a critical objective for international cooperation. This could involve establishing common standards for when and how governments can request data from foreign entities, ensuring due process protections are afforded to individuals regardless of their nationality, and creating mechanisms for mutual legal assistance that are transparent and accountable. Without such harmonization, surveillance powers can be exported and imported, allowing states to circumvent their own domestic privacy protections by leveraging the capabilities and legal systems of allied nations.

Achieving international consensus on privacy standards is, however, fraught with challenges. Nations often have divergent approaches to privacy, shaped by their unique historical experiences, political systems, and economic interests. Some countries, like those in the European Union, have enshrined strong data protection rights in their fundamental laws, viewing privacy as an inalienable human right. Others, particularly those with a greater emphasis on national security or economic growth driven by data analytics, may adopt more permissive stances on data collection and surveillance. This ideological divide can make it difficult to find common ground on issues such as the definition of personal data, the scope of lawful data access, the necessity of independent judicial oversight for surveillance requests, and the extent of data minimization required.

The economic dimension also plays a significant role. Many leading technology companies are multinational corporations whose business models are heavily reliant on the collection and analysis of vast amounts of personal data. These companies may lobby their respective governments to resist international privacy regulations that could impact their operations or profitability. Conversely, countries that are home to these tech giants might be reluctant to impose stringent regulations that could disadvantage their national champions on the global stage. Striking a balance between fostering innovation and economic competitiveness on one hand, and protecting fundamental privacy rights on the other, is a delicate act of diplomacy.

Despite these obstacles, the imperative for international cooperation is undeniable. A coordinated global approach is essential to effectively protect digital rights in an increasingly interconnected world. One avenue for cooperation lies in the development of multilateral treaties and conventions that set baseline privacy standards for all signatory nations. These agreements could draw inspiration from existing frameworks like the GDPR, adapting and expanding upon its principles to create a global floor for data protection. Such treaties could address key areas such as: the lawful basis for data processing; requirements for data security and breach notification; mechanisms for cross-border data transfers, including adequacy decisions and standard contractual clauses; individual rights such as access, rectification, and erasure; and the establishment of independent supervisory authorities.

Furthermore, international bodies, such as the United Nations and regional organizations like the Council of Europe and ASEAN, can serve as crucial platforms for dialogue and standard-setting. These forums can facilitate the exchange of best practices, promote legal harmonization, and foster a shared understanding of the challenges posed by mass surveillance and data exploitation. The Council of Europe's Convention 108, for instance, provides a foundational set of principles for data protection that has been influential globally. Its modernization, Convention 108+, aims to bring it up to date with contemporary data protection challenges, underscoring the ongoing need for international legal development.

Another critical area for cooperation is the establishment of common principles for government access to data held by foreign entities. This involves negotiating bilateral and multilateral agreements on mutual legal assistance (MLA) that include robust safeguards for privacy and due process. Existing MLA frameworks, often rooted in older conventions, can be slow, opaque, and lacking in privacy protections. Modernizing these agreements to reflect the realities of digital data and to incorporate principles of necessity, proportionality, and judicial authorization for data requests is vital. Moreover, international cooperation can focus on limiting the extraterritorial reach of national surveillance laws, preventing states from unilaterally demanding data from companies located in other jurisdictions without due process or reciprocal agreements.

The development and adoption of privacy-enhancing technologies (PETs) can also be a shared goal for international collaboration. Governments and international organizations can foster research and development into PETs and promote their widespread adoption by both public and private sectors. This could involve shared funding for research initiatives, the creation of open standards for PETs, and incentives for companies to implement these technologies. By working together, nations can accelerate the availability and effectiveness of tools that protect privacy, from end-to-end encryption and differential privacy to federated learning and zero-knowledge proofs.

The role of civil society and non-governmental organizations (NGOs) in advocating for global privacy standards cannot be overstated. These organizations often bridge the gap between governments, industry, and citizens, raising awareness, conducting research, and lobbying for stronger protections. International collaboration among these groups can amplify their impact, allowing them to present a unified front in demanding better privacy safeguards from both states and corporations. Networks like Privacy International and Article 19 play a crucial role in monitoring surveillance practices worldwide and advocating for policy reforms.

Addressing the challenges of data localization mandates also requires international dialogue. While some countries implement data localization laws to assert control over data generated within their borders or to facilitate law enforcement access, these measures can also fragment the internet and hinder global data flows, potentially creating new privacy risks if data is less secure in certain domestic environments. Collaborative efforts are needed to find solutions that respect national sovereignty while minimizing the negative impacts on privacy and the free flow of information. This might involve establishing international standards for data security and governance that allow for data to be stored and processed across borders under mutually agreed-upon protections.

Moreover, international cooperation is essential for holding powerful entities accountable. When data breaches occur, or when surveillance practices violate fundamental rights, individuals often face significant hurdles in seeking redress, especially if the responsible parties are located in different jurisdictions. Establishing international mechanisms for dispute resolution and for enforcing privacy judgments across borders would be a significant step forward. This could involve mutual recognition of legal decisions and cooperative efforts to trace and recover illegally accessed data.

The pursuit of a global privacy standard is not about imposing a single, rigid model on all nations. Rather, it is about establishing a shared understanding of fundamental rights and developing common principles and mechanisms to uphold them in the digital age. It is about recognizing that privacy is a global public good, and its protection requires collective action. This approach acknowledges that while national laws and enforcement mechanisms will remain crucial, they are most effective when they are part of a broader, coordinated international effort. Such cooperation can create a virtuous cycle where stronger domestic protections encourage higher global standards, and where international agreements provide a stable and predictable environment for businesses and individuals alike, fostering trust in the digital ecosystem. Ultimately, a unified global stance on digital rights and privacy protections is not just a matter of policy; it is a prerequisite for maintaining individual autonomy, democratic freedoms, and the very integrity of the digital public square in an era defined by pervasive surveillance.

13: BALANCING SECURITY AND LIBERTY

The notion that a society must choose between the twin pillars of national security and individual liberty – a stark, often emotionally charged, dichotomy – is a persistent, yet ultimately misleading, framing of a complex relationship. This perceived trade-off, frequently invoked in political discourse and policy debates, suggests that enhancing one necessarily diminishes the other. It posits a zero-sum game where every expansion of surveillance powers or curtailment of civil liberties is a necessary sacrifice at the altar of safety, and conversely, any insistence on protecting privacy or freedom is an act of naive endangerment. However, a deeper examination, informed by history, political theory, and practical experience, reveals that this dichotomy is not an immutable law of nature but rather a constructed narrative that can, and should, be dismantled. Indeed, the most enduring and effective national security strategies are not those that suppress liberty, but those that foster it, building a foundation of trust and resilience that is inherently more robust.

The allure of the security-liberty false dichotomy lies in its simplicity. It offers a clear, albeit simplistic, rationale for policy decisions, allowing governments to justify intrusive measures by pointing to an existential threat. This framing can be particularly potent during times of crisis, when fear is palpable and the desire for protection is paramount. It allows for the expansion of state power, often through opaque processes, under the guise of protecting the populace. However, this approach is fundamentally flawed because it misidentifies the source of true security. Genuine national security is not merely the absence of external threats; it is also the internal strength and stability of a society, which is directly undermined when fundamental rights are eroded. When citizens feel constantly monitored, when their communications are routinely scrutinized, or when their right to dissent is curtailed, a subtle but significant erosion of social cohesion occurs. This breeds suspicion, alienates segments of the population, and can inadvertently create fertile ground for the very threats that security measures are intended to combat.

History offers compelling evidence that societies that prioritize and protect civil liberties often exhibit greater long-term security and stability. Consider the foundational principles of democratic states, which, despite their inherent freedoms, have historically proven more resilient and adaptable than authoritarian regimes. The ability of citizens to express grievances, to criticize government policies, and to organize politically, while seemingly antithetical to absolute control, actually serves as a crucial safety valve. It allows for the peaceful resolution of societal tensions and provides early warnings of potential instability. When dissent is suppressed, grievances fester, and the likelihood of more extreme reactions increases. Furthermore, a society that respects individual rights is more likely to foster an environment where innovation, critical thinking, and creativity can flourish – qualities that are indispensable for adapting to evolving threats, whether they be technological, economic, or geopolitical.

The idea that strong privacy protections inherently weaken

national security is a particularly prevalent, and often unsubstantiated, claim in the context of digital surveillance. Proponents of broad surveillance powers often argue that encryption and privacy-enhancing technologies are shields for terrorists and criminals, making it impossible for intelligence agencies to gather the information necessary to thwart attacks. While it is true that encrypted communications can pose challenges, the assumption that sacrificing mass surveillance for access to encrypted data is a necessary trade-off is a flawed premise. Firstly, the effectiveness of mass surveillance in preventing specific attacks is often exaggerated, and its efficacy can be undermined by the sheer volume of data collected, leading to signal-to-noise problems where genuine threats are buried amidst a sea of irrelevant information. Intelligence agencies can become overwhelmed, their resources diverted to sifting through vast amounts of data rather than focusing on targeted, actionable intelligence.

Secondly, and more importantly, a society that trusts its government to respect its privacy is more likely to cooperate with law enforcement and intelligence agencies. When citizens believe that their communications are not being indiscriminately monitored, they are more inclined to report suspicious activity or provide information that could be vital to national security. Conversely, if the public perceives surveillance as pervasive and unchecked, and if they believe their private lives are subject to arbitrary intrusion, they are less likely to trust the very institutions responsible for their protection. This erosion of trust can have a chilling effect on public engagement and cooperation, ultimately hindering, rather than helping, the intelligence gathering process. The chilling effect extends beyond explicit cooperation; it can also stifle open discourse and political activism, the very cornerstones of a healthy democracy, making it harder to identify and counter threats that originate from within society itself.

Moreover, the argument that liberties must be curtailed to achieve security often overlooks the significant financial and societal costs associated with widespread surveillance. The infrastructure required for mass data collection, storage, and analysis is immense, diverting substantial resources that could be allocated to other critical areas, such as education, healthcare, or targeted intelligence operations. Furthermore, the erosion of privacy can lead to a host of social ills, including increased anxiety, self-censorship, and a general decline in civic participation. These are not abstract concepts; they represent tangible detriments to the well-being and resilience of a nation, and therefore, to its underlying security.

Reframing the debate beyond the false dichotomy requires a shift in perspective towards synergistic approaches. Instead of viewing security and liberty as competing interests, we should recognize them as mutually reinforcing. Robust legal frameworks that govern surveillance, ensuring that it is targeted, necessary, and proportionate, not only protect individual rights but also lend legitimacy and credibility to the actions of security agencies. When surveillance is conducted under strict oversight, with clear legal boundaries and accountability mechanisms, it is less likely to be perceived as arbitrary or oppressive by the public. This fosters a greater degree of public confidence, which is a vital component of national resilience.

Consider the historical context of constitutional protections. The development of robust legal systems, including protections against unreasonable search and seizure and guarantees of due process, was not seen as a concession to criminals but as a necessary safeguard for all citizens, thereby strengthening the legitimacy of the state and its law enforcement apparatus. The ability of individuals to challenge government actions through the courts, even when those actions are taken in the name of security, ensures that power is not unchecked and that the rule of law is upheld. This adherence to the rule of law, even in difficult times, is a testament to a society's strength and a key determinant of its long-term security.

Furthermore, investing in human intelligence, linguistic expertise, cultural understanding, and sophisticated analytical capabilities can be far more effective in combating complex threats than relying solely on mass data collection. These human-centric approaches require skilled professionals who can build relationships, gather nuanced information, and interpret complex social and political landscapes. When intelligence agencies are perceived as overreaching and intrusive, it becomes more difficult to recruit and retain the kind of talented individuals needed for these critical functions. A reputation for respecting civil liberties can, conversely, attract individuals who are committed to serving their country without compromising their values.

The notion that we must accept a lesser degree of liberty in exchange for a perceived increase in security is a dangerous and self-defeating proposition. It suggests a fundamental misunderstanding of what constitutes genuine security. True security is not built on a foundation of fear and suspicion, but on a bedrock of trust, shared values, and the robust protection of fundamental rights. Societies that embrace this understanding, that actively work to integrate security measures with a deep and abiding respect for individual liberty, are not only more just but also, paradoxically, more secure. They are better equipped to navigate the complexities of the modern world, to adapt to new challenges, and to foster the internal cohesion and resilience that are the ultimate bulwarks against any threat. The path forward lies not in choosing between security and liberty, but in recognizing that they are, in fact, two sides of the same coin, each essential for the flourishing and enduring strength of any free society. The debate should not be about *if* we can have both, but *how* we can best achieve them in concert.

This synergy between security and liberty is not merely theoretical; it is observable in the functioning of well-governed societies. For instance, independent judiciaries play a crucial role in balancing these interests. When surveillance requests are subjected to rigorous judicial review, ensuring they meet strict legal standards of necessity and proportionality, this process not only safeguards individual privacy but also lends legal and public legitimacy to authorized surveillance operations. This contrasts sharply with systems where executive branches can unilaterally authorize broad surveillance, which often leads to public distrust and accusations of overreach. The transparency and accountability inherent in a system with judicial oversight strengthen both the rule of law and, by extension, societal stability.

Moreover, the active participation of civil society in scrutinizing government actions related to security is a vital component of this balanced approach. Non-governmental organizations, investigative journalists, and academics often act as crucial watchdogs, bringing to light potential abuses of power and advocating for reforms. When these actors can operate freely and their concerns are addressed through open dialogue and legislative action, it reinforces public trust in democratic institutions. The ability to question and challenge security measures, rather than being silenced by them, is a hallmark of a resilient society. Efforts to suppress such scrutiny, often under the guise of national security, paradoxically weaken the very society they claim to protect by creating an environment where wrongdoing can flourish unchecked.

Technological advancements also present opportunities

to enhance both security and liberty simultaneously. The development and implementation of privacy-enhancing technologies (PETs) offer a case in point. Secure communication platforms, end-to-end encryption, and anonymization tools can empower individuals to protect their privacy in the digital sphere. While some argue these technologies hinder intelligence gathering, they can also enable legitimate communication and association that might otherwise be chilled by the fear of surveillance. Furthermore, as these technologies become more sophisticated, they can be designed with features that allow for lawful access under strictly defined circumstances, thereby offering a pathway to reconcile competing interests without resorting to blanket surveillance. The challenge lies in fostering innovation in PETs while developing clear, internationally recognized legal frameworks for data access that respect fundamental rights.

The emphasis on intelligence-led policing, as opposed to mass surveillance, offers another model for achieving both security and liberty. By focusing resources on specific threats and individuals based on credible intelligence, law enforcement agencies can be more effective and less intrusive. This approach requires investment in human capital, analytical capabilities, and international cooperation to gather and share actionable intelligence. It stands in stark contrast to dragnet surveillance, which collects data indiscriminately, potentially infringing on the privacy of countless innocent citizens while often failing to identify genuine threats amidst the noise. The former approach builds trust and focuses resources efficiently; the latter breeds suspicion and can overwhelm agencies with data.

The narrative that pits security against liberty is often used to justify expanding state power in ways that are difficult to roll back once a crisis has passed. This phenomenon, sometimes referred to as "emergency creep," illustrates how powers granted in times of crisis can become normalized and institutionalized, even when the original justification has long since faded. By challenging the false dichotomy, we can advocate for temporary, proportionate measures during emergencies, coupled with robust sunset clauses and independent oversight to ensure that civil liberties are not permanently eroded. This requires a vigilant citizenry and political leadership committed to upholding constitutional principles, even when it is politically expedient to do otherwise.

Ultimately, the most secure societies are those that are most free. Freedom of expression, freedom of association, and the right to privacy are not impediments to security; they are its very foundations. They enable societies to self-correct, to adapt, and to maintain the trust and cooperation necessary to face down threats. The persistent insistence on a trade-off between these fundamental values serves the interests of those who seek to expand state power, but it is a disservice to the principles of democratic governance and the pursuit of genuine, sustainable security. Embracing the understanding that liberty and security are mutually reinforcing, rather than opposing forces, is essential for building a future where both can thrive. This requires a conscious and continuous effort to resist simplistic narratives and to champion policies that recognize the intrinsic value of individual freedom as the bedrock upon which collective safety is built. The strength of a nation is not measured solely by its capacity for coercion, but by its commitment to justice and the rights of its people.

The assertion that national security is inextricably linked to the concept of pervasive, indiscriminate data collection – often termed mass surveillance – is a pervasive narrative that warrants critical re-examination. This perspective suggests that the sheer volume of intercepted communications, digital footprints, and personal information is a necessary prerequisite for identifying and neutralizing nascent threats. However, a closer inspection of intelligence gathering methodologies, historical precedents, and the practical realities of information analysis reveals a compelling counter-argument: effective intelligence can, and indeed must, be achieved through precision-focused strategies that eschew the broad sweep of mass surveillance, thereby safeguarding fundamental civil liberties. This approach prioritizes targeted operations, the cultivation of human intelligence, and sophisticated analytical techniques that aim for depth and accuracy rather than sheer quantity.

The efficacy of mass surveillance is often predicated on the assumption that within a vast ocean of data, the few crucial signals of potential threats will inevitably surface. While this "signal in the noise" paradigm holds a theoretical appeal, its practical application is fraught with significant challenges. The sheer scale of data collected by contemporary surveillance programs can easily overwhelm the analytical capabilities of intelligence agencies. Analysts are often tasked with sifting through terabytes of information, much of which is irrelevant to national security concerns, leading to what is known as the "information overload" problem. This dilution of focus can paradoxically make it *harder* to identify genuine threats, as valuable resources are diverted to processing extraneous data. The signal, rather than being amplified, risks being buried under an avalanche of noise. This can lead to missed opportunities, delayed responses, and an inefficient allocation of critical intelligence assets. Consider, for instance, the extensive data mining operations that might collect communications from millions of individuals who have no connection to any threat actor. The cost, both in terms of financial resources and human analytical effort, is substantial, yet the return on investment in terms of actionable intelligence can be remarkably low.

In contrast, targeted surveillance, conducted under strict legal authorization and with a clear, articulable suspicion, offers a more efficient and rights-respecting alternative. This approach mandates that intelligence agencies demonstrate probable cause or a reasonable suspicion that an individual or group is involved in or planning illicit activities before initiating surveillance. This legal standard, enshrined in many democratic legal systems, acts as a crucial bulwark against arbitrary intrusion into the private lives of citizens. When surveillance is narrowly tailored to specific individuals or groups based on credible intelligence, the data collected is inherently more relevant and actionable. Analysts can then focus their efforts on deciphering meaningful patterns, understanding motives, and identifying specific vulnerabilities or plans, rather than engaging in speculative data dredging. This precision not only enhances the effectiveness of intelligence operations but also significantly reduces the risk of infringing upon the privacy of innocent individuals, thereby fostering greater public trust.

The cultivation of human intelligence (HUMINT) remains a cornerstone of effective intelligence gathering, and its importance is amplified when moving away from mass surveillance. HUMINT involves gathering information through human sources – informants, spies, debriefings of individuals, and cultivated relationships within targeted communities. Unlike signals intelligence (SIGINT) or imagery intelligence (IMINT), which rely on technical means, HUMINT provides insights into intentions, motivations, plans, and organizational structures that technical means often cannot capture. A well-placed human source can offer context, nuance, and foresight that raw data alone cannot provide. For example, understanding the internal dynamics of a terrorist cell, the political motivations behind a cyber-attack, or the social grievances that might fuel radicalization often requires direct human interaction and the building of trust. Agencies that invest heavily in recruiting, training, and managing human sources, particularly those with linguistic and cultural expertise, are better positioned to acquire deep, actionable intelligence. This also necessitates fostering environments where individuals feel safe and willing to cooperate with authorities, a condition that is undermined by widespread suspicion of government monitoring.

Furthermore, advanced analytical techniques play a pivotal role in extracting meaningful intelligence from targeted data sets. Rather than relying on automated keyword searches or broad data correlation across massive datasets, sophisticated analytical methods can focus on identifying behavioral patterns, social networks, financial flows, and communication anomalies that are indicative of illicit activity. Techniques such as link analysis, social network analysis, and predictive modeling, when applied to precisely collected intelligence, can illuminate complex networks of actors and uncover intricate plots. For instance, by analyzing the communication patterns and financial transactions of a small group of suspected individuals, analysts can map their associations, identify key facilitators, and anticipate their next moves. This contrasts sharply with mass surveillance, where the sheer volume of data can obscure these subtle but crucial indicators. The effectiveness of these analytical tools is directly proportional to the quality and relevance of the data they process. High-quality, targeted data allows for more accurate and insightful analysis, leading to a higher probability of preempting threats.

The legal and ethical frameworks surrounding intelligence gathering are also critical in shaping the effectiveness of these methods. Targeted surveillance, by its very nature, requires adherence to robust legal standards and judicial oversight. This means that before surveillance can commence, intelligence agencies must typically obtain warrants based on probable cause, demonstrating a specific nexus between the surveillance target and a legitimate security concern. This process, while adding procedural steps, serves to ensure that intelligence operations are grounded in evidence and are proportionate to the threat. The oversight provided by independent judicial bodies acts as a crucial check on executive power, preventing overreach and safeguarding individual rights. When surveillance is conducted within such a framework, it is more likely to be perceived as legitimate by the public, thereby encouraging cooperation and reducing the "chilling effect" on free expression and association. Conversely, mass surveillance, often authorized through less stringent legal mechanisms, erodes public trust and can lead to a climate of fear and self-censorship, which is detrimental to democratic society and, paradoxically, to long-term security.

The economic argument also favors targeted intelligence over mass surveillance. The infrastructure required for the collection, storage, and analysis of petabytes of data is astronomically expensive. These vast sums could be redirected towards more effective intelligence gathering methods, such as investing in skilled human analysts, developing advanced technological tools for specific threats, improving linguistic capabilities, and fostering international cooperation on intelligence sharing. Prioritizing targeted operations allows for a more efficient and strategic allocation of resources, ensuring that the limited budgets of intelligence agencies are used to maximum effect. Focusing on precision means fewer resources are spent on data that will never yield actionable intelligence, and more can be invested in areas that are proven to deliver results.

Historical successes in intelligence operations often point to the efficacy of targeted approaches. Many of the most significant disruptions of terrorist plots, criminal enterprises, and espionage operations have been the result of meticulous investigation and intelligence gathering focused on specific individuals and groups. These operations typically involve a combination of human sources, technical surveillance of specific targets, and sophisticated analysis, all undertaken with clear legal authorization. For example, the dismantling of major drug trafficking networks or the foiling of significant terrorist attacks have rarely been attributed to simply sifting through indiscriminately collected data. Instead, they are typically the product of dedicated intelligence work, often spanning years, involving deep infiltration, careful observation, and the piecing together of fragmented information into a coherent picture of criminal intent and capability.

The argument for precision-based intelligence gathering is not merely a theoretical construct; it is a practical necessity for democracies that seek to uphold both security and liberty. The erosion of privacy through indiscriminate surveillance can foster a climate of distrust, alienate communities, and ultimately undermine the social cohesion upon which national security depends. When citizens feel they are under constant, unwarranted scrutiny, their willingness to engage in civic life, to voice dissent, or to cooperate with authorities can be significantly diminished. This creates a less resilient society, more vulnerable to the very threats that surveillance is intended to combat. Conversely, intelligence agencies that operate within clear legal boundaries, demonstrating transparency and accountability, can build the trust necessary for effective partnerships with the public.

Moreover, the focus on targeted intelligence gathering encourages innovation in analytical techniques and human intelligence methodologies. Instead of relying on the brute force of data collection, agencies are incentivized to develop more sophisticated methods for identifying threats, understanding their operational methods, and anticipating their actions. This can lead to breakthroughs in areas such as behavioral analysis, pattern recognition, and the exploitation of niche intelligence sources. The pursuit of precision fosters a more adaptive and intelligent approach to national security, one that is better equipped to confront the evolving nature of threats in the 21st century. This includes developing expertise in areas like cybersecurity, where understanding the modus operandi of sophisticated state and non-state actors requires detailed, targeted analysis rather than broad data sweeps.

The notion that mass surveillance is an indispensable tool for security also overlooks the inherent limitations of relying solely on technical data. Technical intercepts can provide information about *what* is being said, but they often struggle to convey the nuances of *why* it is being said. Human intelligence, on the other hand, can provide invaluable context regarding motivations, intentions, and the cultural or political factors that drive behavior. A well-cultivated informant can offer insights into the mindset of an adversary that no amount of intercepted data can replicate. Similarly, the ability to conduct thorough background investigations, psychological profiling, and expert analysis of adversary capabilities are critical components of effective intelligence that are best served by targeted rather than indiscriminate data collection.

Furthermore, the argument for precision intelligence gathering aligns with the principles of proportionality and necessity, which are fundamental to international human rights law. Surveillance measures should be necessary to achieve a legitimate aim and proportionate to that aim. Mass surveillance, by its very nature, is difficult to justify as either necessary or proportionate, as it sweeps up vast amounts of data from individuals who pose no threat. Targeted surveillance, when conducted with clear legal authorization and oversight, is far more likely to meet these criteria, ensuring that intelligence operations are both effective and rights-respecting. This adherence to legal and ethical standards is not a weakness but a source of strength, lending legitimacy and public confidence to the work of intelligence agencies.

In conclusion, the transition from mass surveillance to precision-based intelligence gathering represents not a compromise in security, but an enhancement of its effectiveness while simultaneously upholding vital civil liberties. By prioritizing targeted operations, investing in human intelligence, employing sophisticated analytical techniques, and adhering to robust legal frameworks and oversight, intelligence agencies can achieve their mission more efficiently and ethically. This approach fosters public trust, ensures the judicious use of resources, and ultimately builds a more resilient and secure society. The narrative that security is only achievable through pervasive monitoring is a false dichotomy that undermines both our safety and our freedoms. The true path to enduring security lies in the intelligent, targeted, and rights-respecting pursuit of actionable intelligence.

Privacy, far from being a mere personal preference or a cloak for illicit activities, stands as a foundational pillar upon which the edifice of a robust democracy is built. It is not simply the right to keep secrets, but rather the essential precondition for individual autonomy, the unfettered development of thought, and the courage to engage in the public discourse that animates a free society. Without the sanctuary of privacy, the very mechanisms that allow citizens to question authority, to organize for collective action, and to express dissent without fear of retribution are critically compromised. When individuals know, or even suspect, that their every communication, their every online interaction, their every association is subject to observation and potential scrutiny by the state, a chilling effect descends upon civic life. This pervasive awareness of being watched can stifle the spontaneous exchange of ideas, discourage participation in controversial discussions, and foster a climate of self-censorship that is antithetical to democratic ideals. The freedom to explore unconventional ideas, to criticize government policies, or to engage in political activism is fundamentally dependent on the assurance that such activities will not be cataloged, analyzed, and potentially used against one in the future.

The erosion of privacy directly impacts individual autonomy, which is intrinsically linked to the capacity for self-governance, a core tenet of democracy. Autonomy requires the space to make choices free from external coercion or undue influence. When personal information is collected and weaponized, or even just held in reserve by powerful entities, it can be used to subtly or overtly shape behavior, to manipulate public opinion, and to predetermine outcomes. Imagine a society where a government, armed with intimate knowledge of an individual's political leanings, their financial vulnerabilities, or their personal associations, could selectively target them with disinformation campaigns, deny them opportunities, or even fabricate evidence to discredit them. Such a scenario transforms citizens from active participants in a democratic process into passive subjects vulnerable to constant, invisible manipulation. The ability to form one's own opinions, to develop a personal ideology, and to express oneself authentically is a vital component of what it means to be a free citizen, and privacy is the essential enabler of this process. Without it, the marketplace of ideas, a crucial democratic institution, becomes distorted and polluted, rendering genuine deliberation and informed consent nearly impossible.

Furthermore, privacy is inextricably linked to freedom of thought and expression. The capacity to explore ideas, to test hypotheses, and to develop nuanced perspectives often happens in private, through personal reflection, confidential conversations, or the consumption of information without the immediate pressure of public accountability. If individuals believe that their private explorations of ideas – even those that might be considered unpopular or on the fringes of mainstream thought – are being monitored, they may be disinclined to engage in such exploration at all. This can lead to a homogenization of thought, a stifling of intellectual diversity, and a narrowing of the spectrum of acceptable discourse. Democracy thrives on the robust exchange of a wide range of viewpoints, including those that challenge the status quo. When the private space for intellectual development is compromised, the public expression of diverse ideas is inevitably curtailed. The fear of being misinterpreted, or of having nascent or undeveloped thoughts used out of context, can prevent individuals from contributing their unique insights to public debate.

The ability to engage in political and social life without undue fear of reprisal is another critical dimension that privacy safeguards. Democratic societies rely on the active participation of their citizens in various forms of collective action, from peaceful protests and community organizing to forming political associations and lobbying elected officials. These activities often require open communication, the sharing of sensitive information among like-minded individuals, and the building of trust within groups. When the state engages in mass surveillance or pervasive data collection, it can have a profound chilling effect on these vital democratic processes. Individuals might hesitate to join a particular organization, contribute to a campaign, or even express solidarity with a group perceived as controversial, for fear that their involvement will be recorded and potentially used against them, either by the government directly or by other actors who gain access to this data. This fear can disproportionately affect marginalized communities or those who hold minority viewpoints, making it more difficult for them to organize and advocate for their rights. The consequence is a less inclusive and less representative democracy, where the voices of those who are most vulnerable to surveillance are effectively silenced.

The connection between privacy and the weakening of democratic institutions is not theoretical; it is a tangible threat that has manifested in various historical and contemporary contexts. When governments possess vast amounts of personal data on their citizens, they gain an unprecedented capacity for social control. This control can range from subtle nudges and targeted propaganda to more overt forms of intimidation and repression. The aggregation of data allows for the creation of detailed profiles of individuals, which can then be used to predict behavior, identify potential dissenters, and preemptively neutralize perceived threats. This can lead to a society where the state's power is concentrated in the hands of a few, operating with significant opacity and unchecked authority, while the citizenry lives under a constant state of surveillance, their freedoms gradually circumscribed in the name of security or stability. Such a dynamic erodes the very foundations of democratic governance, which are predicated on accountability, transparency, and the protection of individual liberties.

The potential for manipulation and control that stems from the erosion of privacy is particularly alarming in the digital age. The vast datasets collected through online activities – browsing history, social media interactions, geolocation data, purchase records – provide an incredibly detailed and granular picture of individuals' lives, preferences, and vulnerabilities. This information can be exploited for political purposes, such as micro-targeting voters with tailored messages designed to elicit specific emotional responses, often playing on fears or biases. It can also be used to create deepfakes or to spread disinformation that is so personalized that it bypasses rational skepticism. When citizens cannot be sure whether the information they are receiving or the interactions they are having are genuine or orchestrated by unseen forces leveraging their personal data, the integrity of the public sphere is fundamentally undermined. This creates fertile ground for authoritarianism, where the ability to control the flow of information and to manipulate public perception becomes a primary tool of governance.

Protecting privacy, therefore, is not an abstract ideal but a concrete necessity for maintaining a free and open society. It is about safeguarding the conditions that allow for meaningful civic participation. When citizens are confident that their private lives are respected and their communications are not being indiscriminately monitored, they are more likely to engage in public affairs, to voice their opinions, to hold their leaders accountable, and to participate in the democratic process with conviction. This active and informed citizenry is the lifeblood of any healthy democracy. Conversely, a populace living under the shadow of surveillance is a populace disengaged, fearful, and ultimately less capable of self-governance. The ongoing struggle to balance national security imperatives with the protection of individual liberties must therefore recognize the indispensable role of privacy. Without a strong commitment to privacy, the very democracy that national security is meant to protect is at risk of being hollowed out from within, transformed into a system that prioritizes control over freedom, and conformity over genuine democratic engagement. The historical record and contemporary evidence consistently demonstrate that societies that erode privacy, even with the stated intention of enhancing security, invariably end up diminishing both their freedom and, in the long run, their true security. The capacity for citizens to trust their government, and to trust each other, is foundational, and that trust is built, in no small part, upon the mutual understanding and respect for personal privacy.

The very mandate of intelligence agencies, to safeguard national security and protect citizens from a spectrum of threats, inherently places them at a critical juncture where the exercise of immense power must be tempered by profound ethical considerations. The collection, analysis, and dissemination of intelligence, particularly in the digital age, involves the acquisition of vast quantities of personal data, often without the direct, informed consent of the individuals involved. This necessitates a deep examination of the moral responsibilities that accompany such capabilities. The potential for misuse, whether intentional or through negligence, is a constant and significant concern. Data gathered for legitimate national security purposes could, in theory or in practice, be diverted for political advantage, personal vendettas, or to suppress legitimate dissent. This inherent risk demands that intelligence agencies operate within robust ethical frameworks that guide every decision, from the initiation of surveillance to the ultimate disposition of the collected information.

At the heart of these ethical considerations lie fundamental principles that must govern the actions of any agency entrusted with such sensitive powers. The principle of necessity dictates that intelligence collection and surveillance activities should only be undertaken when strictly required to address a genuine and significant threat. This means that speculative or exploratory data gathering, undertaken without a clear and demonstrable need, must be avoided. The question to be asked is not merely whether data *can* be collected, but whether it *must* be collected to achieve a legitimate and vital security objective. This principle acts as a crucial bulwark against the expansion of surveillance into areas that do not pose a demonstrable risk, thereby preserving a zone of privacy for law-abiding citizens. It requires a rigorous and ongoing assessment of threats and the effectiveness of intelligence collection methods in mitigating those threats, ensuring that the intrusion into privacy is a last resort, not a default setting. The temptation to cast a wide net, simply because the technology allows it, must be resisted in favor of a more targeted and judicious approach.

Closely intertwined with necessity is the principle of proportionality. This principle demands that the intrusiveness of any intelligence operation must be commensurate with the gravity of the threat being addressed. In simpler terms, the response must fit the crime, or rather, the threat. Gathering highly sensitive personal information on an entire population to address a minor, isolated security concern would be a clear violation of proportionality. Conversely, a highly intrusive measure might be justifiable if it were the only viable means to prevent a catastrophic attack. This requires a careful balancing act, weighing the potential harm to individual privacy and civil liberties against the potential harm to national security. It necessitates a continuous evaluation of whether less intrusive means could achieve the same security objectives. The adoption of advanced analytical techniques that can extract meaningful intelligence from less sensitive data, or the use of targeted surveillance methods rather than mass data collection, are examples of how proportionality can be upheld. Without this principle, the very notion of proportionate response would be lost, leading to an escalating and unchecked intrusion into private lives under the guise of security.

Transparency, while inherently challenging for intelligence agencies operating in secrecy, remains a critical ethical imperative. While the operational details of specific missions may require confidentiality, the overarching legal frameworks, oversight mechanisms, and the general scope of intelligence activities should be, as far as possible, transparent to the public and their elected representatives. This does not imply revealing classified information that could compromise operations or endanger individuals, but rather fostering an understanding of *what* is being done, *why* it is being done, and *how* it is being overseen. Transparency builds public trust, which is essential for the legitimacy of intelligence operations in a democratic society. When the public understands the necessity and the limits of surveillance, they are more likely to support the legitimate functions of these agencies. Conversely, a complete lack of transparency breeds suspicion, fear, and resentment, undermining the very security the agencies are meant to protect. Establishing clear lines of accountability and robust oversight bodies, composed of individuals with diverse expertise and a commitment to public service, is a vital component of achieving this necessary transparency. These bodies must have the power to scrutinize operations, demand explanations, and recommend or enforce corrective actions.

Beyond these overarching principles, fostering a robust culture of ethical conduct within intelligence agencies is paramount. This is not merely about compliance with regulations, but about embedding ethical reasoning into the daily work and decision-making processes of every individual. This requires comprehensive training that goes beyond legal requirements, delving into the philosophical underpinnings of privacy, liberty, and the social contract. It involves creating an environment where ethical dilemmas can be openly discussed, where whistleblowers are protected, and where ethical lapses are addressed with fairness and seriousness. Leaders within these agencies bear a significant responsibility in setting the tone and demonstrating a commitment to ethical behavior. When ethical conduct is valued and rewarded, and when transgressions are met with appropriate consequences, it reinforces the importance of these values throughout the organization. This proactive approach to ethical development helps to prevent misconduct before it occurs, ensuring that the pursuit of security never eclipses the fundamental commitment to human rights and democratic principles.

The evolution of technology presents a continuous ethical challenge for intelligence agencies. The advent of artificial intelligence, big data analytics, and increasingly sophisticated surveillance tools has amplified both the capabilities and the potential for ethical missteps. For instance, the use of AI in predictive policing or threat assessment, while promising, carries the risk of embedding existing societal biases into algorithms, leading to disproportionate surveillance or suspicion of certain communities. The ethical imperative here is to ensure that these technologies are developed and deployed with a keen awareness of these risks, incorporating safeguards and human oversight to mitigate bias and ensure fairness. Furthermore, the interconnectedness of data means that a breach in one system could have far-reaching consequences for individuals whose information is stored across multiple databases. This amplifies the need for stringent data security measures and clear protocols for data handling and retention.

The question of accountability is central to the ethical landscape of intelligence agencies. In democratic societies, all state power must ultimately be accountable to the people, either directly through elections or indirectly through their representatives. For intelligence agencies, which often operate with a degree of necessary secrecy, establishing effective lines of accountability is particularly complex. This accountability must extend beyond internal disciplinary measures and encompass independent review by legislative bodies, judicial oversight, and, where appropriate, public reporting. Mechanisms such as parliamentary committees with the authority to question intelligence officials, judicial review of surveillance warrants, and independent ombudsman offices can all play a crucial role in ensuring that intelligence agencies remain within their legal and ethical boundaries. The absence of robust accountability mechanisms can create a vacuum where power can be unchecked, leading to a drift away from democratic norms and a disregard for individual liberties.

The ethical considerations also extend to the international dimension of intelligence work. When intelligence agencies collect data on individuals in foreign countries, they must grapple with the differing legal and ethical standards that may apply. Respecting the sovereignty of other nations and adhering to international human rights norms is crucial. The potential for intelligence gathering to be used for economic espionage or to undermine foreign political processes raises significant ethical questions about the conduct of international relations. Ensuring that intelligence activities are conducted in a manner that is consistent with established international law and ethical principles is vital for maintaining global stability and fostering trust between nations.

Furthermore, the concept of "innocent until proven guilty" is a cornerstone of many legal systems and has direct ethical implications for intelligence work. While intelligence agencies may identify individuals as persons of interest based on patterns or associations, the ethical imperative is to avoid treating these individuals as criminals before any wrongdoing has been established. This means that surveillance should be proportionate and targeted, and that the information gathered should not be used to prejudicively label or penalize individuals without due process. The risk of "pre-crime" thinking, where individuals are targeted based on the potential for future actions rather than present evidence, is a significant ethical minefield that intelligence agencies must navigate with extreme caution.

The ethical frameworks governing intelligence agencies must also be dynamic and adaptable, evolving in response to new technologies, emerging threats, and changing societal expectations. What might have been considered ethically acceptable in the past may no longer be so in the current context. This requires a continuous process of reflection, review, and revision of policies and practices. It also necessitates an open dialogue between intelligence agencies, policymakers, legal experts, civil society organizations, and the public to ensure that the balance between security and liberty is constantly being re-evaluated and recalibrated in a way that upholds democratic values. The creation of independent ethics review boards, comprising individuals with diverse backgrounds and expertise, can provide a vital mechanism for this ongoing ethical assessment. These boards could offer non-binding advice, conduct audits of ethical practices, and raise awareness of potential ethical challenges.

The very collection of metadata – information about communications rather than their content – presents a complex ethical landscape. While seemingly innocuous, the aggregation of metadata can reveal intimate details about an individual's life, including their social networks, patterns of movement, and even their political or religious affiliations. The ethical debate here centers on whether such extensive collection, even without accessing content, constitutes an unwarranted intrusion into privacy and whether it can be used to infer sensitive personal information. Proportionality and necessity become particularly important in this context, requiring agencies to demonstrate a clear and compelling need for such data and to ensure that its collection and use are strictly limited. The potential for "function creep," where data collected for one purpose is subsequently used for another, unrelated purpose, is a constant ethical concern that must be actively managed.

Ultimately, the ethical considerations for intelligence agencies are not merely abstract philosophical debates;

they have tangible consequences for individual lives and the health of democratic societies. When intelligence agencies operate ethically, respecting the principles of necessity, proportionality, transparency, and accountability, they can effectively contribute to national security while simultaneously upholding the fundamental rights and liberties of citizens. Conversely, when these ethical considerations are neglected, the pursuit of security can inadvertently lead to the erosion of the very freedoms that these agencies are meant to protect, creating a society that is both less secure and less free. The ongoing commitment to developing and adhering to rigorous ethical standards is therefore not an optional add-on, but an indispensable component of responsible intelligence work in a democratic state. It is a continuous process of vigilance, reflection, and a steadfast dedication to the principles that underpin a just and free society.

The management of sensitive information also brings forth ethical obligations regarding data retention and destruction. Intelligence agencies often hold vast amounts of data, some of which may pertain to individuals who are not and have never been suspected of any wrongdoing. Ethical considerations demand that data be retained only for as long as it is necessary for its intended purpose. Once that purpose has been fulfilled, or the legal retention period has expired, the data should be securely and irrevocably destroyed. This principle helps to mitigate the risks associated with data breaches and prevents the creation of permanent, potentially compromising, dossiers on individuals. Establishing clear policies and audited procedures for data deletion is an essential ethical safeguard, ensuring that individuals' past associations or communications do not remain indefinitely in the hands of state agencies. The absence of such protocols can lead to a situation where individuals are perpetually under a form of digital scrutiny, even long after any legitimate security interest has passed.

Moreover, the ethical implications of intelligence sharing, both domestically and internationally, are profound. When intelligence is shared between agencies, there is a risk that it may be used for purposes beyond its original intent or that it may be shared with entities that do not adhere to the same ethical standards. Establishing clear guidelines and robust safeguards for intelligence sharing is therefore crucial. This includes ensuring that the receiving entity has a legitimate need for the information and that they have the capacity to protect it and use it ethically. Agreements on data handling, limitations on further dissemination, and mechanisms for oversight are essential components of ethical intelligence sharing. Failing to implement these safeguards can lead to the unintended consequences of even well-intentioned intelligence operations, potentially undermining privacy and civil liberties across borders or between different branches of government.

The internal culture of an intelligence agency plays a pivotal role in shaping its ethical behavior. Leaders must actively cultivate an environment where ethical conduct is not just expected but is intrinsically valued. This involves fostering open communication channels where individuals feel empowered to raise concerns about potential ethical breaches without fear of reprisal. It requires leadership that consistently models ethical decision-making, demonstrating a commitment to principles even when faced with difficult choices or external pressures. Training programs should not only focus on legal compliance but also on developing critical thinking skills and a deep understanding of the ethical dimensions of intelligence work. Encouraging a sense of public service and a recognition of the profound responsibility that comes with wielding state power can help to instill a strong ethical compass within the organization. Without this ingrained ethical culture, even the most well-intentioned regulations can become mere bureaucratic hurdles, rather than guiding principles.

The notion of "chilling effect," previously discussed in the context of privacy, is also an ethical consideration for intelligence agencies. By engaging in widespread surveillance, even if legally permissible, agencies can inadvertently stifle legitimate political discourse, activism, and freedom of expression. The ethical responsibility lies in recognizing this potential harm and implementing measures to minimize it. This might involve clearly articulating the specific threats being addressed by surveillance programs, ensuring that data collection is narrowly tailored, and providing assurances that legitimate dissent or investigative journalism will not be targeted. The ethical imperative is to ensure that the pursuit of security does not inadvertently undermine the very democratic freedoms that make society worth protecting.

Finally, the ethical considerations surrounding intelligence agencies are not static; they are in constant dialogue with technological advancements, evolving societal norms, and the ever-changing threat landscape. This necessitates a commitment to continuous learning, adaptation, and a willingness to critically examine existing practices. A proactive approach to ethics, rather than a reactive one, is essential. This involves anticipating potential ethical challenges and developing frameworks to address them before they manifest as misconduct or public distrust. It requires a recognition that the power vested in intelligence agencies is a sacred trust, and that the ethical stewardship of this power is as critical to national security as any operational success. The ultimate goal is to ensure that the pursuit of safety and security remains firmly tethered to the bedrock principles of justice, liberty, and human dignity.

The ongoing quest to harmonize the imperatives of national

security with the fundamental rights and liberties of citizens is not merely a theoretical exercise but a critical, ongoing process that demands concrete, actionable strategies. Having explored the inherent tensions and the ethical underpinnings of intelligence operations, the focus now shifts to identifying practical pathways toward a sustainable equilibrium. This equilibrium is not a fixed point, but rather a dynamic state of balance that requires continuous recalibration in response to evolving threats, technological advancements, and societal expectations. Achieving this delicate balance necessitates a multifaceted approach, weaving together legal reforms, robust oversight mechanisms, technological safeguards, and meaningful public engagement. It is a commitment to a proactive and adaptive governance of security, ensuring that the measures taken to protect society do not inadvertently erode the very freedoms that define it.

A cornerstone of this sustainable equilibrium lies in the realm of legal reform. Existing legislation, often crafted in response to past threats or in a pre-digital era, may no longer adequately address the complexities of contemporary intelligence gathering. There is a pressing need for a comprehensive review and, where necessary, an update of laws governing surveillance, data collection, and the use of intelligence. These reforms should be guided by principles that reflect contemporary understanding of privacy and civil liberties, while still enabling intelligence agencies to effectively counter threats. This could involve clearer definitions of what constitutes a "legitimate" national security interest, stricter limitations on the scope and duration of surveillance, and more explicit prohibitions against the use of intelligence for purposes unrelated to national security, such as political targeting or the suppression of dissent. Furthermore, legislation should establish clear parameters for the collection and retention of data, particularly metadata, ensuring that it is proportionate to the identified threat and that it is securely handled and promptly destroyed when no longer necessary. The legal framework must also provide clear avenues for redress for individuals who believe their rights have been violated, ensuring that accountability is not an abstract concept but a tangible reality. This might involve establishing specialized tribunals or independent bodies capable of adjudicating complaints related to intelligence activities, offering a judicial check on executive power in this sensitive domain.

Complementing legal reforms, the enhancement of oversight mechanisms is paramount. While legislative and judicial oversight are indispensable, they must be sufficiently empowered and resourced to effectively scrutinize the activities of intelligence agencies. This includes ensuring that parliamentary committees or congressional oversight bodies have unfettered access to information, the authority to compel testimony from intelligence officials, and the independence to conduct thorough and unbiased investigations. The composition of these oversight bodies should reflect a diversity of perspectives, including legal experts, civil libertarians, and representatives from different political backgrounds, to ensure a comprehensive and balanced approach to oversight. Moreover, the effectiveness of oversight is contingent on the transparency with which it operates. While classified information must naturally be protected, the processes of oversight, the findings of investigations, and the recommendations for improvement should be made public to the greatest extent possible, fostering public confidence and accountability. Independent ombudsman offices, tasked with reviewing individual complaints and investigating systemic issues within intelligence agencies, can also serve as a vital layer of oversight, providing an accessible point of contact for citizens and acting as an impartial arbiter. The independence of such bodies is critical, requiring their appointment and funding to be insulated from direct political influence.

The rapid advancement of technology presents both unprecedented opportunities for intelligence gathering and significant challenges for the protection of civil liberties. Therefore, a crucial aspect of achieving a sustainable equilibrium involves the implementation of robust technological safeguards. This includes the development and deployment of technologies that minimize the collection of personal data, favor anonymization and aggregation, and incorporate privacy-preserving features by design. Encryption, differential privacy techniques, and secure multi-party computation are examples of technological approaches that can allow for data analysis and threat detection without compromising individual privacy. Furthermore, there is a need for rigorous auditing and ethical review of the algorithms and artificial intelligence systems used by intelligence agencies. These systems must be assessed for bias, accuracy, and adherence to legal and ethical standards, ensuring that they do not inadvertently lead to discriminatory outcomes or unwarranted surveillance of specific communities. The principle of "privacy by design" should be embedded into the development lifecycle of all new surveillance technologies, ensuring that privacy considerations are addressed from the outset rather than as an afterthought. This requires close collaboration between technologists, legal experts, ethicists, and civil society organizations to ensure that technological solutions are both effective for security and respectful of fundamental rights. The challenge lies in ensuring that these safeguards are not merely theoretical but are rigorously implemented and independently verified.

Public engagement and education are equally vital components of this endeavor. A well-informed citizenry is better equipped to understand the complexities of national security and the trade-offs involved in balancing security and liberty. Intelligence agencies and policymakers have a responsibility to foster a greater public understanding of their mission, their methods (within the bounds of necessary secrecy), and the oversight mechanisms in place. This can be achieved through clear, accessible communication, public consultations on proposed policies, and educational initiatives that explain the legal and ethical frameworks governing intelligence activities. Building public trust is essential, and this trust can only be earned through transparency, accountability, and a demonstrable commitment to upholding democratic values. When the public understands the necessity of certain measures and the safeguards in place to protect their rights, they are more likely to support the legitimate work of intelligence agencies. Conversely, a lack of transparency and a perception of unchecked power can breed suspicion and erode public confidence, which is ultimately detrimental to national security. Moreover, fostering a culture of public discourse on these issues, involving diverse stakeholders, can lead to more informed policy decisions and a greater collective ownership of the balance being struck. This engagement should not be a one-off event but an ongoing dialogue, adapting to new challenges and societal concerns.

The concept of "proportionality" needs to be continuously re-examined and rigorously applied in practice. This means that any intrusive surveillance or data collection measure must be strictly necessary to achieve a specific, legitimate national security objective, and that the intrusiveness of the measure must be commensurate with the gravity of the threat. This requires a constant evaluation of whether less intrusive means could achieve the same security outcome. For instance, instead of mass surveillance, intelligence agencies should prioritize targeted collection based on specific intelligence leads. The "all-or-nothing" approach to data collection, often enabled by technology, must be resisted in favor of a more judicious and evidence-based approach. This principle also extends to the type of data collected; efforts should be made to collect the minimum amount of personal information necessary to achieve the objective, and to avoid collecting highly sensitive data unless absolutely essential. The "necessity and proportionality" test should be the guiding principle for all intelligence operations, subject to regular review by independent bodies.

Furthermore, the development of robust internal ethics frameworks within intelligence agencies is not a substitute for external oversight and legal constraints, but a necessary complement. These internal frameworks should go beyond mere compliance with regulations and foster a deep-seated ethical culture. This involves comprehensive training for all personnel, emphasizing the importance of privacy, civil liberties, and democratic accountability. It also requires leadership that consistently models ethical behavior, creates channels for reporting ethical concerns without fear of retaliation, and ensures that ethical lapses are addressed promptly and appropriately. Establishing internal ethics committees or appointing ethics officers within agencies can provide dedicated resources for guidance and review. This proactive approach to cultivating an ethical mindset is crucial in preventing misconduct and ensuring that the pursuit of security remains aligned with fundamental values. The promotion of a "speak-up" culture, where individuals feel empowered to question decisions or raise concerns without fear of reprisal, is a critical indicator of a healthy ethical environment.

The international dimension of intelligence sharing and operations also demands careful consideration. As intelligence networks become increasingly globalized, agreements on data sharing and operational conduct must be guided by a shared commitment to human rights and democratic principles. When intelligence is shared with foreign partners, it is imperative to ensure that those partners adhere to similar standards of oversight, accountability, and respect for privacy. This requires clear agreements on the conditions under which intelligence can be shared, limitations on its use, and mechanisms for verifying compliance. The potential for intelligence sharing to inadvertently circumvent domestic legal protections or to be used for purposes that violate international human rights norms is a significant concern that requires constant vigilance. Ethical guidelines must extend to how intelligence is gathered in foreign jurisdictions, respecting national sovereignty and international law.

The concept of "chilling effect" warrants continued attention within this framework. While intelligence activities are designed to prevent threats, their very existence and visibility can inadvertently discourage legitimate forms of expression, association, and political dissent. A sustainable equilibrium requires intelligence agencies to actively mitigate this chilling effect. This can be achieved through clear communication about the specific threats being addressed, ensuring that surveillance is narrowly tailored and not broadly applied to entire populations or communities without a clear justification, and providing assurances that lawful dissent and investigative journalism will not be targeted. The principle of "innocent until proven guilty" must remain a guiding star, ensuring that individuals are not treated as suspects based on speculative associations or pre-emptive judgments. The focus must remain on preventing demonstrable threats, rather than on pre-empting potential future actions based on profiles or predictions without concrete evidence.

Finally, the dynamic nature of the threat landscape and technological evolution necessitates an adaptive approach to maintaining the equilibrium between security and liberty. This means that policies, legal frameworks, and oversight mechanisms cannot remain static; they must be subject to continuous review and revision. The establishment of independent commissions or expert panels tasked with periodically assessing the effectiveness and ethical implications of intelligence practices can provide valuable input for policy adjustments. This ongoing process of reassessment and adaptation is not a sign of weakness but a testament to a commitment to upholding democratic values in a changing world. It requires a willingness to learn from mistakes, to engage with criticism, and to continuously seek ways to improve the balance between the vital need for security and the fundamental imperative to protect individual freedoms. The resilience of a democratic society is, in part, measured by its ability to adapt its security apparatus without sacrificing its core liberties, thereby fostering a system that is both safe and free, a true testament to the enduring power of its foundational principles. This constant calibration ensures that the pursuit of safety never comes at the cost of the very values that make a society worth defending, creating a robust and trustworthy security apparatus that serves, rather than subverts, the democratic order. The integration of technology should be viewed not as an inevitability that dictates the erosion of privacy, but as a tool that, when guided by ethical principles and robust oversight, can enhance security without compromising the foundational rights of citizens. This proactive, iterative approach is the only path to a truly sustainable equilibrium.

14: THE CITIZEN'S ROLE IN SURVEILLANCE REFORM

O ur digital existence, often perceived as a collection of scattered online interactions, is in reality a meticulously constructed mosaic – our digital footprint. Every click, every search, every shared photograph, and every message sent contributes to an ever-expanding record of our lives. This seemingly innocuous accumulation of data paints a remarkably detailed portrait, one that extends far beyond what most individuals consciously comprehend. Understanding the nature and scope of this footprint is not merely an exercise in digital hygiene; it is a fundamental prerequisite for engaging meaningfully in the ongoing discourse surrounding surveillance reform and for reclaiming a degree of agency over our personal information. The digital world, while offering unparalleled convenience and connectivity, operates on a foundation of data, and it is crucial for every citizen to grasp how this data is generated, collected, and subsequently utilized.

At its core, a digital footprint comprises all the traces of digital activity that a person leaves behind. This can be broadly categorized into two types: active and passive. Active footprints are those we intentionally create. This includes posting on social media, sending emails, writing blog posts, uploading photos and videos, and filling out online forms with personal details. Whenever you sign up for a new service, create a profile, or share your thoughts in a public forum, you are actively contributing to your digital footprint. Think of it as the digital equivalent of speaking in a public square; your words and actions are recorded and can be observed by others. Each piece of information willingly shared, from your date of birth to your favorite hobbies, becomes a data point.

The passive footprint, on the other hand, is less obvious and often goes unnoticed. This includes data collected through your online browsing habits, such as the websites you visit, the search queries you make, the links you click, and the amount of time you spend on particular pages. Every time you connect to the internet, your device is assigned an IP address, which can reveal your general geographical location. Websites use cookies – small text files stored on your browser – to track your activity, remember your preferences, and sometimes to serve targeted advertising. Your interaction with applications, even when you're not actively using them, can also generate passive data. For instance, location services on your smartphone, even if seemingly dormant, can be transmitting your whereabouts. The sheer volume of this passively collected data is staggering, and it often goes unscrutinized by the user, forming a rich tapestry of personal habits, interests, and associations.

The implications of this extensive digital footprint are profound, especially when considering the purposes to which this data is put. Primarily, data is collected to enhance user experience and provide personalized services. Search engines tailor results based on your past searches, streaming services recommend movies and music based on your viewing history, and e-commerce sites suggest products you might like. This personalization, while often beneficial, is predicated on the detailed profiling of individuals. Beyond personalization, data is a valuable commodity for marketing and advertising. Companies analyze browsing patterns, purchase histories, and demographic information to target specific consumer groups with tailored advertisements, aiming to influence purchasing decisions. This is the engine that drives much of the free online services we enjoy; in essence, we are often paying with our data.

However, the collection and utilization of data extend beyond commercial interests. Governments and law enforcement agencies, citing national security and public safety concerns, also collect and analyze vast amounts of data. This can range from metadata associated with communications – such as who called whom, when, and for how long – to the content of communications themselves, often obtained through warrants or, in some jurisdictions, through broader legal frameworks that allow for bulk data collection. The capacity of modern technology to collect, store, and analyze data at an unprecedented scale means that even seemingly innocuous digital activities can be aggregated and analyzed to infer sensitive personal information, such as political leanings, religious beliefs, health conditions, or personal relationships.

In this landscape of pervasive data collection, understanding privacy policies and managing online settings becomes an act of digital self-defense. Privacy policies, often lengthy and filled with legal jargon, are the agreements between users and online service providers regarding the collection, use, and sharing of personal data. While rarely read in full, they are crucial documents that outline what information is collected, why it is collected, how it is protected, and with whom it might be shared. Many users click "agree" without truly understanding the implications, effectively granting broad permissions for data usage. It is incumbent upon citizens to make a concerted effort to understand these policies, or at least to seek out summaries and analyses provided by consumer advocacy groups. Tools exist that can help simplify these policies, making the information more accessible.

Managing online settings is another critical aspect of controlling one's digital footprint. Most social media platforms, search engines, and web browsers offer a range of privacy controls that allow users to limit the data collected about them and how it is shared. On social media, this can involve adjusting who can see your posts, your friend list, or your personal information. It may also include opting out of targeted advertising or limiting the use of your data for facial recognition or location tracking. Search engines often provide options to review and delete your search history, and to disable personalized search results. Web browsers can be configured to block third-party cookies, limit tracking, and send "Do Not Track" requests to websites. While these settings may not always be foolproof, and some services may function less effectively without certain data, actively engaging with them is a vital step in asserting control.

Furthermore, recognizing the implications of our digital activities requires a conscious effort to reflect on the potential downstream effects of our online behavior. Every piece of information shared, every website visited, and every interaction recorded can become part of a permanent record. This data can be used to build profiles that influence decisions made about us by entities ranging from potential employers and insurance providers to credit bureaus and even criminal justice systems. For instance, a publicly available social media post expressing dissatisfaction with a former employer could be interpreted negatively by a prospective new employer. Similarly, browsing for health-related information could, in some contexts, lead to higher insurance premiums if such data were accessible and utilized in actuarial calculations. The permanence of digital information means that even casual online activities can have long-lasting consequences.

The concept of "digital literacy" is central to empowering citizens in this environment. Digital literacy extends beyond simply knowing how to use a computer or navigate the internet. It encompasses the ability to critically evaluate online information, understand the privacy implications of digital activities, and employ strategies to protect personal data. Educational initiatives play a crucial role in fostering this literacy. Schools, community organizations, and government bodies can all contribute to educating the public about their digital rights and responsibilities. This includes demystifying complex topics such as encryption, metadata, and data analytics, and explaining in clear, accessible terms how these technologies impact individual privacy.

The notion that "if you have nothing to hide, you have nothing to fear" is a dangerous oversimplification that fails to acknowledge the broader societal implications of pervasive surveillance and data collection. Even for individuals with no illicit activities, the constant monitoring and profiling can create a chilling effect on freedom of expression, association, and thought. The knowledge that one's communications, associations, and even private browsing habits are being recorded can lead to self-censorship, discouraging individuals from exploring controversial ideas, engaging in political dissent, or seeking out sensitive personal information. This is why the fight for digital privacy is intrinsically linked to the broader struggle for civil liberties.

Moreover, the aggregation of data about large populations can reveal patterns and insights that are then used to influence public opinion, shape political discourse, and even predict and manipulate behavior. Algorithms trained on vast datasets can inadvertently perpetuate and amplify existing societal biases, leading to discriminatory outcomes in areas such as loan applications, hiring decisions, and even criminal sentencing. Understanding how these systems work, and advocating for their transparency and fairness, is therefore a civic responsibility.

Practical steps individuals can take to manage their digital footprint include:

Reviewing and cleaning up online profiles: Regularly audit social media accounts, deleting old posts, photos, and information that is no longer relevant or that you are uncomfortable sharing.

Using strong, unique passwords and enabling two-factor authentication: This is a fundamental security measure to protect accounts from unauthorized access.

Being mindful of app permissions: When installing new apps, carefully review the permissions they request and consider whether they are necessary for the app's functionality.

Utilizing privacy-focused browsers and search engines: Consider using alternatives like DuckDuckGo, Brave, or Firefox with enhanced privacy settings.

Using virtual private networks (VPNs): VPNs can mask your IP address and encrypt your internet traffic, providing an additional layer of privacy.

Limiting location sharing: Turn off location services for apps and devices when they are not actively needed, or set them to only share location while the app is in use.

Being cautious about public Wi-Fi: Avoid conducting sensitive transactions or accessing confidential information on unsecured public networks.

Practicing selective sharing: Think critically before

posting personal information online, especially on public platforms. Consider the audience and the potential long-term implications.

Regularly checking privacy settings: Most online services update their policies and settings periodically. Make it a habit to review and adjust your privacy preferences.

Understanding data brokers: Be aware that numerous companies specialize in collecting and selling personal data. While direct opt-out can be challenging, being aware of their existence is the first step.

The fight for surveillance reform is not solely the domain of legislators and intelligence agencies; it begins with an empowered and informed citizenry. By understanding the nature of our digital footprint, the mechanisms of data collection, and the potential implications of our online activities, we can make more informed decisions about our digital lives. This heightened awareness allows us to engage more effectively in advocating for stronger privacy protections, demanding greater transparency from both corporations and governments, and ultimately, shaping a digital future that respects and safeguards our fundamental rights and freedoms. The ability to navigate the digital landscape with knowledge and intentionality is a cornerstone of modern citizenship, enabling individuals to participate fully in society without sacrificing their privacy or their autonomy. It transforms passive consumption of digital services into active participation in the governance of one's own data, a vital component of a healthy democracy in the digital age.

The digital age has undeniably amplified the capabilities of surveillance, presenting a complex challenge to the very fabric of civil liberties. While technological advancements have offered unprecedented benefits, they have also created an environment where the potential for unwarranted monitoring is pervasive. Yet, within this complex landscape lies a potent counter-force: the engaged and informed citizen. The power of individual and collective advocacy, coupled with robust political engagement, forms the bedrock upon which meaningful surveillance reform can be built. It is through these avenues that citizens can actively participate in shaping the digital public square and ensuring that technological progress aligns with democratic values and fundamental human rights.

One of the most direct and impactful ways citizens can engage in surveillance reform is through direct communication with their elected representatives. Legislators, by their very nature, are responsive to the concerns of their constituents. When a significant number of individuals voice their opinions on issues like data privacy, government surveillance programs, or the need for stronger oversight, these concerns are more likely to be translated into legislative action. This communication can take many forms, from well-crafted letters and emails to phone calls and in-person meetings. The key is to be informed, articulate, and persistent. Instead of generic pleas, constituents should clearly state their concerns, reference specific policies or proposed legislation, and propose concrete solutions. For example, an individual might write to their representative advocating for stronger warrant requirements for accessing digital communications, citing concerns about the erosion of Fourth Amendment protections in the digital realm. They might also point to specific pieces of legislation, either existing or proposed, and explain why they support or oppose them, offering reasoned arguments. Town hall meetings, while often boisterous, provide valuable opportunities to pose questions directly to representatives and to make one's views known publicly, creating a record of citizen engagement.

Beyond direct outreach, a crucial avenue for impact lies in supporting and amplifying the work of dedicated privacy advocacy groups. Organizations such as the Electronic Frontier Foundation (EFF), the American Civil Liberties Union (ACLU), the Center for Democracy & Technology (CDT), and numerous smaller, specialized groups dedicate themselves to researching, litigating, and lobbying for digital rights and privacy protections. These organizations possess the expertise, resources, and established networks necessary to effect significant change. By becoming a member, donating, or volunteering time, individuals can provide essential support for these efforts. Advocacy groups often lead campaigns to educate the public, draft model legislation, challenge surveillance practices in court, and engage directly with policymakers. Supporting them is akin to pooling resources and expertise, allowing for a more strategic and effective approach to complex issues. For instance, when a new surveillance technology is proposed or deployed, these organizations are often at the forefront of analyzing its implications and mobilizing public opposition or support for regulatory measures. Their efforts in filing amicus briefs in key court cases, or in pushing for legislation like the USA FREEDOM Act or the proposed American Privacy Rights Act, demonstrate the tangible impact of organized advocacy.

Public demonstrations and civil disobedience, when employed thoughtfully and strategically, can also serve as powerful tools for raising awareness and demanding action on surveillance reform. While often perceived as radical, peaceful protests can effectively capture public attention and signal the depth of citizen concern. These events can bring issues that might otherwise remain obscure into the mainstream media and public discourse. Whether it's a march on a government building, a silent vigil, or a coordinated online campaign, these actions demonstrate a collective will for change. However, the effectiveness of such actions is often amplified when they are integrated with other forms of engagement, such as lobbying efforts or public education campaigns. For example, a protest against a particular surveillance program might be timed to coincide with a legislative debate, thereby increasing the pressure on lawmakers. The history of civil rights and other social movements is replete with examples of how public demonstration, when combined with sustained political pressure and legal challenges, has led to transformative societal change. The digital realm itself can also be a site for protest, through coordinated online actions, viral awareness campaigns, or even acts of digital civil disobedience aimed at highlighting the invasiveness of certain technologies.

Engaging in informed debate and fostering public understanding is another critical component of citizen advocacy. The complex nature of surveillance technologies and their legal frameworks can often be opaque to the general public. Citizens have a role to play in demystifying these issues, sharing information, and encouraging critical thinking within their own communities and social networks. This can involve writing letters to the editor of local newspapers, participating in online forums, hosting community discussions, or simply having conversations with friends and family. By sharing articles, analyses, and personal reflections, individuals can help to build a more informed public opinion, which in turn can influence policy. The narrative around surveillance is often shaped by powerful interests; therefore, it is vital for citizens to contribute to this discourse by providing alternative perspectives grounded in privacy and civil liberties principles. Educational initiatives, whether formal or informal, are essential for equipping citizens with the knowledge they need to participate meaningfully in these debates.

Furthermore, citizens can actively participate in the

legislative process itself. Many legislative bodies offer opportunities for public comment on proposed rules and regulations. Submitting detailed, well-reasoned comments can directly influence the final shape of these policies. Similarly, engaging with legislative staff, attending committee hearings, and even testifying on behalf of organizations or as private citizens can provide valuable insights to policymakers. The process of drafting legislation is often iterative, and public input at various stages can lead to significant improvements in the protection of privacy and civil liberties. For example, during the drafting of data privacy bills, public comments might highlight loopholes or unintended consequences that lawmakers had not considered, leading to amendments that strengthen protections.

The power of collective action cannot be overstated. While individual actions are important, it is often the aggregation of these actions, coordinated through organized efforts, that yields the most significant results. When citizens unite under a common cause, their voices are amplified, making it impossible for policymakers to ignore their demands. This can manifest in various ways: signing petitions, joining organized campaigns, participating in consumer boycotts of companies with problematic data practices, or forming local advocacy groups. The digital landscape itself provides new platforms for organizing and mobilizing. Social media can be used to rapidly disseminate information, coordinate actions, and build broad coalitions. However, it is also important to be aware of the potential for surveillance and manipulation within these digital spaces, necessitating a strategic approach to online organizing.

The success of surveillance reform hinges on sustained pressure. Policy changes do not typically occur overnight. It requires persistent effort, a willingness to adapt strategies, and a long-term commitment to the cause. This means continuing to engage with elected officials even after initial requests, supporting advocacy groups consistently, and remaining vigilant against new threats to privacy. The landscape of surveillance technology is constantly evolving, and so too must the efforts to reform it. This ongoing commitment ensures that the issue remains on the political agenda and that progress, however incremental, is made.

Moreover, citizens can contribute to surveillance reform by advocating for greater transparency and accountability in government and corporate surveillance practices. This includes demanding access to information about how data is collected and used, pushing for independent oversight mechanisms, and holding individuals and institutions accountable when privacy violations occur. Transparency is not merely about disclosing information; it is about fostering a culture of openness and responsibility. When citizens demand to know how their data is being handled, they are asserting their right to digital self-determination. This can involve supporting legislation that mandates greater transparency in government surveillance programs, such as the publication of statistics on data requests or the disclosure of surveillance technologies employed. Similarly, advocating for stronger enforcement of existing privacy laws and pushing for new legislation that creates meaningful penalties for data misuse are crucial steps.

The notion of "digital citizenship" has evolved to encompass not just responsible online behavior but also active participation in shaping the digital environment. Surveillance reform is a core component of this evolving concept. By embracing their roles as active participants rather than passive consumers of digital services, citizens can reclaim agency over their digital lives and contribute to a more just and equitable digital society. This empowerment is not just about protecting personal data; it is about safeguarding democratic principles, freedom of expression, and the very notion of a private sphere in an increasingly interconnected world. Each informed communication to a legislator, each donation to a privacy advocacy group, each public discussion about digital rights, contributes to a larger movement for surveillance reform. It is through this multifaceted approach, combining individual action with collective organization and sustained political engagement, that citizens can effectively advocate for a digital future that respects their fundamental rights and freedoms. The ongoing struggle for surveillance reform is a testament to the enduring power of an informed and mobilized citizenry to shape policy and protect the essential liberties that define a free society. This continuous engagement ensures that the rapid pace of technological innovation does not outstrip our commitment to privacy and civil liberties, creating a dynamic and responsive system of governance in the digital age. The ability to challenge, to question, and to advocate for change is not merely a privilege but a responsibility for all who value a democratic society.

The proactive adoption of privacy-enhancing technologies (PETs) represents a tangible and increasingly vital pathway for citizens to assert control over their digital lives and directly contribute to the broader movement for surveillance reform. In an era where data is a commodity and digital footprints are constantly being tracked, utilizing tools designed to shield personal information is not merely a matter of preference, but a fundamental act of self-defense and a powerful statement of intent to those who seek to monitor our activities. These technologies act as digital bulwarks, empowering individuals to reclaim a degree of anonymity and security in an increasingly transparent digital world.

At the forefront of this technological empowerment are encrypted messaging applications. Platforms like Signal, Telegram (with its end-to-end encrypted "secret chats"), and WhatsApp (which uses the Signal protocol for end-to-end encryption) offer a vital layer of protection for personal communications. Unlike standard SMS messages or unencrypted digital conversations, end-to-end encryption ensures that only the sender and the intended recipient can read the messages. Even the service provider, government agencies, or malicious actors who might intercept the data are rendered unable to decipher its content. This is achieved through complex cryptographic algorithms that scramble the message at the sender's device and only unscramble it on the recipient's device, using unique cryptographic keys held by each end-user. The widespread adoption of these platforms not only protects individual conversations but also signals to telecommunications companies and governments that there is a significant public demand for secure and private communication channels. When millions opt for encrypted services, it creates a de facto standard, pushing the entire market towards more secure infrastructure. Consider the network effect: as more people use Signal, for instance, it becomes more advantageous for their contacts to also adopt it, creating a growing ecosystem of secure communication. This shift in user behavior can influence product development, forcing companies to compete on privacy features rather than solely on functionality or cost.

Beyond messaging, the very act of browsing the internet can be a window into our lives. Secure browsers and browser extensions play a crucial role in mitigating the pervasive tracking that occurs online. Browsers like Brave, for instance, are built with privacy as a core feature, automatically blocking trackers and ads by default. Tools like the Tor Browser, on the other hand, anonymize internet traffic by routing it through a series of volunteer-operated servers, making it exceedingly difficult to trace online activity back to its source. While Tor is often associated with illicit activities, its primary function is to protect the privacy of all users, including journalists, activists, and ordinary citizens who wish to avoid targeted advertising, censorship, or surveillance. Furthermore, browser extensions such as Privacy Badger (from the Electronic Frontier Foundation), uBlock Origin, and DuckDuckGo Privacy Essentials can be added to more conventional browsers like Firefox or Chrome to block trackers, cookies, and malicious scripts that collect user data without explicit consent. The cumulative effect of millions of users deploying these tools is a significant reduction in the amount of data available to be collected by websites, advertisers, and potentially surveillance agencies. Each blocked tracker, each anonymized session, contributes to a less surveilled internet. This increasing demand for privacy-conscious browsing compels web developers and platform providers to reconsider their data collection practices, as a significant portion of their user base is actively opting out of invasive tracking.

Virtual Private Networks (VPNs) offer another powerful tool for enhancing online privacy and security. A VPN encrypts an internet connection and routes it through a server located elsewhere, masking the user's IP address and making it appear as though they are browsing from the VPN server's location. This not only shields users from potential surveillance by their Internet Service Provider (ISP) but also protects them when using public Wi-Fi networks, which are often unsecured and vulnerable to man-in-the-middle attacks. The growing popularity of VPNs has made them a mainstream privacy tool. When users choose VPN providers, they are often faced with a spectrum of services, some with robust no-logging policies and others that are less trustworthy. Educating oneself about reputable VPN providers and choosing services that prioritize user privacy over data monetization is essential. The increased market share for VPNs demonstrates a clear consumer preference for secure and private internet access, pushing the industry towards greater transparency and stronger privacy commitments from service providers. This consumer-driven demand can influence how ISPs and mobile carriers handle user data, as they face competition from services that offer a more private alternative.

The concept extends to operating systems and device security as well. Citizens can opt for operating systems that are designed with privacy and open-source principles in mind, such as Linux distributions like Tails (The Amnesic Incognito Live System), which is specifically designed for anonymity and can be run from a USB stick without leaving traces on the host computer. While not as mainstream as Windows or macOS, the increasing availability and usability of privacy-focused alternatives empower users to make more informed choices about the fundamental software that underpins their digital interactions. Furthermore, understanding and implementing basic security practices, such as using strong, unique passwords managed by password managers, enabling two-factor authentication (2FA) whenever possible, and regularly updating software to patch vulnerabilities, are crucial components of a comprehensive privacy strategy. These are not exotic technologies but fundamental digital hygiene practices that significantly reduce the attack surface available to both casual snoopers and sophisticated surveillance operations.

The collective impact of citizens embracing these privacy-enhancing technologies is profound. It shifts the market dynamics. As more individuals demand tools that protect their data, companies are incentivized to innovate and invest in privacy features. This creates a virtuous cycle: increased demand leads to better products, which in turn attracts more users. This is particularly evident in the competitive landscape of consumer electronics and software. Companies that are perceived as prioritizing user privacy can gain a significant competitive advantage. Conversely, companies with a reputation for lax data protection or extensive data sharing practices can suffer reputational damage and loss of market share. For example, the introduction of features like Apple's App Tracking Transparency, which requires apps to ask users for permission before tracking them across other companies' apps and websites, was partly a response to growing public and regulatory pressure for greater privacy. While controversial for some advertisers, it signaled a significant shift in how user data can be handled.

Furthermore, citizen adoption of PETs serves as a powerful educational tool. As individuals become more familiar with tools like VPNs or encrypted messaging, they are more likely to discuss these issues with their peers, family, and colleagues. This demystifies complex technologies and raises general awareness about digital privacy threats. When people experience firsthand the benefits of secure browsing or private communication, they become natural advocates, spreading knowledge and encouraging others to adopt similar practices. This grassroots education is essential for building a broad-based movement for surveillance reform, as it moves the conversation from technical jargon to tangible personal benefits.

The development and popularization of privacy-focused services also put pressure on legislative bodies. When a significant portion of the population actively uses and advocates for privacy tools, it becomes more difficult for governments to implement overly broad surveillance measures. Legislators are more likely to consider the privacy implications of new technologies and policies when they know that their constituents are equipped with and utilizing tools that allow them to bypass or resist intrusive surveillance. This creates a feedback loop where technological solutions can inform and drive policy changes. For instance, if widespread use of encrypted communications makes it harder for law enforcement to intercept messages, it might lead to legislative debates about the necessity of backdoors or lawful access mechanisms, forcing a public discussion about the trade-offs between security and privacy.

Moreover, the open-source nature of many PETs fosters transparency and allows for community scrutiny. Projects like Signal, Tor, and many Linux distributions are built on publicly available code. This means that security experts and privacy advocates worldwide can examine the code for vulnerabilities or backdoors, building trust and ensuring that the technologies perform as advertised. By supporting and using these open-source tools, citizens are actively participating in a decentralized, community-driven approach to privacy protection. This contrasts sharply with proprietary systems where the inner workings are hidden, making it difficult to assess their true privacy implications.

The economic incentives for companies are also shifting. While data collection has traditionally been a lucrative business model, the increasing consumer demand for privacy is forcing companies to explore alternative revenue streams, such as subscription services, privacy-preserving advertising, or value-added services that do not rely on extensive personal data harvesting. This transition is not without its challenges, but the growing market for privacy-conscious products and services indicates a fundamental shift in consumer expectations. By consciously choosing products and services that respect their privacy, citizens are actively shaping the market and rewarding companies that align with their values. This includes everything from choosing search engines like DuckDuckGo over those that aggressively track user queries, to opting for email services that offer end-to-end encryption.

However, it is crucial to acknowledge that no technology is foolproof. PETs can be complex to implement correctly, and even the most robust tools can have vulnerabilities or be circumvented under certain circumstances. For example, metadata – information about communications, such as who communicated with whom and when – can still be collected even with encrypted content. Furthermore, the effectiveness of a VPN depends heavily on the trustworthiness of the VPN provider itself. It is therefore essential for citizens to remain informed, to research the tools they use, and to understand their limitations. Continuous education and vigilance are paramount. The journey towards a more private digital life is an ongoing process, not a destination, and it requires a commitment to staying informed about emerging threats and evolving solutions.

In conclusion, the proactive and widespread adoption of privacy-enhancing technologies by citizens is a cornerstone of effective surveillance reform. It empowers individuals, shifts market dynamics, fosters public education, and puts pressure on policymakers. By making informed choices about the digital tools they use – from encrypted messaging apps and secure browsers to VPNs and privacy-focused operating systems – citizens can actively build a more private and secure digital environment for themselves and contribute to a broader cultural and political movement that prioritizes fundamental rights in the digital age. Each user who chooses a privacy-respecting service, each conversation that raises awareness about digital security, adds a crucial building block to the edifice of digital liberty. This active engagement with technology is not just about personal protection; it is a powerful form of civic participation, a direct assertion of agency in an era defined by data and digital oversight.

The journey toward meaningful surveillance reform is not solely paved with technological solutions; it is equally, if not more, reliant on the persistent and informed engagement of citizens with the institutions that wield significant power over their data and privacy. While the adoption of privacy-enhancing technologies (PETs) provides a crucial layer of individual defense, a truly robust system of accountability and democratic oversight necessitates a deep dive into the operational realities of both corporations and government entities that collect, process, and utilize our personal information. This involves a proactive demand for transparency – a fundamental right that, when exercised, empowers citizens to understand the scope of surveillance, scrutinize its justification, and ultimately hold those responsible accountable for their actions. Without transparency, the intricate machinery of data collection and surveillance operates in the shadows, leaving individuals vulnerable and democratic processes weakened.

One of the most direct and impactful avenues for demanding transparency lies in exercising one's right to access personal data held by technology companies. Virtually every online service, from social media platforms and search engines to e-commerce sites and cloud storage providers, amasses vast quantities of information about its users. This data can include browsing history, search queries, location data, purchase records, communication logs, and even biometric information, depending on the services used. Many jurisdictions, particularly in regions with strong data protection laws like the European Union (with its General Data Protection Regulation - GDPR) or California (with the California Consumer Privacy Act - CCPA), grant individuals the legal right to request a copy of the personal data that a company holds about them. This process, often referred to as a "data subject access request" (DSAR) or a "right to access," is a powerful tool for illuminating the extent of corporate data collection.

To effectively utilize this right, citizens must first identify which companies hold their data and then navigate the often-complex procedures for submitting these requests. Many companies provide dedicated portals or contact points for

DSARs, though finding them can sometimes require diligent searching through their privacy policies or terms of service. It is crucial to be specific in the request, clearly stating what information is being sought, such as "all data collected about my account since [date]" or "records of my location history." The response from the company should ideally be comprehensive, providing the requested data in a readable and usable format. However, companies may sometimes attempt to limit the scope of the disclosure or provide data in an obfuscated manner. In such instances, persistence and a clear understanding of one's rights are vital. The information received through a DSAR can be eye-opening, revealing patterns of behavior, inferred characteristics, and data points that an individual may not have even realized were being collected. This firsthand knowledge is invaluable in understanding the landscape of digital surveillance and the commercial data economy.

Beyond simply requesting data, scrutinizing the privacy policies of these companies is another critical element of demanding transparency. These documents, often lengthy and couched in legalistic language, are intended to inform users about how their data is collected, used, shared, and protected. However, they are frequently opaque, vague, and designed to obscure rather than clarify. Citizens must dedicate the time to read and understand these policies, paying close attention to clauses related to data sharing with third parties, the purposes for which data is processed, the duration of data retention, and the security measures in place. Websites and organizations dedicated to privacy, such as the Electronic Frontier Foundation (EFF) or the Future of Privacy Forum, often provide resources that help demystify privacy policies and highlight key areas of concern. By understanding these policies, citizens can make more informed decisions about which services to use and can identify potential areas where companies are falling short of their stated commitments.

Furthermore, a collective demand for clearer, more accessible, and more user-friendly privacy policies can pressure companies to adopt more transparent practices. Companies that are willing to explain their data practices in plain language, with clear examples, build trust and demonstrate a commitment to user privacy that can be a significant competitive advantage.

The demand for transparency must extend forcefully to government agencies and intelligence services. While the collection of data by governments is often justified in the name of national security, public safety, or law enforcement, the scope and methods of these activities are frequently shrouded in secrecy. Citizens have a right to know how their government is monitoring them, what powers it is exercising, and what safeguards are in place to prevent abuse. This is where the concept of "accountability" truly comes into play. Without transparency regarding government surveillance programs, it becomes nearly impossible to ensure that these powers are being used appropriately, that they are proportionate to the threats they aim to address, and that they are not infringing upon fundamental civil liberties.

One crucial aspect of this demand is advocating for the public disclosure of details surrounding government surveillance programs. This can involve a variety of actions. For instance, when news breaks about a new surveillance initiative or a court ruling related to government data access, citizens can demand that the government provide more context and justification. This might involve writing to elected officials, participating in public consultations where available, or supporting journalistic investigations that seek to shed light on classified or proprietary government operations. Freedom of Information Act (FOIA) requests, or their equivalents in other countries, are vital legal tools that citizens can use to compel government agencies to release non-classified information about their activities. While governments often cite national security to withhold information, persistent and well-crafted FOIA requests can sometimes yield valuable insights into the types of data being collected, the legal frameworks governing surveillance, and the agencies involved.

Furthermore, citizens should actively support legislative efforts that promote greater transparency in government surveillance. This includes advocating for laws that require greater public reporting on the number of data requests made to telecommunications companies, the types of data sought, and the success rates of these requests. It also involves pushing for more oversight from legislative bodies and independent review boards, ensuring that these bodies have the necessary access and authority to scrutinize surveillance activities effectively. When citizens make their voices heard through contacting their representatives, participating in town hall meetings, or supporting advocacy groups, they can influence the legislative agenda and push for reforms that prioritize openness and accountability.

The relationship between government and the private technology sector in the context of surveillance is a critical area where transparency is paramount. Companies often cooperate with government agencies by providing access to user data, either voluntarily or through legal mandates. Understanding the extent and nature of this cooperation is essential for a complete picture of surveillance. This includes demanding clarity on how companies respond to government data requests, whether they challenge overly broad requests, and what policies they have in place regarding data retention and disclosure to government entities. Publicly accessible transparency reports, published by technology companies, are a step in the right direction, but they often lack the detail needed for a thorough understanding. Citizens should demand that these reports provide more granular data, such as the specific types of data disclosed, the legal authorities under which disclosure was made, and the number of requests blocked or modified.

The ultimate goal of demanding transparency is to foster a culture of accountability. When both companies and governments are more open about their data practices, they are more likely to be held responsible for any breaches of privacy or abuses of power. An informed public, armed with knowledge about how their data is being handled, is better equipped to challenge questionable practices, demand stronger protections, and elect representatives who prioritize privacy rights. This is not a passive endeavor; it requires ongoing engagement, a willingness to ask difficult questions, and a persistent effort to pry open the curtains of secrecy that often surround digital surveillance.

Moreover, transparency is not a static concept; it must evolve alongside the technologies and legal frameworks that govern them. As new surveillance capabilities emerge, citizens must continue to press for disclosure and oversight. This might involve understanding how artificial intelligence and machine learning are being used for surveillance purposes, how biometric data is being collected and analyzed, or how encrypted communications are being targeted. Each new frontier of surveillance demands a corresponding commitment to transparency.

The role of investigative journalism and civil society organizations in demanding and facilitating transparency cannot be overstated. These entities often serve as crucial conduits of information, uncovering and disseminating details about surveillance programs that would otherwise remain hidden. Citizens can support these efforts by consuming their reporting, donating to their causes, and participating in campaigns that advocate for greater openness.

By empowering these watchdogs, citizens amplify their own voices and contribute to a more robust system of checks and balances.

Ultimately, the demand for transparency from companies and governments is an exercise of fundamental democratic citizenship. It is about ensuring that power is exercised responsibly, that rights are protected, and that the digital world we inhabit is one where individuals are informed and empowered, not merely subjects of opaque systems of observation and control. Each DSAR filed, each privacy policy scrutinized, each letter to a representative, each piece of investigative journalism supported, is a step towards building a more transparent, accountable, and rights-respecting digital society. It is a continuous process of engagement, demanding that the invisible become visible, and that the unaccountable be held to account. Without this relentless pursuit of openness, the promise of digital liberty remains elusive, overshadowed by the specter of pervasive, unchecked surveillance. The power to demand and achieve transparency lies with an informed and active citizenry, ready to illuminate the dark corners of data collection and surveillance.

The journey towards dismantling pervasive surveillance and securing digital privacy is not a solitary quest, nor is it confined to the halls of power or the pronouncements of legal scholars. At its heart, it is a fundamentally human endeavor, rooted in shared understanding and collective action. The power to effect meaningful change, to shift the landscape of surveillance from one of opaque control to one of transparent accountability, lies within the interconnectedness of individuals and their ability to educate and mobilize one another. This subsection delves into the profound impact of peer-to-peer education and community organizing as the essential building blocks for a robust and sustainable movement dedicated to surveillance reform. It's about igniting awareness, fostering dialogue, and transforming passive concern into active participation, thereby building a formidable grassroots force capable of challenging entrenched practices and championing our digital rights.

Consider the simple act of conversation. How often do we find ourselves lamenting the constant barrage of targeted ads, the uncanny accuracy of recommendations, or the nagging feeling that our online activities are being meticulously tracked? These shared experiences, often dismissed as minor annoyances, are potent entry points for discussing the broader implications of digital surveillance. The first step in educating others is to demystify the often-abstract concepts of data collection and privacy. Instead of abstract discussions about algorithms and metadata, ground the conversation in tangible examples that resonate with everyday life. Talk about how a casual search for a medical condition might lead to persistent, sometimes intrusive, advertising related to that condition, or how seemingly innocuous app permissions can grant access to a wealth of personal data. Frame these discussions not as technical lectures, but as relatable concerns about personal autonomy and the right to live without constant observation.

When engaging with friends, family, and colleagues, authenticity and empathy are your most valuable assets. Share your own journey of discovery, the moments when you realized the extent of data collection and its potential consequences. This personal narrative can be far more persuasive than any data point or statistic. Explain, in simple terms, what kind of data is being collected, who is collecting it, and for what purposes. For instance, you might explain that every website visited, every search query entered, and every message sent contributes to a vast digital profile that can be used to influence purchasing decisions, shape opinions, or even determine eligibility for services. When discussing with parents or older relatives who might be less familiar with digital technologies, draw parallels to real-world privacy concerns they already understand, such as protecting personal financial information or maintaining confidentiality in personal conversations. The goal is to foster a sense of shared vulnerability and a recognition that privacy is not merely a technical issue, but a fundamental human right that affects everyone.

Reliable information is the bedrock of any informed movement. In an era of misinformation, it is crucial to share accurate and well-sourced material. This can take many forms. Curate a list of reputable organizations that are actively working on digital rights and privacy, such as the Electronic Frontier Foundation (EFF), the American Civil Liberties Union (ACLU), or local civil liberties groups. Share their articles, reports, and explainer videos through social media, email, or messaging apps. Highlight key findings from investigative journalism that have exposed surveillance practices. When sharing information, be discerning. Prioritize resources that break down complex topics into easily digestible formats, such as infographics, short explainer videos, or clear and concise articles. Avoid sensationalism and focus on factual reporting and well-reasoned arguments. For example, instead of simply stating that "governments are spying on everyone," explain the legal frameworks that permit certain types of surveillance, the specific technologies being employed, and the potential for overreach. By providing credible information, you empower your network to form their own informed opinions and become advocates in their own right.

Building a movement also necessitates creating spaces for collective engagement. Local community organizing offers a powerful pathway to amplify individual voices. This could involve starting a local chapter of a national privacy advocacy group, or simply organizing informal meetups to discuss digital rights. These gatherings can be a forum for sharing personal experiences, learning from each other, and strategizing about local actions. For instance, a group might decide to write letters to local elected officials regarding data privacy policies in public services, or organize a presentation at a community center to educate a wider audience. Even small-scale actions, like collectively contacting a local internet service provider to inquire about their data retention policies, can demonstrate a unified community interest and build momentum.

Online advocacy groups and forums also play a crucial role in connecting individuals across geographical boundaries. Platforms like Reddit, Discord, or dedicated privacy forums can be invaluable for sharing information, coordinating actions, and finding like-minded individuals. Participating in these online communities allows for the exchange of ideas, the identification of emerging issues, and the rapid dissemination of urgent calls to action. For example, during legislative debates surrounding new surveillance powers, online groups can quickly mobilize to contact lawmakers, organize petition drives, or share talking points for public commentary. The interconnectedness of the internet, ironically, can be harnessed to foster deeper, more meaningful human connection and collective purpose.

A key aspect of building a movement is fostering a shared understanding of the principles at stake. Privacy is not just about hiding personal information; it's about autonomy, dignity, and the freedom to explore ideas and express oneself without fear of reprisal or judgment. When discussing surveillance, connect it to these broader democratic values. Explain how constant surveillance can have a chilling effect on free speech, association, and even thought. Consider the implications for marginalized communities, who may be disproportionately targeted by surveillance technologies, and emphasize how privacy is intrinsically linked to social justice. By framing the issue in terms of fundamental rights and societal well-being, you can broaden the appeal of the movement beyond a niche tech-savvy audience to a wider segment of the population.

Encourage active participation by making it easy for people to get involved. Provide clear calls to action. This could be as simple as signing a petition, contacting a representative, or supporting a specific piece of legislation. When asking people to take action, explain the impact of their contribution. For example, if you're organizing a letter-writing campaign to a senator, explain how a flood of constituent letters can influence their voting decisions. Offer resources like pre-written email templates or contact information for elected officials to lower the barrier to entry for participation. The more accessible and impactful the actions you propose, the more likely people are to engage.

The long-term success of surveillance reform hinges on its ability to become a mainstream concern, embedded in our collective consciousness. This requires persistent and consistent effort in educating and mobilizing others. It's about creating a culture where digital privacy is not an afterthought, but a fundamental expectation. This involves celebrating small victories, sharing stories of impact, and continuously adapting strategies as the landscape of surveillance evolves. For instance, when a particular transparency report is published by a tech company, or when a legislative amendment is proposed, use these as opportunities to re-engage your network and discuss the latest developments. Maintain the momentum by continuing to share information, foster dialogue, and provide avenues for collective action. The strength of the movement lies not just in its numbers, but in the depth of its members' commitment and their shared understanding of why this fight is so critical. It is through this dedicated, grassroots effort – through countless conversations, shared resources, and organized actions – that the foundations for meaningful and lasting surveillance reform are truly laid. The power resides in our collective voice, amplified by education and united by a common purpose to safeguard our digital lives.

15: THE PATH FORWARD: A CALL FOR ACCOUNTABLE GOVERNANCE

T he current architecture of surveillance in the United States, largely a legacy of post-9/11 security imperatives, has ossified into a system characterized by its breadth, depth, and often, its opacity. It is a system built on the premise that vast amounts of digital information, collected indiscriminately from virtually all citizens, are a necessary bulwark against potential threats. This approach, however, has demonstrably outstripped its original justification, evolving into a pervasive, suspicionless dragnet that erodes fundamental civil liberties and democratic norms. Reimagining this apparatus requires not merely tinkering at the edges, but a fundamental architectural redesign – a shift from a model predicated on ubiquitous data hoarding to one grounded in necessity, proportionality, and genuine accountability. This is not a call for the abolition of all intelligence gathering, but for its rigorous recalibration, ensuring that national security interests are pursued without sacrificing the privacy and freedom that define a just society.

The core of this reimagining lies in dismantling the infrastructure of mass surveillance. The indiscriminate collection of metadata, the bulk acquisition of phone records, email content, and internet browsing histories, represents the antithesis of targeted investigation. Instead, we must champion a paradigm where data collection is an exception, not the rule, initiated only when specific, credible suspicion of wrongdoing exists, and authorized through stringent legal processes. This means moving away from programs that Hoover up the digital lives of millions to programs that seek out specific individuals or communications demonstrably linked to illicit activities. Such a shift necessitates a re-evaluation of the legal frameworks that currently permit such broad data acquisition. Legislation must be strengthened to explicitly prohibit suspicionless collection and to mandate that any authorized surveillance be narrowly tailored to the specific investigation at hand, minimizing the collection of incidental data pertaining to innocent individuals. This is not a rhetorical flourish; it is a practical necessity for restoring constitutional protections in the digital age.

Furthermore, the concept of "necessity" must be rigorously defined and enforced. In the context of surveillance, necessity implies that the proposed action is the least intrusive means available to achieve a legitimate government objective. This requires agencies to exhaust all alternative, less intrusive methods of investigation before resorting to surveillance that impacts privacy. For instance, if a suspect can be identified and apprehended through conventional policing methods or by seeking voluntarily provided information, these avenues should be pursued before any form of electronic surveillance is considered. The presumption should always be in favor of privacy, with the burden of proof resting squarely on the government to demonstrate why a particular surveillance measure is absolutely indispensable and why less invasive alternatives are insufficient. This principle of proportionality, central to many international human rights frameworks, is regrettably underdeveloped in the current U.S. approach.

The cornerstone of any reimagined surveillance state must be robust, independent oversight. The current oversight mechanisms, while present, have proven insufficient to curb the excesses of intelligence agencies. This is due in part to inherent conflicts of interest, with oversight often residing within the executive branch or legislative committees that are themselves deeply intertwined with national security agencies. To rectify this, we need to cultivate a more empowered and independent oversight architecture. This could involve establishing an independent civilian oversight board, composed of individuals with expertise in civil liberties, technology, and law, who are insulated from political pressures and possess the authority to conduct thorough audits, investigate complaints, and recommend concrete remedial actions. This board should have unfettered access to information, including classified data, and its findings and recommendations should be made public to the greatest extent possible, fostering transparency and accountability.

Moreover, the judicial oversight role needs to be strengthened. The Foreign Intelligence Surveillance Court (FISC), while intended as a safeguard, has been criticized for its ex parte proceedings and its reliance on government-provided information, leading to what some have termed a "rubber-stamping" of surveillance requests. Reimagining the judicial role would involve increasing transparency in FISC proceedings, allowing for amicus curiae briefs from civil liberties advocates, and ensuring that judges are equipped with sufficient technical expertise to critically evaluate surveillance methodologies. The presumption of probable cause, a bedrock of Fourth Amendment protections, must be strictly applied in all surveillance authorizations, eliminating loopholes that allow for broad, dragnet-style surveillance. The judiciary's role should not be to facilitate surveillance, but to act as a genuine check on government power, protecting individual rights from potential overreach.

Transparency is another critical pillar for reimagining the surveillance state. The current opacity surrounding government surveillance practices breeds distrust and hinders democratic accountability. While certain aspects of intelligence gathering must, by necessity, remain classified, this classification should not be used as a blanket shield to conceal illegal or unconstitutional activities. A more accountable system would involve significantly increasing public disclosure regarding the types of surveillance conducted, the legal authorities invoked, the aggregate data collected (without revealing sensitive operational details), and the metrics of success and failure. This could manifest in regular, declassified reports from intelligence agencies detailing their surveillance activities, oversight findings, and any instances of abuse or misconduct. Such transparency would empower citizens to engage in informed debate about the scope and necessity of surveillance, and to hold their elected officials accountable for the policies they enact.

The notion of "accountability" itself needs to be redefined. Currently, when surveillance abuses come to light, the consequences for those responsible are often minimal. Reimagining accountability requires establishing clear lines of responsibility within intelligence agencies and implementing meaningful penalties for violations of privacy rights and legal statutes. This could include robust disciplinary measures for individuals who engage in unlawful surveillance, as well as mechanisms for individuals whose rights have been violated to seek redress, whether through civil litigation or dedicated administrative processes. The principle that no one is above the law, including those tasked with national security, must be unequivocally upheld.

The technological landscape of surveillance is also in constant flux, and any reimagining of the state apparatus must grapple with this dynamic reality. The proliferation of artificial intelligence, facial recognition technology, predictive policing algorithms, and the increasing interconnectedness of the Internet of Things (IoT) present new and evolving challenges to privacy and civil liberties. A forward-looking approach would involve proactively developing ethical guidelines and regulatory frameworks for these emerging technologies *before* they become deeply embedded in surveillance systems. This requires a commitment to foresight, anticipating potential harms and establishing safeguards in advance, rather than reacting to abuses after the fact. For example, strict limitations or outright bans on certain applications of facial recognition technology in public spaces, or mandates for algorithmic transparency and bias mitigation in AI-driven surveillance systems, are crucial steps.

Furthermore, the privatization of surveillance capabilities presents a complex challenge. Much of the data collected and analyzed by government agencies is facilitated by private sector technology companies. Reimagining the surveillance state must therefore extend to regulating the relationship between government and private industry in the realm of data collection and analysis. This includes demanding greater transparency from tech companies about their data-sharing practices with governments, and establishing clear legal boundaries that prevent private entities from circumventing public privacy protections. The outsourcing of surveillance functions to private contractors, often with less stringent oversight, also needs careful scrutiny and regulation.

The economic incentives driving the surveillance economy are also a critical factor. The business model of many tech companies relies on the collection and monetization of user data. This creates a inherent conflict of interest, where the desire for profit can overshadow concerns for user privacy. A reimagined approach would involve exploring alternative economic models that do not rely on pervasive data collection, and potentially imposing data minimization requirements on companies that process personal information. This is a complex undertaking, requiring a nuanced understanding of digital economies, but it is essential for disrupting the ongoing cycle of data exploitation that fuels much of modern surveillance.

Crucially, the reimagining of the surveillance state must be an inclusive and democratic process. The decisions about the scope and nature of surveillance directly impact every citizen, and therefore, the public must be actively involved in shaping these policies. This involves fostering greater public understanding of surveillance technologies and their implications, as well as creating accessible avenues for public input and participation in the policy-making process. Open forums, public consultations, and educational initiatives are vital for ensuring that surveillance policies reflect the values and priorities of the society they are meant to serve. The current trend towards technocratic decision-making, where policies are shaped by a select group of experts and officials with limited public input, must be reversed.

In essence, the path forward is not about finding a perfect balance between security and privacy, a perpetually elusive equilibrium. Instead, it is about fundamentally altering the default settings of our digital society. We must move from a posture of mass, suspicionless surveillance, justified by abstract threats and opaque processes, to one that is rooted in specific, demonstrable suspicion, narrowly tailored interventions, and unwavering, transparent accountability. This requires a courageous re-examination of our laws, our institutions, and our technological practices. It demands that we prioritize human rights and civil liberties, not as secondary considerations to national security, but as integral components of a truly secure and democratic society. This is a long and challenging road, but one that is essential for preserving the freedoms that define the American experiment. It is a call to build a surveillance apparatus that serves the public good, rather than one that undermines it through its very design. This requires a sustained commitment to reform, driven by an informed and engaged citizenry, demanding that government power be exercised with respect for individual dignity and liberty. The future of privacy and freedom in the digital age hinges on our collective willingness to undertake this vital reimagining.

The transition to a targeted, necessary, and accountable surveillance model necessitates a significant recalibration of intelligence agency mandates and operational doctrines. Currently, many agencies are structured and funded to collect and analyze vast quantities of data, with significant resources allocated to processing metadata and identifying patterns across broad populations. To shift towards a suspicion-based approach, these agencies would need to reorient their priorities, investing more heavily in traditional investigative techniques, human intelligence, and the development of highly specific, technology-driven tools that can be deployed only under strict legal authorization. This transition would likely involve a cultural shift within these organizations, emphasizing precision and restraint over sheer volume of data. It would also require a reassessment of performance metrics, moving away from measures of data collected or analyzed, towards measures of successful, targeted investigations that yield concrete results with minimal collateral impact on innocent individuals.

Legislation must be the bedrock of this reform. Existing statutes, such as the PATRIOT Act and FISA, need to be fundamentally reformed or, in some cases, repealed and replaced. The concept of "reasonable suspicion" must be clearly defined and rigorously enforced in the context of electronic surveillance, distinguishing it clearly from the lower standards that have, in practice, allowed for suspicionless data collection. Provisions that permit the bulk collection of data, regardless of whether it pertains to individuals suspected of any wrongdoing, must be eliminated. New legislative frameworks should explicitly codify the principles of data minimization, purpose limitation, and proportionality, ensuring that any data collected is strictly necessary for a specific, authorized purpose and is not retained for longer than required. Furthermore, mechanisms for legislative oversight need to be strengthened, granting congressional committees more robust powers to scrutinize intelligence activities, conduct independent investigations, and hold agencies accountable for compliance with privacy laws. This includes ensuring that agencies provide timely and complete information to oversight bodies, and that those bodies have the resources and expertise to effectively fulfill their oversight responsibilities.

The international dimension of surveillance also demands consideration. The U.S. surveillance apparatus operates within a global network, often exchanging data and intelligence with foreign governments. Reimagining the U.S. system must also involve a review of these international data-sharing agreements and practices. This means ensuring that U.S. allies adhere to comparable privacy standards and that data shared with foreign partners is not used in ways that would violate U.S. constitutional protections. The extraterritorial reach of U.S. surveillance laws and practices also raises complex questions about the rights of non-U.S. citizens, and any reform effort must grapple with these extraterritorial implications, striving for a framework that respects fundamental human rights globally, even as it prioritizes the rights of U.S. citizens.

Education and public discourse are not merely steps towards reform; they are ongoing necessities for sustaining it. As technology evolves and new surveillance capabilities emerge, the public must remain informed and engaged. This requires a commitment to ongoing civic education, empowering citizens with the knowledge and tools to understand the implications of surveillance and to advocate for their digital rights. It also means fostering a media environment that is committed to investigative journalism on surveillance issues, holding powerful institutions accountable and providing the public with accurate and critical information. Without a vigilant and informed populace, any reforms, however well-intentioned, are vulnerable to erosion over time.

Ultimately, the reimagining of the surveillance state is a profound challenge that touches upon the core values of a democratic society. It is about reclaiming a balance where national security does not necessitate the erosion of fundamental freedoms. It is about building a system that is not only effective in identifying threats but is also respectful of individual dignity and liberty. This requires a courageous commitment to change, a willingness to challenge established practices, and a steadfast dedication to the principles of justice and accountability. The goal is not to eliminate intelligence gathering, but to ensure it is conducted responsibly, transparently, and in full accordance with the rights enshrined in our Constitution. The current trajectory of ever-expanding, opaque surveillance is unsustainable and ultimately self-defeating, undermining the very freedoms it purports to protect. A move towards a targeted, necessary, and accountable model is not just a policy preference; it is an imperative for the preservation of a free and democratic society in the digital age. This reimagining is an investment in trust, in liberty, and in the future of democratic governance.

The trajectory toward a future where technology serves democratic ideals rather than undermines them hinges on the implementation of concrete, actionable proposals. This requires a conscious, collective effort to embed fundamental rights into the very architecture of our digital lives and governance structures. Foremost among these is the urgent need to codify digital rights, establishing a clear and robust legal framework that explicitly enumerates and protects the privacy, freedom of expression, and associational rights of individuals in the digital sphere. Such a charter would serve as a bulwark against the pervasive overreach that has characterized recent decades, providing a clear mandate for how technology should be developed and deployed within a democratic society. This goes beyond simply adapting existing legal principles to new technologies; it necessitates a proactive creation of rights that anticipate future challenges. For instance, this could include the right to data minimization, ensuring that personal information is collected only when strictly necessary and not in vast, indiscriminate troves. It would also encompass the right to algorithmic transparency and accountability, allowing individuals to understand how decisions affecting them are made by automated systems and to challenge biased or erroneous outcomes. The right to be free from warrantless digital searches and seizures, and the right to secure and private communications, must be unequivocally affirmed and legally enforceable, with clear penalties for violations.

Building upon a foundation of codified digital rights, the strengthening of independent oversight bodies is paramount. The current mechanisms, often criticized for their insularity and lack of genuine independence, require a significant overhaul. Proposals must focus on empowering these bodies with the resources, authority, and insulation from political influence necessary to effectively scrutinize government surveillance activities. This could involve establishing truly independent civilian review boards, comprised of experts in technology, law, civil liberties, and ethics, who are appointed through a transparent and bipartisan process. These boards should possess the authority to conduct unannounced audits, access all relevant information without undue redaction, and initiate investigations into alleged abuses. Furthermore, their findings and recommendations should be made public, subject only to narrowly defined national security exceptions, fostering public trust and accountability. The legislative branch also plays a crucial role. Congressional committees tasked with oversight must be equipped with greater investigative powers and access to expert staff who can independently analyze the technical and legal complexities of surveillance programs. Mechanisms to prevent undue influence from the executive branch, such as requiring regular, unclassified briefings and providing robust whistleblower protections, are also vital components of this strengthened oversight architecture. The judiciary, as the ultimate arbiter of constitutional rights, must also be empowered. This could involve reforms to the Foreign Intelligence Surveillance Court (FISC), such as allowing for the appointment of independent technical experts to assist judges, increasing transparency in its proceedings by permitting meaningful amicus curiae briefs from civil liberties organizations, and ensuring that the standard of probable cause, a cornerstone of the Fourth Amendment, is rigorously applied to all surveillance requests.

The advancement of technological innovation must be consciously steered towards privacy-enhancing solutions. Instead of treating privacy as an afterthought or a regulatory burden, it should be a guiding principle in the design and development of new technologies. This approach, often termed "privacy by design," requires developers and engineers to proactively incorporate privacy protections from the outset of any project. This can manifest in various ways, such as developing encryption technologies that are more robust and accessible to the public, creating decentralized communication platforms that limit the ability of any single entity to monitor conversations, and fostering the development of anonymization techniques that allow for data analysis without compromising individual identities. Governments can incentivize this shift through research and development grants, tax credits for companies that prioritize privacy-enhancing technologies, and by setting clear procurement standards that favor privacy-conscious solutions. Furthermore, public investment in open-source privacy tools and technologies can democratize access to these protective measures, ensuring that they are not solely the domain of affluent individuals or corporations. This also extends to promoting the development of "explainable AI" and auditable algorithms, allowing for a deeper understanding and scrutiny of how these systems operate and make decisions, particularly when they intersect with government surveillance.

Fostering greater public participation in decisions about surveillance is not merely a desirable democratic ideal; it is an essential safeguard against unchecked governmental power. The current model, where critical decisions about the scope and nature of surveillance are often made in opaque, technocratic settings with limited public input, is inherently undemocratic. To rectify this, we must cultivate a more informed and engaged citizenry. This involves significant investment in public education initiatives that demystify surveillance technologies and explain their implications for civil liberties and democratic processes. Accessible online resources, public forums, and collaborations with educational institutions can play a vital role in raising public awareness. Beyond education, creating meaningful channels for public input into policy-making is crucial. This could involve establishing citizen assemblies or deliberative panels tasked with studying surveillance issues and providing recommendations to policymakers. Online platforms for public comment on proposed surveillance regulations, coupled with a commitment by government agencies to genuinely consider and respond to public feedback, can also foster a more participatory approach. Furthermore, supporting independent investigative journalism and civil society organizations that monitor government surveillance activities is critical. These entities act as essential watchdogs, holding power accountable and providing the public with the information necessary to engage in informed debate. Ensuring robust legal protections for whistleblowers who expose surveillance abuses is also a non-negotiable element of this participatory framework.

Ultimately, the proposals for a more democratic future in the digital age coalesce around a rights-centric approach to technology and governance. This means that every decision regarding the development, deployment, and oversight of surveillance capabilities must be grounded in an unwavering commitment to protecting and upholding fundamental human rights. It requires a paradigm shift from viewing privacy and civil liberties as obstacles to security to recognizing them as essential components of a secure and just society. When national security is invoked as a justification for intrusive surveillance, the burden of proof must lie with the government to demonstrate that the proposed measures are not only necessary but also proportionate, targeted, and subject to rigorous, independent oversight. This rights-centric approach demands that we ask not just "Can we do this?" but "Should we do this?" and "How can we do this in a way that respects fundamental freedoms?" It means prioritizing human dignity and autonomy in the face of increasingly powerful technological capabilities. It requires a constant vigilance, a willingness to challenge assumptions, and a commitment to adapt our legal and ethical frameworks as technology evolves. This is not a static endeavor; it is an ongoing process of societal deliberation and democratic self-governance, ensuring that the tools of the digital age are wielded to empower citizens and strengthen democratic institutions, rather than to erode them. This commitment to a rights-centric future is not merely an idealistic aspiration; it is a pragmatic necessity for ensuring the long-term health and resilience of democratic societies in an increasingly digitized world. It is about actively shaping a future where technology serves humanity, rather than one where humanity is subservient to technology. This requires a proactive, rather than reactive, approach to policy-making, anticipating potential harms and establishing safeguards before they become entrenched realities.

The practical implementation of these proposals will undoubtedly face resistance, often framed in terms of national security imperatives or the inherent complexities of technological change. However, history teaches us that perceived threats to security are often used as pretexts for the expansion of state power, and that technological advancement does not, by itself, dictate a surrender of fundamental liberties. The challenge before us is to embrace innovation while simultaneously fortifying the democratic guardrails that protect individual freedoms. This means fostering a culture within government and the private sector that values transparency, accountability, and respect for human rights. It requires an ongoing dialogue between policymakers, technologists, civil liberties advocates, and the public to ensure that the digital future we are building is one that reflects our deepest democratic values. This dialogue must be inclusive, ensuring that diverse voices and perspectives are heard, particularly those from communities that have historically been disproportionately affected by surveillance and other forms of state overreach.

Furthermore, international cooperation will be essential in shaping a global digital future that uphms democratic norms. As data flows across borders with unprecedented ease, and as surveillance technologies are adopted and adapted by nations worldwide, a fragmented approach to digital rights and oversight is insufficient. Collaborative efforts to establish international norms and standards for data protection, cybersecurity, and government surveillance are crucial. This involves engaging in diplomatic efforts to promote privacy-respecting technologies, advocate for transparency in international data-sharing agreements, and push for the adoption of human-rights-centered approaches to digital governance on a global scale. The lessons learned from domestic reforms can and should inform these international discussions, contributing to a more just and equitable digital landscape for all. The ethical considerations surrounding the export of surveillance technologies also warrant close attention, ensuring that these powerful tools are not deployed in ways that suppress dissent or violate human rights in other jurisdictions.

The economic underpinnings of the digital age also demand scrutiny within the context of democratic governance. The business models of many technology companies, reliant on the pervasive collection and monetization of personal data, create a powerful incentive structure that can conflict with the public interest in privacy. Reimagining governance in this space may necessitate exploring alternative economic models for digital services that do not depend on mass data collection. This could include promoting subscription-based services, supporting open-source and community-driven platforms, or even exploring forms of public investment in digital infrastructure that prioritizes user privacy. Regulatory frameworks may also need to evolve to address the concentration of power in the digital economy, ensuring that market competition does not lead to a race to the bottom in terms of privacy standards. The concept of data as a public good, rather than solely a private commodity, is a notion worth exploring further in this context.

The ongoing evolution of artificial intelligence and its integration into surveillance systems presents a particularly pressing challenge. Without robust ethical guidelines and legal frameworks, AI-driven surveillance could exacerbate existing societal biases, automate discrimination, and create unprecedented levels of predictive control. Proposals must therefore include mandates for algorithmic transparency, independent auditing of AI systems for bias and accuracy, and clear lines of accountability for the outcomes produced by these technologies. The development of "red lines" for AI in surveillance, such as prohibitions on autonomous lethal weapons or facial recognition systems used for mass surveillance in public spaces without strict justification, is also a critical component of safeguarding democratic futures. The very definition of "suspicion" in the context of AI-driven analysis needs careful consideration, ensuring that predictive models do not criminalize individuals based on statistical correlations rather than concrete evidence of wrongdoing.

In conclusion, the path forward towards a more democratic digital future is paved with concrete proposals that center on codifying digital rights, strengthening independent oversight, fostering privacy-enhancing innovation, and prioritizing public participation. This is not merely about regulating technology; it is about shaping the fundamental relationship between citizens, the state, and the digital tools that increasingly mediate our lives. It requires a sustained commitment to democratic principles, a willingness to challenge entrenched power structures, and a proactive embrace of a future where technology serves the public good, rather than undermining the very freedoms it is intended to enhance. The ongoing dialogue and action on these fronts are crucial for ensuring that the digital age ushers in an era of expanded liberty and democratic accountability, rather than one of pervasive surveillance and diminished freedoms. This is a call to actively build a digital society that reflects our highest ideals, ensuring that the innovations of today empower the democracies of tomorrow.

The ongoing debate surrounding national security and civil liberties often frames the two as inherently antagonistic, a zero-sum game where gains in one inevitably come at the expense of the other. This dichotomy, however, is a false one, particularly when examining the operational realities of intelligence gathering. A robust national security apparatus, capable of effectively anticipating and mitigating threats, does not necessitate the abandonment of fundamental rights. In fact, the opposite is true: a security framework that respects and upholds civil liberties can be significantly more effective, sustainable, and legitimate. The pursuit of comprehensive, all-encompassing surveillance, often justified by the nebulous concept of "all-source intelligence," can paradoxically dilute an agency's ability to focus on genuine threats. When intelligence professionals are tasked with sifting through unmanageable volumes of data, much of it irrelevant or benign, their capacity to identify and analyze critical information is diminished. This deluge of data can obscure the signal within the noise, leading to missed opportunities or, worse, the misallocation of valuable resources.

Consider the historical context: the era preceding widespread digital surveillance was not an era of unchecked vulnerability. Intelligence agencies operated effectively, albeit with different tools and methodologies, to counter significant threats. Their effectiveness was not predicated on the indiscriminate collection of every communication, every online interaction, or every movement of every citizen. Instead, it relied on targeted intelligence gathering, human sources, and sophisticated analytical techniques that focused on specific threats and individuals of interest. The advent of advanced computing power and the proliferation of digital data have indeed presented new opportunities for intelligence collection, but they have also amplified the potential for systemic error and overreach when not tempered by strict legal and ethical boundaries. Reforms aimed at protecting civil liberties, therefore, are not merely about safeguarding individual freedoms; they are also about optimizing the effectiveness of intelligence operations themselves. By shifting away from mass surveillance and towards more precise, targeted collection methods, intelligence agencies can concentrate their efforts on genuine threats, fostering a more efficient and productive security posture. This focus allows for deeper analysis, better source development, and a more nuanced understanding of complex geopolitical landscapes, ultimately contributing to more accurate threat assessments and more effective counter-terrorism and national security strategies.

Moreover, the erosion of public trust, a direct consequence of perceived or actual overreach in surveillance activities, represents a significant operational handicap for any intelligence agency. When citizens believe their communications are routinely monitored without cause, or that their digital lives are subject to constant, unchecked scrutiny, a climate of suspicion and fear can emerge. This can have a chilling effect on legitimate public discourse, discourage whistleblowers from coming forward with vital information about wrongdoing, and foster a general distrust of government institutions. In contrast, an intelligence framework that is demonstrably accountable and respectful of civil liberties can cultivate a more cooperative environment. When the public understands that surveillance is targeted, legally authorized, and subject to rigorous oversight, they are more likely to support necessary security measures and may even be more willing to provide information that aids in threat detection. This symbiotic relationship between security and trust is fundamental. Agencies that operate transparently within established legal bounds, demonstrating a commitment to due process and individual rights, are better positioned to garner the support and cooperation necessary to achieve their objectives. This is not to suggest that intelligence agencies should operate in the public spotlight; their very nature requires a degree of discretion. However, the principles guiding their operations—legality, necessity, proportionality, and accountability—must be visible and demonstrably upheld.

The argument that more data equals more security is a simplistic and often misleading one. The quality of intelligence, not merely its quantity, is what truly matters. Mass surveillance, by its nature, generates vast quantities of data that is often of low value, requiring immense resources to process and analyze, with a high probability of false positives and an equally high probability of missing crucial signals amidst the noise. This approach can be akin to searching for a specific needle in an ever-growing haystack, a task that becomes exponentially more difficult and inefficient as the haystack expands. Targeted intelligence collection, on the other hand, operates on the principle of precision. It involves identifying specific threats, individuals, or networks of interest and then employing legal and technically sound methods to gather information relevant to those specific targets. This approach not only conserves resources but also significantly increases the likelihood of obtaining actionable intelligence. For example, instead of collecting the metadata of millions of phone calls, an agency might seek legal authorization to monitor the communications of a specific individual or group credibly suspected of planning an attack, based on prior intelligence or law enforcement leads. This focused approach allows analysts to dedicate more time and expertise to understanding the context, intent, and capabilities of the subjects of their investigation, leading to a deeper and more accurate intelligence picture.

Furthermore, the legal frameworks governing intelligence gathering must be robust enough to ensure that any collection is narrowly tailored and proportional to the identified threat. This means that the scope of surveillance must be defined by the specific intelligence requirements, not by the technological capacity to collect data indiscriminately. The Foreign Intelligence Surveillance Act (FISA) in the United States, for instance, was intended to provide a legal basis for targeted surveillance of foreign powers and their agents. However, interpretations and implementations of FISA, particularly concerning the collection of "incidental" data pertaining to U.S. persons, have often blurred the lines between targeted collection and mass surveillance. Reforms should aim to reinforce the original intent of such legislation, ensuring that any collection impacting the privacy of citizens is based on individualized suspicion and subjected to judicial review based on probable cause, as enshrined in constitutional protections. This means that data collection should not be a fishing expedition but a carefully directed effort to gather specific information to verify or refute a concrete suspicion of wrongdoing.

The shift towards more effective intelligence capabilities, therefore, hinges on a recalibration of priorities and methodologies. It involves investing in sophisticated analytical tools that can process and interpret targeted data more effectively, rather than simply acquiring more data. It necessitates the development of advanced human intelligence capabilities, fostering relationships with informants and sources that can provide context and insights that raw data alone cannot offer. It also requires a commitment to ongoing training and professional development for intelligence analysts and officers, equipping them with the skills to understand complex threats, identify patterns, and make informed judgments. These are the building blocks of effective intelligence, and they are not diminished by a respect for civil liberties; rather, they are enhanced by it.

The notion that privacy protections inherently weaken intelligence capabilities is a red herring, often used to resist necessary reforms. In reality, privacy-enhancing technologies and practices can be leveraged to improve security. For instance, end-to-end encryption, while sometimes presenting challenges for surveillance, also makes communications more secure against malicious actors and foreign adversaries who might seek to intercept them. Intelligence agencies can adapt by developing lawful access mechanisms that do not require backdoors or the weakening of encryption for all users. This could involve working with technology providers to establish secure, legally authorized channels for accessing data when necessary, under strict judicial oversight. Such an approach preserves the privacy of the general population while still allowing for legitimate intelligence gathering. Moreover, the development and deployment of privacy-preserving analytics techniques can enable agencies to glean insights from large datasets without compromising individual identities. Techniques like differential privacy or federated learning can allow for statistical analysis and the identification of patterns without exposing sensitive personal information, thereby achieving security goals while minimizing privacy intrusions.

The sustainability of intelligence operations is also intrinsically linked to their legitimacy, which in turn is derived from public trust and adherence to legal and ethical norms. Surveillance programs that are perceived as overly intrusive or unwarranted are vulnerable to legal challenges, public outcry, and ultimately, political pressure that can lead to their curtailment or complete dismantling. This cyclical pattern of overreach followed by backlash is counterproductive to long-term security objectives. A more measured, rights-respecting approach, conversely, builds a more stable and enduring foundation for intelligence activities. By operating within clearly defined legal boundaries and demonstrating accountability, intelligence agencies can foster an environment where their work is understood and, where necessary, supported by the public. This long-term perspective is crucial. Security is not a static endpoint but an ongoing process, and the effectiveness of intelligence agencies depends on their ability to adapt and maintain their legitimacy over time.

The argument for maintaining essential intelligence capabilities must therefore be understood within the broader context of democratic governance. In a democratic society, the power of the state, particularly the power to surveil its citizens, must be constrained and accountable. This is not an impediment to security but a fundamental requirement of a free society. The pursuit of security must not come at the cost of the very freedoms it is meant to protect. By embracing reforms that prioritize targeted collection, judicial oversight, and transparency, intelligence agencies can not only enhance their effectiveness but also reaffirm their commitment to democratic values. This dual objective is not only achievable but essential for building a resilient and secure nation in the digital age. It is about recognizing that true security is not merely the absence of threats but the presence of a just and lawful society where individual liberties are respected and protected. Intelligence agencies, as instruments of the state, have a critical role to play in upholding these principles, ensuring that the tools they employ serve to protect, rather than diminish, the democratic fabric of society. This requires a continuous dialogue and a commitment to adapting legal and operational frameworks to meet the evolving challenges of the 21st century, always with an eye toward balancing legitimate security needs with the unassailable rights of individuals. The efficacy of intelligence operations should be measured not only by the threats they thwart but also by their fidelity to the democratic principles they are sworn to uphold.

The preceding discussion has illuminated the intricate tension between national security imperatives and the preservation of civil liberties, arguing that these are not mutually exclusive but rather interdependent. A truly effective national security posture is one that is deeply rooted in democratic principles, respecting the rights and freedoms of the citizenry. This foundation of trust and legitimacy is not a hindrance to intelligence operations but a crucial enabler, fostering the cooperation and understanding necessary for long-term security. It is within this framework that we must now consider the indispensable roles of those who build the digital tools and those who govern their use: the technologists and the policymakers.

Technologists, by their very nature, are architects of the future, crafting the systems and platforms that increasingly define our social, economic, and political lives. The power they wield, often unseen and sometimes unintended, is profound. The algorithms that curate our news feeds, the encryption that secures our communications, the data storage solutions that house vast repositories of personal information – these are not neutral creations. Each line of code, each design choice, carries with it inherent biases, capabilities, and potential consequences for privacy, autonomy, and democratic discourse. Consequently, technologists bear a significant ethical responsibility that extends beyond mere functionality and efficiency. They must engage with the societal implications of their creations, contemplating how their innovations might be wielded by governments, corporations, or malicious actors, and considering the impact on fundamental human rights. This often requires a proactive, rather than reactive, approach to ethics, embedding considerations of privacy, fairness, and accountability into the very design process.

The call for technologists to be more socially conscious is not an indictment of innovation, but an plea for mindful innovation. It is about fostering a culture within the tech industry where the potential for misuse, particularly by state actors, is a primary consideration. This involves moving beyond a purely utilitarian view of technology to one that recognizes its embedded social and political values. For instance, the development of sophisticated data mining tools, while offering immense potential for analysis, also presents significant risks of enabling mass surveillance and profiling. Technologists involved in creating such tools have a duty to consider how these capabilities can be constrained or regulated, perhaps by building in safeguards, transparency mechanisms, or by actively engaging in public discourse about their deployment. Similarly, the continuous advancements in areas like facial recognition, behavioral analysis, and predictive policing software demand a critical examination of their potential for discrimination and unwarranted intrusion into private lives. The developers of these technologies are not merely engineers; they are also de facto policymakers, shaping the boundaries of what is technically feasible and, by extension, what might become permissible.

Policymakers, on the other hand, bear the solemn responsibility of safeguarding the public interest, which inherently includes both security and liberty. In the rapidly evolving landscape of digital technologies, this task is particularly challenging. The speed at which technological advancements occur often outpaces the deliberative processes of governance. Laws and regulations, designed for a bygone era, can quickly become obsolete, leaving critical gaps that allow for unchecked surveillance or the erosion of privacy. Therefore, policymakers must make a concerted effort to bridge this knowledge gap. This means actively seeking to understand the technical realities of modern surveillance –

how data is collected, stored, processed, and analyzed; the capabilities and limitations of various technologies; and the potential for both unintended consequences and deliberate misuse. It requires moving beyond abstract principles to a granular understanding of the digital infrastructure that underpins contemporary intelligence gathering.

The effective collaboration between technologists and policymakers is not merely desirable; it is essential for building a system of accountable governance in the digital age. This partnership must be founded on mutual respect, a shared commitment to democratic values, and a willingness to engage in continuous, interdisciplinary dialogue. Policymakers need to be educated on the intricacies of the technologies they seek to regulate, and technologists need to understand the legal and ethical frameworks that govern their creations. This exchange can take many forms: public consultations, expert advisory panels, legislative hearings where technical experts are given ample opportunity to explain complex issues, and even internships or secondments where individuals can gain firsthand experience in both domains. The goal is to foster a shared understanding that allows for the development of smart, effective, and rights-respecting policies.

Consider the area of data retention. Policymakers grapple with questions of how long telecommunications data should be stored, what types of data should be retained, and who should have access to it. Without a deep understanding of how data is generated, the technical feasibility of various retention periods, the costs involved, and the analytical value at different stages of retention, policymakers are essentially making decisions in a vacuum. Technologists, conversely, may possess the knowledge to suggest more efficient or less intrusive data management practices that could meet security needs while minimizing privacy risks. For example, advances in data anonymization and differential privacy could offer alternative approaches to raw data retention, allowing for statistical analysis without compromising individual identities. A collaborative approach would involve policymakers understanding these technical possibilities and technologists providing clear, accessible explanations of their benefits and limitations.

Another critical area is the implementation of oversight mechanisms for surveillance programs. While legal frameworks may mandate oversight, the effectiveness of that oversight hinges on the ability of those conducting it – often legislators or independent bodies – to comprehend the technical operations they are scrutinizing. This includes understanding metadata analysis, the use of specialized software, and the processes for obtaining warrants or legal authorization. Technologists can play a vital role in educating oversight bodies, demystifying complex technical procedures, and helping to identify potential loopholes or areas where existing oversight mechanisms are insufficient. This knowledge transfer is crucial for ensuring that oversight is not a mere formality but a substantive check on power.

The challenge is to move beyond adversarial relationships, where policymakers might view technologists with suspicion for creating tools that enable surveillance, and where technologists might view policymakers as ignorant or obstructive. Instead, the focus must shift to a constructive partnership aimed at achieving shared goals. This means creating platforms and forums where these discussions can happen openly and productively. It requires recognizing that technologists are not monolithic; there is a diversity of opinions and ethical stances within the technology community, and policymakers should actively engage with this spectrum of views. Similarly, policymakers must be willing to adapt their thinking as technology evolves, rather than rigidly adhering to outdated assumptions.

The concept of "lawful access" provides a concrete example of this necessary collaboration. As encryption technologies become more robust, the debate intensifies around how law enforcement and intelligence agencies can access encrypted communications for legitimate national security purposes. Technologists have developed sophisticated encryption methods, while policymakers are tasked with balancing the need for access with the imperative to protect user privacy. A collaborative approach could explore technical solutions that provide targeted access under strict legal authorization, without requiring the creation of "backdoors" that would weaken security for all users. This might involve working with technology companies to develop secure, auditable mechanisms for data disclosure when legally compelled, ensuring that such access is rare, targeted, and subject to robust judicial oversight. The development of such solutions requires a deep understanding of cryptography, network architecture, and legal frameworks, necessitating joint efforts from both technologists and policymakers.

Furthermore, the increasing use of artificial intelligence (AI) in surveillance operations presents both immense opportunities and significant ethical quandaries. AI can be used to analyze vast datasets, identify patterns, and predict potential threats with unprecedented speed and scale. However, it also raises concerns about algorithmic bias, the potential for autonomous decision-making in surveillance, and the lack of transparency in how AI systems arrive at their conclusions. Policymakers need to understand how these AI systems work, what data they are trained on, and what safeguards are in place to prevent bias and ensure accountability. Technologists involved in developing these AI systems have a duty to proactively address these concerns, embedding ethical considerations into their design and development processes, and providing clear explanations of their capabilities and limitations. The creation of robust AI governance frameworks, which clearly define the permissible uses of AI in surveillance, the required transparency, and the mechanisms for redress, can only be achieved through this interdisciplinary collaboration.

This collaboration is also vital for fostering public trust. When the public understands that the technologies used for national security are developed and deployed with careful consideration for privacy and civil liberties, and that policymakers are actively engaged in ensuring accountability, trust in government institutions is strengthened. Conversely, a lack of transparency and understanding can breed suspicion and erode public confidence, making it harder to implement necessary security measures. By involving technologists in the policymaking process and educating policymakers about technology, we can create a more transparent and accountable system that enjoys broader public support. This transparency need not compromise operational security; rather, it can focus on the principles, processes, and oversight mechanisms governing the use of technology.

The very definition of "security" is also being reshaped by technology. In an era of cyber warfare, digital espionage, and the weaponization of information, national security is no longer solely about physical borders and military might. It also encompasses the integrity of digital infrastructure, the security of personal data, and the resilience of democratic processes against digital interference. Technologists are at the forefront of understanding these evolving threats, and policymakers must actively solicit their insights to develop effective strategies. This requires a shift in how national security is conceptualized, moving from a purely state-centric view to one that recognizes the interconnectedness of digital systems and the vulnerability of societies to technological disruption.

Ultimately, the path forward requires a commitment to continuous learning and adaptation from both technologists and policymakers. The digital realm is not static; it is in a constant state of flux, with new technologies emerging and existing ones being refined at an astonishing pace. This necessitates an ongoing dialogue and a willingness to revisit and revise policies and practices as circumstances change. It means establishing durable mechanisms for consultation and collaboration that are not dependent on specific individuals or events. This enduring partnership is the bedrock upon which a truly accountable and democratic surveillance system can be built, ensuring that technological advancement serves the public good and upholds the fundamental rights that define a free society. The challenge is immense, but the stakes – the future of privacy, security, and democratic governance in the digital age – demand nothing less than our most concerted and collaborative efforts.

The journey through the complex terrain of national security and civil liberties, particularly as shaped by the evolving digital landscape, has revealed a stark imperative: the need for a proactive, engaged citizenry and responsive, accountable lawmakers. We stand at a critical juncture, where the very architecture of our societies is increasingly mediated by technology, and where the balance between collective safety and individual freedoms is constantly being renegotiated. To simply observe these shifts without participation is to cede control over our future, allowing policies and practices to be determined by default rather than by deliberate, democratic choice. Therefore, this moment calls for a robust and sustained call to action, directed at every level of our society.

For citizens, the path forward is one of informed vigilance and active participation. The digital age, while presenting unprecedented challenges to privacy, also offers unparalleled opportunities for connection, information dissemination, and collective action. It is incumbent upon each of us to understand the scope and implications of surveillance technologies, to question the rationale behind their deployment, and to advocate for policies that safeguard our fundamental rights. This begins with education – seeking out reliable sources of information, engaging in thoughtful dialogue, and resisting the siren song of apathy or overwhelm. The erosion of liberties is rarely a sudden, dramatic event; it is often a gradual, incremental process that begins with small compromises. Recognizing these early warning signs and speaking out against them is a crucial civic duty.

Participation in democratic processes is not limited to casting a ballot every few years. It extends to contacting elected representatives, attending town hall meetings, signing petitions, supporting organizations that champion civil liberties, and even engaging in peaceful protest. When we see legislation that threatens privacy, we must make our voices heard. When we witness the overreach of government agencies in the digital sphere, we must demand accountability. This might involve scrutinizing the budgets allocated to intelligence agencies, questioning the legal justifications for expanded surveillance powers, and advocating for independent oversight mechanisms that possess real teeth. Furthermore, fostering a culture of digital literacy within our communities, ensuring that individuals of all ages and backgrounds understand their digital rights and how to protect them, is a vital long-term strategy. This could manifest in community workshops, educational programs in schools, and accessible online resources.

The power of collective action cannot be overstated. History is replete with examples of how organized citizens, united by a common cause, have successfully challenged powerful interests and brought about meaningful change. In the context of surveillance and civil liberties, this means building coalitions, sharing information, and amplifying the voices of those most affected by intrusive practices. It means supporting investigative journalism that exposes governmental overreach and holding accountable those who violate public trust. It means demanding transparency not as an abstract ideal, but as a concrete requirement for the legitimate exercise of state power. When lawmakers feel the sustained pressure of an informed and engaged electorate, they are far more likely to prioritize the protection of fundamental rights.

For our lawmakers, the call to action is equally urgent, demanding a renewed commitment to their oath of office and a deep understanding of the digital realities facing their constituents. The era of assuming that security concerns can always justify the infringement of liberties is over. Lawmakers must proactively engage with the technological advancements that shape our world, seeking not to stifle innovation, but to channel it in ways that are compatible with democratic values and human rights. This requires moving beyond partisan divides and engaging in genuine, evidence-based policymaking.

Firstly, lawmakers must prioritize comprehensive and ongoing education about the technologies used by intelligence and law enforcement agencies. This means actively seeking out the expertise of technologists, ethicists, and civil liberties advocates, and creating forums where these diverse perspectives can be heard and debated without undue political influence. Understanding the capabilities of data mining, facial recognition, predictive analytics, and encryption is not merely a technical exercise; it is a foundational requirement for crafting effective and rights-respecting legislation. Without this understanding, laws risk being either irrelevant or actively harmful, creating loopholes that enable abuse or stifle legitimate activity.

Secondly, there is a critical need to strengthen oversight mechanisms. Existing oversight bodies, whether legislative committees or internal review boards, must be granted sufficient resources, access, and authority to effectively monitor surveillance activities. This includes ensuring that oversight is truly independent, free from undue influence by the very agencies it is meant to scrutinize. Lawmakers must champion legislation that mandates greater transparency regarding surveillance programs, including clear reporting requirements, public audits, and robust avenues for redress when rights are violated. The principle of "innocent until proven guilty" must extend into the digital realm, meaning that individuals should not be subjected to mass surveillance or data collection without a clear, legally defined justification.

Thirdly, lawmakers have a responsibility to update and adapt legal frameworks to address the realities of the digital age. This means revisiting laws related to data privacy, search and seizure, and freedom of expression in the online space. It means resisting the temptation to enact broad, sweeping legislation that grants excessive powers to the state in the name of security, and instead focusing on targeted, proportionate measures that are subject to strict judicial review. The concept of "lawful access" in encrypted communications, for example, requires careful consideration of technical feasibility, privacy implications, and the potential for creating vulnerabilities that could be exploited by malicious actors. Policymakers must engage with these complexities, seeking solutions that balance legitimate law enforcement needs with the imperative to protect the security and privacy of all citizens.

Furthermore, lawmakers must foster a culture of accountability within government agencies. This involves establishing clear ethical guidelines for the use of surveillance technologies, implementing strong internal controls, and ensuring that individuals who abuse their authority are held responsible. Transparency in the use of public funds, including those allocated to intelligence and national security, is also paramount. Citizens have a right to know how their tax dollars are being spent, especially when those expenditures involve the potential infringement of their fundamental rights.

The collaborative approach, as discussed previously, is not merely an option but a necessity. Lawmakers should actively seek partnerships with the technology sector, academia, and civil society to develop innovative solutions that enhance security without compromising liberty. This could involve supporting the development of privacy-preserving technologies, promoting best practices in data security, and fostering public-private dialogues on emerging threats and challenges.

The ultimate goal is to build a society where security and liberty are not seen as competing interests, but as mutually reinforcing pillars of a strong and resilient democracy. A government that respects the rights and privacy of its citizens is a government that earns their trust and cooperation. This trust is the most valuable asset in any national security strategy. When citizens feel that their government acts with integrity and accountability, they are more likely to support necessary security measures and to contribute to the collective safety of the nation. Conversely, a pervasive sense of being monitored, judged, or controlled breeds fear, distrust, and ultimately, a weakening of the social fabric.

This call to action is not a suggestion; it is an imperative. The decisions made today regarding surveillance, data privacy, and governmental accountability will shape the kind of society we inhabit for generations to come. It requires sustained effort, unwavering commitment, and a willingness to challenge the status quo. Citizens must remain vigilant, informed, and engaged, holding their elected officials to account. Lawmakers must rise to the occasion, embracing their responsibility to govern with both wisdom and integrity, ensuring that the pursuit of security never comes at the expense of the fundamental freedoms that define a just and democratic society. The path forward is not an easy one, but it is a path we must walk together, united in our commitment to a future where technology serves humanity, and where the rights and dignity of every individual are protected. The ongoing struggle for privacy and civil liberties demands our sustained attention and our collective resolve. It is a continuous process, requiring constant adaptation and a steadfast defense of democratic principles in the face of evolving technological and geopolitical landscapes. By working together, citizens and lawmakers can forge a future where security and liberty are not mutually exclusive, but rather synergistic elements of a truly thriving and resilient society. This commitment to collective responsibility is the bedrock upon which we can build a future that upholds democratic ideals and vigorously protects fundamental human rights in the digital age, ensuring that the tools designed to protect us do not become the instruments of our subjugation.

GLOSSARY

Bulk Data Collection: The acquisition of large quantities of digital information, often without individualized suspicion, typically by government agencies. This can include communication metadata, internet browsing history, and location data.

Chilling Effect: The discouragement of the exercise of constitutional rights, such as freedom of speech or association, due to fear of governmental surveillance or retaliation.

Encryption: A process of encoding information so that only authorized parties can access it. End-to-end encryption ensures that messages are unreadable by third parties, including the service provider.

Metadata: Data that describes other data. In telecommunications, metadata can include information about who communicated with whom, when, for how long, and from where, but not the content of the communication itself.

National Security Letter (NSL): A type of administrative subpoena used by the FBI to obtain certain types of information from third parties, such as financial institutions and telecommunications companies, without prior judicial approval.

Oversight: The process by which legislative bodies or independent bodies monitor and review the activities of executive agencies, particularly in relation to surveillance and intelligence gathering.

Predictive Policing: The use of data analysis and algorithms to identify individuals or locations likely to be involved in future criminal activity.

Section 702 of the FISA Amendments Act: A US law that allows the government to conduct warrantless surveillance on non-US persons located outside the United States, even if they are communicating with US persons.

Surveillance: The close observation of a person or area, especially by government or law enforcement agencies, often involving electronic means.

Warrant: A legal document issued by a judge or magistrate authorizing law enforcement to conduct a search or seizure, or to make an arrest.

www.ingramcontent.com/pod-product-compliance
Lightning Source LLC
Chambersburg PA
CBHW050545270326
41926CB00012B/1915